Elements of Continuum Mechanics and Conservation Laws

Elements of Continuum Mechanics and Conservation Laws

Sergei K. Godunov and Evgenii I. Romenskii

S. L. Sobolev Institute of Mathematics SB RAS
Novosibirsk, Russia

Kluwer Academic / Plenum Publishers
New York, Boston, Dordrecht, London, Moscow

Library of Congress Cataloging-in-Publication Data

С. К. Годунов, Е. И. Роменский
ЭЛЕМЕНТЫ МЕХАНИКИ СПЛОШНЫХ СРЕД
И ЗАКОНЫ СОХРАНЕНИЯ
Научная книга (ИДМИ), Новосибирск, 1998

Translated from the Russian with new material added by the authors by Tamara Rozhkovskaya, Novosibirsk, Russia

ISBN: 0-306-47735-1

233 Spring Street, New York, New York 10013

http://www.wkap.nl/

10 9 8 7 6 5 4 3 2 1

A C.I.P. record for this book is available from the Library of Congress

Printed in United Kingdom by Biddles/IBT Global.

Preface

This English edition draws on the original work of S. Godunov [5]. This book was published in Russia in 1978 and received the M. A. Lavrent'ev prize of the Russian Academy of Science in 1993. Later, the material of [5] was included in the book by both authors [7, Chapters I-IV] published in 1998. This edition also presents a new concept developed by S. Godunov on the basis of examples due to E. Romenskii. This concept was essentially developed while preparing the English translation of [7] and was intensively discussed with specialists in France, Russia, and USA last year. Four years ago we could only conjecture as to the existence of relationships between the structure of thermodynamical conservation laws and representations of the rotation group. Now, we can say more owing to the very recent results of S. Godunov and V. Gordienko (cf. Appendix and references there).

Chapters I-IV were based on lectures given by S. Godunov at the Physical Department of the Novosibirsk State University. This course was based on the original results of both authors and concerned formulations of differential equations describing high-rate deformations of metals. To study such processes, equations of elasticity should be stated in one zone, whereas equations of hydrodynamics should be used in other zones where, for example, cumulative jets appear. Plastic deformations can occur in transition zones, which causes the residual stresses. The suggested model differs from the equations of elasticity by relaxation terms simulating plastic deformations and transistion of an elastic medium to a fluid. This model was often used for numerical purposes in simulating high-rate deformations of metals [2, 6].

Using thermodynamics laws and the so-called generalized thermodynamical potentials, we can write differential equations describing physical processes in a special form. The simplest variant of such a model suggested by S. Godunov in 1961 and discussed in Sections 19–22 is well known and widely used at present [1, 3, 4, 8–10].

It should be noted that approaches to systematization of equations simulating processes in continuous media are extensively discussed in the literature. This topic is attractive for many researchers. However, the study is still far from a complete theory and there exist many different and sometimes opposite points of view, which was observed even between the authors here.

Naturally, the concept presented in the Appendix cannot be regarded as a final scheme of formalization of differential equations describing processes in continuous media but the ideas and methods expressed there could be helpful for the further study of these problems and hopefully attract the attention of not only specialists in mechanics and physics but also mathematicians studying solutions to partial differential equations and their properties.

In my opinion, the further development of the theory of thermodynamically compatible conservation laws will lead to very interesting results of general applicability. In addition, an appropriate systematization of various models of physical processes could be also useful in numerical methods

Sergei Godunov

November, 2002
Novosibirsk, Academtown

Contents

Chapter I

Elementary Properties of Deformations and Stresses

§ 1. Deformations and Strain Rate

Elements of a moving continuous medium change in various ways. Some of them change their size and shape (i.e., a deformation occurs), whereas other elements move as a rigid body (i.e. the distance between any two particles remains unchanged). The change of elements of the medium may be characterized on the basis of a description of the motion.

1.1. The velocity of a material particle with Cartesian coordinates x_1, x_2, x_3 at a time t is a vector with components

$$u_1 = u_1(x_1, x_2, x_3, t), \quad u_2 = u_2(x_1, x_2, x_3, t), \quad u_3 = u_3(x_1, x_2, x_3, t).$$

The trajectory of a particle passing through a point (x_{10}, x_{20}, x_{30}) at a time t_0 can be found as a solution (the integral curve) $x_i = x_i(t)$, $i = 1, 2, 3$, of the ordinary differential equations

$$\frac{dx_i}{dt} = u_i(x_1, x_2, x_3, t) \tag{1.1}$$

with the initial condition

$$x_i\Big|_{t=t_0} = x_{i0}, \quad i = 1, 2, 3. \tag{1.2}$$

For the sake of convenience, we indicate the dependence of x_i on the initial point (x_{10}, x_{20}, x_{30}) and write

$$\begin{aligned} x_1 &= x_1(x_{10}, x_{20}, x_{30}, t), \\ x_2 &= x_2(x_{10}, x_{20}, x_{30}, t), \\ x_3 &= x_3(x_{10}, x_{20}, x_{30}, t). \end{aligned} \tag{1.3}$$

Three functions $x_i = x_i(x_{10}, x_{20}, x_{30}, t)$, $i = 1, 2, 3$, describe the change of the Cartesian coordinates of material particles of a moving medium during some time-interval $t_0 < t < t_1$. It is natural to assume that the functions x_1, x_2, x_3 are sufficiently smooth (piecewise continuously differentiable) and $x_i(x_{10}, x_{20}, x_{30}, t_0) = x_{i0}$. In this case, using the Taylor formula, for a fixed t we can write the approximate formulas

$$\begin{aligned} x_1 &= x_1(x_{10}^0, x_{20}^0, x_{30}^0, t) + \frac{\partial x_1}{\partial x_{10}}(x_{10} - x_{10}^0) + \frac{\partial x_1}{\partial x_{20}}(x_{20} - x_{20}^0) \\ &\quad + \frac{\partial x_1}{\partial x_{30}}(x_{30} - x_{30}^0) + \dots, \\ x_2 &= x_2(x_{10}^0, x_{20}^0, x_{30}^0, t) + \frac{\partial x_2}{\partial x_{10}}(x_{10} - x_{10}^0) + \frac{\partial x_2}{\partial x_{20}}(x_{20} - x_{20}^0) \\ &\quad + \frac{\partial x_2}{\partial x_{30}}(x_{30} - x_{30}^0) + \dots, \\ x_3 &= x_3(x_{10}^0, x_{20}^0, x_{30}^0, t) + \frac{\partial x_3}{\partial x_{10}}(x_{10} - x_{10}^0) + \frac{\partial x_3}{\partial x_{20}}(x_{20} - x_{20}^0) \\ &\quad + \frac{\partial x_3}{\partial x_{30}}(x_{30} - x_{30}^0) + \dots, \end{aligned} \tag{1.4}$$

where the partial derivatives $\frac{\partial x_i}{\partial x_{j0}}$ are taken at the point $(x_{10}^0, x_{20}^0, x_{30}^0, t)$. Formulas (1.4) describe a transformation of a neighborhood of $(x_{10}^0, x_{20}^0, x_{30}^0)$ during the time-interval (t_0, t). We see that it is approximated by the affine transformation

$$\begin{aligned} x_1 &= l_1 + c_{11}(x_{10} - x_{10}^0) + c_{12}(x_{20} - x_{20}^0) + c_{13}(x_{30} - x_{30}^0), \\ x_2 &= l_2 + c_{21}(x_{10} - x_{10}^0) + c_{22}(x_{20} - x_{20}^0) + c_{23}(x_{30} - x_{30}^0), \\ x_3 &= l_3 + c_{31}(x_{10} - x_{10}^0) + c_{32}(x_{20} - x_{20}^0) + c_{33}(x_{30} - x_{30}^0), \end{aligned} \tag{1.5}$$

where $l_j = x_j(x_{10}^0, x_{20}^0, x_{30}^0, t)$ and $c_{ij} = \frac{\partial x_i}{\partial x_{j0}}$. Without loss of generality, we can assume that $(x_{10}^0, x_{20}^0, x_{30}^0)$ is the origin: $x_{i0}^0 = 0$, $i = 1, 2, 3$. It is intuitively clear that parallel translations do not affect shapes of elements and the distance between particles and cannot generate any internal stresses. Therefore, we omit the vector (l_1, l_2, l_3) in (1.5) and consider linear transformations

$$\begin{aligned} x_1 &= c_{11}x_{10} + c_{12}x_{20} + c_{13}x_{30}, \\ x_2 &= c_{21}x_{10} + c_{22}x_{20} + c_{23}x_{30}, \\ x_3 &= c_{31}x_{10} + c_{32}x_{20} + c_{33}x_{30} \end{aligned} \tag{1.6}$$

or, in matrix form,

$$\begin{pmatrix} x_1 \\ x_2 \\ x_3 \end{pmatrix} = C \begin{pmatrix} x_{10} \\ x_{20} \\ x_{30} \end{pmatrix}, \qquad C = ||c_{ij}|| = \begin{bmatrix} c_{11} & c_{12} & c_{13} \\ c_{21} & c_{22} & c_{23} \\ c_{31} & c_{32} & c_{33} \end{bmatrix},$$

or, for brevity,

$$x = Cx_0.$$

Using linear transformations, we can consider displacements of particles in some neighborhood as though we observed them under the microscope. In continuum mechanics, it is reasonable to consider only linear transformations (1.6) with $\det C > 0$.

• A motion of a medium, or a linear change of coordinates, is called a *deformation* of the medium if there are material particles such that the distance between them changes during this motion.

The affine transformation (1.5) is a deformation if the matrix C is not orthogonal.

1.2. A continuous medium always resists any deformation in one way or another. Therefore, if a medium moves with deformation, internal stresses necessarily appear or (if such stresses already exist) change. It is natural to expect that the stresses in a neighborhood of some material particle depend only on the history of deformation of this neighborhood. But this history is kept by the functions $c_{ij}(t)$ defining the linear transformation (1.6) which approximates the actual deformation of the neighborhood. We note that, in order to determine the stressed state of the medium at some time t^*, we sometimes need to know only the values $c_{ij}(t^*)$, whereas in some cases the knowledge about the whole functions c_{ij} or the values $c_{ij}(t)$ for all t sufficiently close to t^* can be required. In this section we use an intuitive notion of stresses. To formalize it, we need to introduce some mathematical formalism, which will be done later.

1.3. We return to deformations and consider the linear transformation (1.6) in more detail. As is known from linear algebra, any nonsingular matrix $C = ||c_{ij}||$, $\det C > 0$, admits the *Cayley polar decomposition*

$$C = SW \quad (S = S^*,\ WW^* = I), \tag{1.7}$$

where S is a *symmetric matrix* ($S = S^*$) and W is an *orthogonal matrix* ($WW^* = I$). In turn, the symmetric matrix S can be represented as the product

$$S = UKU^* \quad (UU^* = I) \tag{1.8}$$

of an orthogonal matrix U and a diagonal matrix

$$K = \begin{bmatrix} \pm k_1 & 0 & 0 \\ 0 & \pm k_2 & 0 \\ 0 & 0 & \pm k_3 \end{bmatrix}.$$

Introducing the notation $V = U^*W$ and taking into account (1.8), we reduce the representation (1.7) to the form

$$C = UKV \quad (UU^* = I,\ VV^* = I). \tag{1.9}$$

Further, the diagonal matrix K can be written as the product of a diagonal matrix with nonnegative entries and an orthogonal matrix. For example,

$$\begin{bmatrix} +k_1 & 0 & 0 \\ 0 & -k_2 & 0 \\ 0 & 0 & +k_3 \end{bmatrix} = \begin{bmatrix} +k_1 & 0 & 0 \\ 0 & +k_2 & 0 \\ 0 & 0 & +k_3 \end{bmatrix} \begin{bmatrix} +1 & 0 & 0 \\ 0 & -1 & 0 \\ 0 & 0 & +1 \end{bmatrix}.$$

Let $K = K_+R$, where $RR^* = I$ and K_+ is a diagonal matrix with nonnegative entries. Then the decomposition (1.7) takes the form

$$C = UKV = UK_+RV = UK_+\widetilde{V} \quad (\widetilde{V} = RV,\ \widetilde{V}\widetilde{V}^* = I). \tag{1.10}$$

Since $\det C = \det U \det K_+ \det \widetilde{V} > 0$, each diagonal element of the matrix K_+ is strictly positive and the determinants of the matrices U and $\widetilde{V}$ are simultaneously equal to $+1$ or -1. Setting

$$\widetilde{U} = \begin{cases} +U & \text{if } \det U = +1, \\ -U & \text{if } \det U = -1, \end{cases} \qquad \widetilde{\widetilde{V}} = \begin{cases} +\widetilde{V} & \text{if } \det \widetilde{V} = \det U = +1, \\ -\widetilde{V} & \text{if } \det \widetilde{V} = \det U = -1, \end{cases}$$

we obtain the representation $C = \widetilde{U}K_+\widetilde{\widetilde{V}}$ with $\det \widetilde{U} = +1$ and $\det \widetilde{\widetilde{V}} = +1$.

• *Any nonsingular matrix C with* $\det C > 0$ *can be represented in the form*

$$C = UKV, \tag{1.11}$$

where

$$UU^* = VV^* = I, \quad \det U = \det V = +1,$$
$$K = \begin{bmatrix} k_1 & 0 & 0 \\ 0 & k_2 & 0 \\ 0 & 0 & k_3 \end{bmatrix}, \quad k_i > 0. \tag{1.12}$$

• An orthogonal matrix U $(UU^* = I)$ with $\det U = 1$ is called a *rotation matrix* or simply a *rotation* for brevity.

• *The representation* (1.11) *shows that any linear transformation* (1.6) *with a nonsingular matrix C,* $\det C > 0$, *consists in the following successive* *operations: a rotation V, an extension with coefficients k_1, k_2, k_3 along the coordinate axes, and a rotation U. In the case $k_1 = k_2 = k_3 = 1$, the medium moves as a rigid body; otherwise, if $k_i \neq 1$ for some i, there exist points such that the distance between them changes under the linear transformation* (1.6) *with the matrix C, i.e., the medium is deformed.*

• A medium is said to be *elastic* if the stresses depend only on the temperature and on the transformation C of the initial state into the deformed state of the medium.

1.4. We will often consider stresses, called *viscous stresses*, that arise when some elements of a medium move relative to other elements. Viscous stresses depend on the rate of deformation expressed by the so-called strain rate tensor (the definition is given below). To study viscous stresses, it is more convenient to describe a moving medium in terms of the velocity field $u_1 = u_1(x_1, x_2, x_3, t)$, $u_2 = u_2(x_1, x_2, x_3, t)$, $u_3 = u_3(x_1, x_2, x_3, t)$ instead of the position vector (1.3). In the case of a rigid-body motion, no stresses should arise. In particular, this happens if u_i, $i = 1, 2, 3$, are independent of x_1, x_2, x_3. Therefore, it is natural to assume that the viscous stresses depend only on the velocity gradient

$$\left\| \frac{\partial u_i}{\partial x_j} \right\| = \begin{bmatrix} \dfrac{\partial u_1}{\partial x_1} & \dfrac{\partial u_1}{\partial x_2} & \dfrac{\partial u_1}{\partial x_3} \\ \dfrac{\partial u_2}{\partial x_1} & \dfrac{\partial u_2}{\partial x_2} & \dfrac{\partial u_2}{\partial x_3} \\ \dfrac{\partial u_3}{\partial x_1} & \dfrac{\partial u_3}{\partial x_2} & \dfrac{\partial u_3}{\partial x_3} \end{bmatrix}. \tag{1.13}$$

Naturally, the viscous stresses may also depend on properties and parameters of the medium such as density, temperature, etc. We write the matrix (1.13) as the sum of the symmetric matrix with entries

$$u_{ik} = u_{ki} = \frac{1}{2}\left(\frac{\partial u_i}{\partial x_k} + \frac{\partial u_k}{\partial x_i}\right) \tag{1.14}$$

and the skew-symmetric matrix

$$\begin{bmatrix} \dfrac{\partial u_1}{\partial x_1} & \dfrac{\partial u_1}{\partial x_2} & \dfrac{\partial u_1}{\partial x_3} \\ \dfrac{\partial u_2}{\partial x_1} & \dfrac{\partial u_2}{\partial x_2} & \dfrac{\partial u_2}{\partial x_3} \\ \dfrac{\partial u_3}{\partial x_1} & \dfrac{\partial u_3}{\partial x_2} & \dfrac{\partial u_3}{\partial x_3} \end{bmatrix} = \begin{bmatrix} u_{11} & u_{12} & u_{13} \\ u_{21} & u_{22} & u_{23} \\ u_{31} & u_{32} & u_{33} \end{bmatrix} + \begin{bmatrix} 0 & -\omega_3 & \omega_2 \\ \omega_3 & 0 & -\omega_1 \\ -\omega_2 & \omega_1 & 0 \end{bmatrix}. \tag{1.15}$$

• The symmetric matrix $\|u_{ik}\|$ with entries (1.14) is called the *strain rate tensor*, and the nonzero components ω_1, ω_2, ω_3 of the skew-symmetric matrix (1.15) form the so-called *vorticity*.

We clarify the meaning of the formal decomposition (1.15) of the velocity gradient $\left\|\frac{\partial u)i}{\partial x_j}\right\|$ into symmetric and skew-symmetric parts.

Particles in a neighborhood of a point (x_1, x_2, x_3) at a time t have coordinates $x_1 + \delta x_1$, $x_2 + \delta x_2$, $x_3 + \delta x_3$ and the velocities with components

$$u_i(x_1, x_2, x_3, t) + \frac{\partial u_i}{\partial x_1}\delta x_1 + \frac{\partial u_i}{\partial x_2}\delta x_2 + \frac{\partial u_i}{\partial x_3}\delta x_3 + \dots, \qquad i = 1, 2, 3.$$

In a time-interval Δt, the point $x_i + \delta x_i$ goes to a point $x_i' + \delta x_i'$:

$$x_i' + \delta x_i' = (x_i + u_i \Delta t) + \delta x_i + \left[\frac{\partial u_i}{\partial x_1}\delta x_1 + \frac{\partial u_i}{\partial x_2}\delta x_2 + \frac{\partial u_i}{\partial x_3}\delta x_3\right]\Delta t + \dots.$$

Since we do not take into account a rigid-body motion (i.e., we neglect the translation $x_i \to x_i' = x_i + u_i \Delta t + \dots$ of the center of the neighborhood), the transformation during the time-interval Δt is expressed as follows:

$$\begin{pmatrix} \delta x_1' \\ \delta x_2' \\ \delta x_3' \end{pmatrix} = \left\|\delta_{ij} + \Delta t \frac{\partial u_i}{\partial x_j}\right\| \begin{pmatrix} \delta x_1 \\ \delta x_2 \\ \delta x_3 \end{pmatrix} = \left(I + \Delta t \left\|\frac{\partial u_i}{\partial x_j}\right\|\right) \begin{pmatrix} \delta x_1 \\ \delta x_2 \\ \delta x_3 \end{pmatrix}, \tag{1.16}$$

where $\delta_{ij} = 0$ if $i \neq j$ and $\delta_{ii} = 1$.

Using (1.15), we represent the matrix $I + \Delta t \left\|\frac{\partial u_i}{\partial x_j}\right\|$ as follows:

$$I + \Delta t \left\|\frac{\partial u_i}{\partial x_j}\right\| = (I + \Delta t \|u_{ij}\|) \begin{bmatrix} 1 & -\omega_3 \Delta t & \omega_2 \Delta t \\ \omega_3 \Delta t & 1 & -\omega_1 \Delta t \\ -\omega_2 \Delta t & \omega_1 \Delta t & 1 \end{bmatrix} + O(\Delta t^2). \tag{1.17}$$

The linear transformation with the matrix

$$\begin{bmatrix} 1 & -\omega_3 \Delta t & \omega_2 \Delta t \\ \omega_3 \Delta t & 1 & -\omega_1 \Delta t \\ -\omega_2 \Delta t & \omega_1 \Delta t & 1 \end{bmatrix} = I + \Delta t \begin{bmatrix} 0 & -\omega_3 & \omega_2 \\ \omega_3 & 0 & -\omega_1 \\ -\omega_2 & \omega_1 & 0 \end{bmatrix}$$

can be regarded as an infinitesimal rotation since

$$\begin{bmatrix} 1 & -\omega_3 \Delta t & \omega_2 \Delta t \\ \omega_3 \Delta t & 1 & -\omega_1 \Delta t \\ -\omega_2 \Delta t & \omega_1 \Delta t & 1 \end{bmatrix}^* \begin{bmatrix} 1 & -\omega_3 \Delta t & \omega_2 \Delta t \\ \omega_3 \Delta t & 1 & -\omega_1 \Delta t \\ -\omega_2 \Delta t & \omega_1 \Delta t & 1 \end{bmatrix}$$

$$= \begin{bmatrix} 1 & 0 & 0 \\ 0 & 1 & 0 \\ 0 & 0 & 1 \end{bmatrix} + O(\Delta t^2).$$

This is the rotation by the angle $\Delta t \sqrt{\omega_1^2 + \omega_2^2 + \omega_3^2}$ about the axis directed along the vector $(\omega_1, \omega_2, \omega_3)$. Since rotations do not lead to the appearance of internal friction, it is natural to assume that the viscous stresses depend only on the strain rate tensor $\|u_{ik}\|$.

With the same accuracy, we have

$$I + \Delta t \left\| \frac{\partial u_i}{\partial x_j} \right\| = \begin{bmatrix} 1 & -\omega_3 \Delta t & \omega_2 \Delta t \\ \omega_3 \Delta t & 1 & -\omega_1 \Delta t \\ -\omega_2 \Delta t & \omega_1 \Delta t & 1 \end{bmatrix} (I + \Delta t \|u_{ij}\|) + O(\Delta t^2), \tag{1.18}$$

i.e., the order of factors on the right-hand side of (1.17) is of no importance.

Using the representation of a symmetric matrix (cf. (1.8)), we represent the strain rate tensor $\|u_{ij}\|$ in the form

$$\|u_{ij}\| = W \begin{bmatrix} \lambda_1 & 0 & 0 \\ 0 & \lambda_2 & 0 \\ 0 & 0 & \lambda_3 \end{bmatrix} W^*. \tag{1.19}$$

We can assume that $\det W = +1$, i.e., W is a rotation matrix. Using (1.19), we can write

$$I + \Delta t \|u_{ij}\| = W \begin{bmatrix} 1 + \lambda_1 \Delta t & 0 & 0 \\ 0 & 1 + \lambda_2 \Delta t & 0 \\ 0 & 0 & 1 + \lambda_3 \Delta t \end{bmatrix} W^*.$$

Hence the linear transformation with the matrix $I + \Delta t \|u_{ij}\|$ consists of a rotation W^* that transforms some three orthogonal axes into the coordinate axes, the extension with coefficients $1 + \lambda_k \Delta t$, $k = 1, 2, 3$, along the coordinate axes, and the inverse rotation W which transform the coordinate axes into their previous location. In other words, this transformation extends with coefficients $1 + \lambda_k \Delta t$ along the orthogonal axes obtained from the coordinate axes by the rotation

$$W = \begin{bmatrix} w_{11} & w_{12} & w_{13} \\ w_{21} & w_{22} & w_{23} \\ w_{31} & w_{32} & w_{33} \end{bmatrix}.$$

The unit directing vectors of these axes are columns of the matrix W and are obtained as follows:

$$W \begin{pmatrix} 1 \\ 0 \\ 0 \end{pmatrix} = \begin{pmatrix} w_{11} \\ w_{21} \\ w_{31} \end{pmatrix}, \quad W \begin{pmatrix} 0 \\ 1 \\ 0 \end{pmatrix} = \begin{pmatrix} w_{12} \\ w_{22} \\ w_{32} \end{pmatrix}, \quad W \begin{pmatrix} 0 \\ 0 \\ 1 \end{pmatrix} = \begin{pmatrix} w_{13} \\ w_{23} \\ w_{33} \end{pmatrix}.$$

• *The linear transformation* (1.16) *with the matrix* $I + \Delta t \left\| \frac{\partial u_i}{\partial x_j} \right\|$ *can be represented as the product of an infinitesimal rotation and an extension* $I + \Delta t \|u_{ij}\|$ *along three orthogonal axes. The strain rate tensor* $\|u_{ij}\|$ *characterizes the rate of deformation.*

1.5. We are interested in transformation of the strain rate tensor $\|u_{ij}\|$ under an orthogonal change of coordinates. Let W be a rotation matrix.

Applying W, we thereby pass to the new components of the velocity vector and the position vector

$$\begin{pmatrix} u_1' \\ u_2' \\ u_3' \end{pmatrix} = W \begin{pmatrix} u_1 \\ u_2 \\ u_3 \end{pmatrix}, \quad \begin{pmatrix} x_1' \\ x_2' \\ x_3' \end{pmatrix} = W \begin{pmatrix} x_1 \\ x_2 \\ x_3 \end{pmatrix}.$$

Then the matrix of the derivatives $\dfrac{\partial u_i'}{\partial x_j'}$ has the form

$$\begin{bmatrix} \dfrac{\partial u_1'}{\partial x_1'} & \dfrac{\partial u_1'}{\partial x_2'} & \dfrac{\partial u_1'}{\partial x_3'} \\ \dfrac{\partial u_2'}{\partial x_1'} & \dfrac{\partial u_2'}{\partial x_2'} & \dfrac{\partial u_2'}{\partial x_3'} \\ \dfrac{\partial u_3'}{\partial x_1'} & \dfrac{\partial u_3'}{\partial x_2'} & \dfrac{\partial u_3'}{\partial x_3'} \end{bmatrix} = W \begin{bmatrix} \dfrac{\partial u_1}{\partial x_1} & \dfrac{\partial u_1}{\partial x_2} & \dfrac{\partial u_1}{\partial x_3} \\ \dfrac{\partial u_2}{\partial x_1} & \dfrac{\partial u_2}{\partial x_2} & \dfrac{\partial u_2}{\partial x_3} \\ \dfrac{\partial u_3}{\partial x_1} & \dfrac{\partial u_3}{\partial x_2} & \dfrac{\partial u_3}{\partial x_3} \end{bmatrix} W^*.$$

Passing to the transposed matrices, we find

$$\begin{bmatrix} \dfrac{\partial u_1'}{\partial x_1'} & \dfrac{\partial u_2'}{\partial x_1'} & \dfrac{\partial u_3'}{\partial x_1'} \\ \dfrac{\partial u_1'}{\partial x_2'} & \dfrac{\partial u_2'}{\partial x_2'} & \dfrac{\partial u_3'}{\partial x_2'} \\ \dfrac{\partial u_1'}{\partial x_3'} & \dfrac{\partial u_2'}{\partial x_3'} & \dfrac{\partial u_3'}{\partial x_3'} \end{bmatrix} = W \begin{bmatrix} \dfrac{\partial u_1}{\partial x_1} & \dfrac{\partial u_2}{\partial x_1} & \dfrac{\partial u_3}{\partial x_1} \\ \dfrac{\partial u_1}{\partial x_2} & \dfrac{\partial u_2}{\partial x_2} & \dfrac{\partial u_3}{\partial x_2} \\ \dfrac{\partial u_1}{\partial x_3} & \dfrac{\partial u_2}{\partial x_3} & \dfrac{\partial u_3}{\partial x_3} \end{bmatrix} W^*$$

which implies

$$||u_{ik}'|| = W\, ||u_{ik}||\, W^*$$

because, by definition,

$$u_{ik} = \frac{1}{2}\left(\frac{\partial u_i}{\partial x_k} + \frac{\partial u_k}{\partial x_i}\right), \quad u_{ik}' = \frac{1}{2}\left(\frac{\partial u_i'}{\partial x_k'} + \frac{\partial u_k'}{\partial x_i'}\right).$$

We note that the matrix C in (1.6) is transformed by the same rule if we make the change of coordinates x_1, x_2, x_3 with an orthogonal matrix W:

$$\begin{pmatrix} x_1' \\ x_2' \\ x_3' \end{pmatrix} = W \begin{pmatrix} x_1 \\ x_2 \\ x_3 \end{pmatrix}, \quad \begin{pmatrix} x_{10}' \\ x_{20}' \\ x_{30}' \end{pmatrix} = W \begin{pmatrix} x_{10} \\ x_{20} \\ x_{30} \end{pmatrix}.$$

From the last equality we find

$$\begin{pmatrix} x_{10} \\ x_{20} \\ x_{30} \end{pmatrix} = W^* \begin{pmatrix} x_{10}' \\ x_{20}' \\ x_{30}' \end{pmatrix}.$$

Multiplying the equalities

$$\begin{pmatrix} x_1 \\ x_2 \\ x_3 \end{pmatrix} = C \begin{pmatrix} x_{10} \\ x_{20} \\ x_{30} \end{pmatrix} = CW^* \begin{pmatrix} x'_{10} \\ x'_{20} \\ x'_{30} \end{pmatrix}$$

by W from the left, we find

$$\begin{pmatrix} x'_1 \\ x'_2 \\ x'_3 \end{pmatrix} = W \begin{pmatrix} x_1 \\ x_2 \\ x_3 \end{pmatrix} = WCW^* \begin{pmatrix} x'_{10} \\ x'_{20} \\ x'_{30} \end{pmatrix} = C' \begin{pmatrix} x'_{10} \\ x'_{20} \\ x'_{30} \end{pmatrix}.$$

Hence

$$C' = WCW^*. \tag{1.20}$$

In the future we will often use matrices that are transformed by the rule (1.20), i.e., a matrix C is multiplied by an orthogonal matrix W from the left and by W^* from the right. Such matrices will be referred to as *second-order tensors.*

- *Under the change of coordinates*

$$x'_k = \sum_i w_{ki} x_i, \quad \sum_i w_{ki} w_{li} = \delta_{kl} = \begin{cases} 0 & \text{if } k \neq l, \\ 1 & \text{if } k = l, \end{cases}$$

the components c_{ij} are transformed by the following rule:

$$c'_{kl} = \sum_{i,j} w_{ki} c_{ij} w_{lj} = \sum_{i,j} w_{ki} w_{lj} c_{ij}. \tag{1.21}$$

§ 2. General Properties of Stresses

In Section 1, we used the intuitive notion of stresses. In this section, we formalize this notion and discuss it in detail.

2.1. We begin by considering a motionless medium in equilibrium under the action of external forces. The forces can be applied to the interior of a region occupied by the medium (the body force $\mathcal{F}(x_1, x_2, x_3)$) or to the boundary surface (the surface force $\mathcal{P}(x_1, x_2, x_3)$). Then $\mathcal{F}(x_1, x_2, x_3)dx_1dx_2dx_3$ is the total force acting on a volume element $dx_1dx_2dx_3$ and $\mathcal{P}(x_1, x_2, x_3, t)ds$ is the total force acting on a surface element ds. We assume that the components of $\mathcal{F}$ and $\mathcal{P}$ are piecewise smooth functions. Thereby we do not consider surface forces applied to a point and body forces concentrated on interior surfaces. It is convenient to represent $\mathcal{F}$ as $\mathcal{F} = \rho\mathbf{F}$, where $\rho = \rho(x_1, x_2, x_3)$ is the density of the medium. By the *density* one usually understand the mass per unit volume. More exactly, it should be said that the *density* is the mass of an infinitesimal element divided by the volume of this element. Thus, $\mathbf{F} = \begin{pmatrix} F_1 \\ F_2 \\ F_3 \end{pmatrix}$ is the specific body force (per unit mass).

We consider a medium in equilibrium and cut it along some plane $n_1x_1 + n_2x_2 + n_3x_3 = \text{const}$. Since the equilibrium state should be preserved, surface forces should be applied to both sides of the cut. We accept the following postulate.

Postulate 2.1. *The surface force applied to one side of the cut is equal to the surface force applied to the other side but point in the opposite direction.*

Postulate 2.1 will be justified under some additional conditions later.

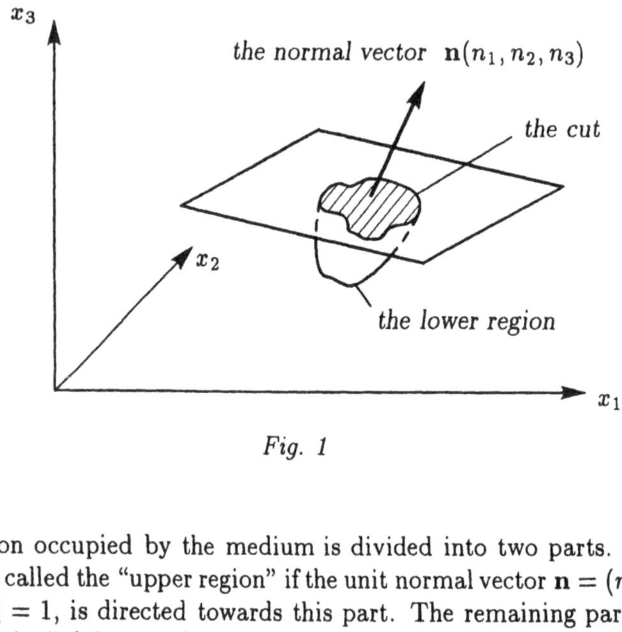

Fig. 1

The region occupied by the medium is divided into two parts. A part is conditionally called the "upper region" if the unit normal vector $\mathbf{n} = (n_1, n_2, n_3)$, $n_1^2 + n_2^2 + n_3^2 = 1$, is directed towards this part. The remaining part is called the "lower region" (cf. Fig. 1).

• The surface force (per unit area) applied to the "lower" side of the cut is called the *stress vector* and is denoted by $\mathbf{\Sigma}$.

The stress vector depends, in general, on a point (x_1, x_2, x_3) and on the direction of the normal $\mathbf{n} = (n_1, n_2, n_3)$, i.e., $\mathbf{\Sigma} = \mathbf{\Sigma}(x_1, x_2, x_3; n_1, n_2, n_3)$. By Postulate 2.1, we have the equality

$$\mathbf{\Sigma}(x_1, x_2, x_3; n_1, n_2, n_3) = -\mathbf{\Sigma}(x_1, x_2, x_3; -n_1, -n_2, -n_3). \tag{2.1}$$

The components of the vector $\mathbf{\Sigma}$ are denoted by $\sigma_1, \sigma_2, \sigma_3$, i.e., we write

$$\mathbf{\Sigma} = \begin{pmatrix} \sigma_1 \\ \sigma_2 \\ \sigma_3 \end{pmatrix}, \qquad \sigma_i = \sigma_i(x_1, x_2, x_3; n_1, n_2, n_3).$$

The stress vectors on area elements perpendicular to the coordinate axes play an important role because they completely determine the stressed state of the medium under consideration. Therefore, we use the special notation for the stress vectors directed towards to the positive direction of the coordinate axes.

If the normal vector to an area element is directed along the x_1-axis ($n_1 = 1$, $n_2 = 0$, $n_3 = 0$), the components of the stress vector are denoted by $\sigma_{11}, \sigma_{21}, \sigma_{31}$:

$$\begin{aligned} \sigma_1(x_1, x_2, x_3; 1, 0, 0) &= \sigma_{11}(x_1, x_2, x_3), \\ \sigma_2(x_1, x_2, x_3; 1, 0, 0) &= \sigma_{21}(x_1, x_2, x_3), \\ \sigma_3(x_1, x_2, x_3; 1, 0, 0) &= \sigma_{31}(x_1, x_2, x_3). \end{aligned}$$

If the normal vector to an area element is directed along the x_2-axis ($n_1 = 0$, $n_2 = 1$, $n_3 = 0$), the components of the stress vector are denoted by $\sigma_{12}, \sigma_{22}, \sigma_{32}$:

$$\begin{aligned} \sigma_1(x_1, x_2, x_3; 0, 1, 0) &= \sigma_{12}(x_1, x_2, x_3), \\ \sigma_2(x_1, x_2, x_3; 0, 1, 0) &= \sigma_{22}(x_1, x_2, x_3), \\ \sigma_3(x_1, x_2, x_3; 0, 1, 0) &= \sigma_{32}(x_1, x_2, x_3). \end{aligned}$$

If the normal vector to an area element is directed along the x_3-axis ($n_1 = 0$, $n_2 = 0$, $n_3 = 3$), the components of the stress vector are denoted by $\sigma_{13}, \sigma_{23}, \sigma_{33}$:

$$\begin{aligned} \sigma_1(x_1, x_2, x_3; 0, 0, 1) &= \sigma_{13}(x_1, x_2, x_3), \\ \sigma_2(x_1, x_2, x_3; 0, 0, 1) &= \sigma_{23}(x_1, x_2, x_3), \\ \sigma_3(x_1, x_2, x_3; 0, 0, 1) &= \sigma_{33}(x_1, x_2, x_3). \end{aligned}$$

• The *stress tensor* is the square matrix whose columns are the stress vectors on area elements perpendicular to the coordinate axes and is denoted by

$$\begin{bmatrix} \sigma_{11} & \sigma_{12} & \sigma_{13} \\ \sigma_{21} & \sigma_{22} & \sigma_{23} \\ \sigma_{31} & \sigma_{32} & \sigma_{33}, \end{bmatrix} = \|\sigma_{ij}\|. \tag{2.2}$$

In Section 3, we show that, the stress tensor $\|\sigma_{ij}\|$ is transformed by the rule (1.21) under an orthogonal change of coordinates.

• *The stress tensor completely determines the stressed state of the medium under consideration, i.e., it is possible to find the stress vector on any area element provided that the stress tensor is known.*

To justify this assertion, we use the fact that the medium is in equilibrium under the action of body and surface forces. We consider an arbitrary region V

with boundary S and write the equation of equilibrium for the medium occupied V:

$$\iint_S \mathbf{\Sigma}(x_1, x_2, x_3; n_1, n_2, n_3)\, ds + \iiint_V \rho(x_1, x_2, x_3)\mathbf{F}(x_1, x_2, x_3)\, dx_1 dx_2 dx_3 = 0, \tag{2.3}$$

where n_1, n_2, n_3 are the components of the unit outward normal to the surface S at (x_1, x_2, x_3). The equality (2.3) means that the sum of all applied forces vanishes. We write (2.3) componentwise ($i = 1, 2, 3$):

$$\iint_S \sigma_i(x_1, x_2, x_3; n_1, n_2, n_3)\, ds + \iiint_V \rho(x_1, x_2, x_3)F_i(x_1, x_2, x_3)\, dx_1 dx_2 dx_3 = 0. \tag{2.4}$$

Lemma 2.1. *If piecewise smooth continuous functions $\sigma(x_1, x_2, x_3; n_1, n_2, n_3)$ and $f(x_1, x_2, x_3)$ are such that for any region V with boundary S*

$$\iint_S \sigma(x_1, x_2, x_3; n_1, n_2, n_3)\, ds + \iiint_V f\, dx_1 dx_2 dx_3 = 0,$$

then

$$\sigma(x_1, x_2, x_3; n_1, n_2, n_3) = n_1\sigma(x_1, x_2, x_3; 1, 0, 0) + n_2\sigma(x_1, x_2, x_3; 0, 1, 0) + n_3\sigma(x_1, x_2, x_3; 0, 0, 1).$$

Remark 2.1. In the formulation of Lemma 2.1, the condition that σ is continuous with respect to n_1, n_2, n_3 can be omitted. However, we keep this condition in order to simplify the proof.

Remark 2.2. By Lemma 2.1, $\sigma_i = \sigma_{i1}n_1 + \sigma_{i2}n_2 + \sigma_{i3}n_3$ or, in matrix form,

$$\mathbf{\Sigma} = \begin{pmatrix} \sigma_1 \\ \sigma_2 \\ \sigma_3 \end{pmatrix} = \begin{bmatrix} \sigma_{11} & \sigma_{12} & \sigma_{13} \\ \sigma_{21} & \sigma_{22} & \sigma_{23} \\ \sigma_{31} & \sigma_{32} & \sigma_{33} \end{bmatrix} \begin{pmatrix} n_1 \\ n_2 \\ n_3 \end{pmatrix}. \tag{2.5}$$

In particular, this implies the equality (2.1), i.e., the assertion of Postulate 2.1 is justified.

PROOF OF LEMMA 2.1. We consider a sequence of similar tetrahedrons T_1, $T_2, \ldots$, $T_N, \ldots$ contracting to some interior point (x_{10}, x_{20}, x_{30}). Let the tetrahedron T_N be given by the inequalities

$$\nu_1(x_1 - x_{10}) + \nu_2(x_2 - x_{20}) + \nu_3(x_3 - x_{30}) \leqslant h_N,$$
$$\nu_1(x_1 - x_{10}) \geqslant 0, \quad \nu_2(x_2 - x_{20}) \geqslant 0, \quad \nu_3(x_3 - x_{30}) \geqslant 0,$$

where $\nu_i \neq 0$ $(\nu_1^2 + \nu_2^2 + \nu_3^2 = 1)$ and $h_N > 0$ $(h_N \to 0$ as $N \to \infty)$. The faces of the tetrahedron T_N are located on the planes $x_1 = x_{10}$, $x_2 = x_{20}$, $x_3 = x_{30}$ and are connected by mutually perpendicular edges that are parallel to the coordinate axes. We write out the lengths of the edges:

$$\frac{h_N}{|\nu_1|}, \quad \frac{h_N}{|\nu_2|}, \quad \frac{h_N}{|\nu_3|}$$

and the area of the faces:

$$s_1 = \frac{h_N^2}{2|\nu_2\nu_3|}, \quad s_2 = \frac{h_N^2}{2|\nu_3\nu_1|}, \quad s_3 = \frac{h_N^2}{2|\nu_1\nu_2|}.$$

The normals to the faces are directed along the coordinate axes in the direction depending on the sign of ν_1, ν_2, ν_3. Thus, the outward normal to the face $x_1 = x_{10}$ has the components $(-\operatorname{sgn}\nu_1 = \pm 1, 0, 0)$. The normals to the faces $x_2 = x_{20}$ and $x_3 = x_{30}$ have the components $(0, -\operatorname{sgn}\nu_2, 0)$ and $(0, 0, -\operatorname{sgn}\nu_3)$ respectively. The face located on the plane $\nu_1(x_1 - x_{10}) + \nu_2(x_2 - x_{20}) + \nu_3(x_3 - x_{30}) = h_N$ has the unit outward normal (ν_1, ν_2, ν_3). The altitude of the tetrahedron from the vertex x_{10}, x_{20}, x_{30} to this face intersects it at the point $x_1 = x_{10} + H\nu_1$, $x_2 = x_{20} + H\nu_2$, $x_3 = x_{30} + H\nu_3$, where H is the length of the altitude and can be found from the relation

$$\begin{aligned} h_N &= \nu_1(x_1 - x_{10}) + \nu_2(x_2 - x_{20}) + \nu_3(x_3 - x_{30}) \\ &= \nu_1 H\nu_1 + \nu_2 H\nu_2 + \nu_3 H\nu_3 = H(\nu_1^2 + \nu_2^2 + \nu_3^2) = H. \end{aligned}$$

Thus, $H = h_N$. The area s_4 multiplied by $\frac{1}{3}h_N$ is equal to the volume of the tetrahedron: $s_4\dfrac{h_N}{3} = \dfrac{1}{6}\dfrac{h_N^3}{|\nu_1\nu_2\nu_3|}$. Therefore, $s_4 = \dfrac{h_N^2}{2|\nu_1\nu_2\nu_3|}$. Assume that $\sigma(x_1, x_2, x_3; n_1, n_2, n_3)$ is a smooth function. Then the value of σ at any point of the tetrahedron differs from the value of σ at the top (x_{10}, x_{20}, x_{30}) (for fixed n_1, n_2, n_3) by a quantity of order $O(h_N)$:

$$\begin{aligned} \sigma(x_1, x_2, x_3; n_1, n_2, n_3) &= \sigma(x_{10}, x_{20}, x_{30}; n_1, n_2, n_3) + O(h_N) \\ &= \sigma^0(n_1, n_2, n_3) + O(h_N), \end{aligned}$$

where $\sigma^0(n_1, n_2, n_3) = \sigma(x_{10}, x_{20}, x_{30}; n_1, n_2, n_3)$. Consider the identity

$$\iint\limits_{S_N} \sigma(x_1, x_2, x_3; n_1, n_2, n_3)\, ds + \iiint\limits_{T_N} f\, dx_1 dx_2 dx_3 = 0,$$

where the first integral is taken over the surface S_N consisting of four faces of area s_1, s_2, s_3, s_4. The second integral is taken over the tetrahedron T_N and is of order $O(h_N^3)$.

It is easy to see that

$$\iint\limits_{S_N} \sigma(x_1, x_2, x_3; n_1, n_2, n_3)\, ds + \iiint\limits_{T_N} f\, dx_1 dx_2 dx_3$$
$$= [\sigma^0(-\operatorname{sgn}\nu_1, 0, 0) + O(h_N)]s_1 + [\sigma^0(0, -\operatorname{sgn}\nu_2, 0) + O(h_N)]s_2$$
$$+ [\sigma^0(0, 0, -\operatorname{sgn}\nu_3) + O(h_N)]s_3 \quad + [\sigma^0(\nu_1, \nu_2, \nu_3) + O(h_N)]s_4 + O(h_N^3)$$
$$= \sigma^0(-\operatorname{sgn}\nu_1, 0, 0)\frac{h_N^2}{2|\nu_2\nu_3|} + \sigma^0(0, -\operatorname{sgn}\nu_2, 0)\frac{h_N^2}{2|\nu_3\nu_1|}$$
$$+ \sigma^0(0, 0, -\operatorname{sgn}\nu_3)\frac{h_N^2}{2|\nu_1\nu_2|} + \sigma^0(\nu_1, \nu_2, \nu_3)\frac{h_N^2}{2|\nu_1\nu_2\nu_3|} + O(h_N^3) = 0.$$

We divide this equality by $\dfrac{h_N^2}{2\nu_1\nu_2\nu_3}$ and pass to the limit as $h_N \to 0$. Then for any $\nu_i \neq 0$, $\nu_1^2 + \nu_2^2 + \nu_3^2 = 1$, we have

$$|\nu_1|\sigma^0(-\operatorname{sgn}\nu_1, 0, 0) + |\nu_2|\sigma^0(0, -\operatorname{sgn}\nu_2, 0) + |\nu_3|\sigma^0(0, 0, -\operatorname{sgn}\nu_3)$$
$$+ \sigma^0(\nu_1, \nu_2, \nu_3) = 0.$$

To simplify this relation, we use the identities

$$|\nu_1|\sigma^0(-\operatorname{sgn}\nu_1, 0, 0) = -\operatorname{sgn}\nu_1|\nu_1|\sigma^0(1, 0, 0) = -\nu_1\sigma^0(1, 0, 0),$$
$$|\nu_2|\sigma^0(0, -\operatorname{sgn}\nu_2, 0) = -\nu_2\sigma^0(0, 1, 0),$$
$$|\nu_3|\sigma^0(0, 0, -\operatorname{sgn}\nu_3) = -\nu_3\sigma^0(0, 0, 1)$$

and write it in the form

$$\nu_1\sigma^0(1, 0, 0) + \nu_2\sigma^0(0, 1, 0) + \nu_3\sigma^0(0, 0, 1) - \sigma^0(\nu_1, \nu_2, \nu_3)$$
$$\equiv \nu_1\sigma(x_{10}, x_{20}, x_{30}; 1, 0, 0) + \nu_2\sigma(x_{10}, x_{20}, x_{30}; 0, 1, 0)$$
$$+ \nu_3\sigma(x_{10}, x_{20}, x_{30}; 0, 0, 1) - \sigma(x_{10}, x_{20}, x_{30}; \nu_1, \nu_2, \nu_3) = 0. \tag{2.6}$$

We see that (2.6) differs from the required relation in Lemma 2.1 only by the notation. The difference is that the parameters n_i in Lemma 2.1 take any values such that $n_1^2 + n_2^2 + n_3^2 = 1$.

The above identities follow from the equalities

$$\sigma^0(-1, 0, 0) = -\sigma^0(1, 0, 0),$$
$$\sigma^0(0, -1, 0) = -\sigma^0(0, 1, 0),$$
$$\sigma^0(0, 0, -1) = -\sigma^0(0, 0, 1)$$

which, in turn, follow from the equality (2.1) and the continuity of σ^0. Thus, letting $\nu_1 \to +1$, $\nu_2 \to 0$, $\nu_3 \to 0$ ($\nu_1^2 + \nu_2^2 + \nu_3^2 = 1$) and passing to the limit, we find $\sigma^0(-1, 0, 0) + \sigma^0(1, 0, 0) = 0$. By the limit passage, we can get rid of the restrictions $n_i \neq 0$ ($\nu_i \neq 0$). The lemma is proved. □

2.2. The above arguments remain valid in the case of a moving medium under the action of a smoothly distributed body and surface forces provided that the velocity and acceleration of particles are bounded piecewise smooth continuous functions. Indeed, we consider a moving medium and repeat the above arguments with some slight modifications.

We cut a stressed moving medium at some time t along a plane $n_1x_1 + n_2x_2 + n_3x_3 = \text{const}$, $n_1^2 + n_2^2 + n_3^2 = 1$. Since this action should not affect the motion, we apply forces to both sides of the cut. Consider an area element ds having the normal (n_1, n_2, n_3) and passing through a point (x_1, x_2, x_3). The surface force applied to the lower side of the cut is $ds\mathbf{\Sigma}(x_1, x_2, x_3; n_1, n_2, n_3)$.

• The surface force (per unit area) is denoted by $\mathbf{\Sigma} = (\sigma_1, \sigma_2, \sigma_3)$ and is called the *stress vector* on this area element.

It turns out that *under rotations of the area element, the stress vector* $\mathbf{\Sigma}$ *is transformed in the same way as in the stationary case.* To justify this assertion, we use the momentum conservation law. We consider a region V occupied by the medium at some time t. Let S be the smooth boundary of V. Denote by $\rho = \rho(x_1, x_2, x_3, t)$ the density of the medium and by the components $u_i = u_i(x_1, x_2, x_3, t)$ of the velocity vector. Then the momentum of the mass in V is the vector with components ($i - 1, 2, 3$)

$$p_i(t) = \iiint\limits_{V(t)} \rho(x_1, x_2, x_3, t) u_i(x_1, x_2, x_3, t)\, dx_1 dx_2 dx_3,$$

where the region $V = V(t)$ changes with time since we deal with a moving medium.

The derivatives $\dfrac{dp_i(t)}{dt}$ are found from the acceleration

$$\frac{du_i}{dt} = \frac{\partial u_i}{\partial t} + u_1 \frac{\partial u_i}{\partial x_1} + u_2 \frac{\partial u_i}{\partial x_2} + u_3 \frac{\partial u_i}{\partial x_3}$$

of particles located in $V(t)$ by the formula

$$\frac{dp_i(t)}{dt} = \iiint\limits_{V(t)} \rho \frac{du_i}{dt}\, dx_1 dx_2 dx_3.$$

The total mass contained in $V(t)$,

$$m = \iiint\limits_{V(t)} \rho(x_1, x_2, x_3)\, dx_1 dx_2 dx_3,$$

is independent of time.

The time-derivative of the momentum is the vector $\frac{d\mathbf{p}}{dt}$ with components $\frac{dp_i}{dt}$ which is equal to the sum of the applied forces:

$$\iint\limits_{S(t)} \boldsymbol{\Sigma}(x_1, x_2, x_3; n_1, n_2, n_3, t)\, ds + \iiint\limits_{V(t)} \rho(x_1, x_2, x_3, t)\mathbf{F}(x_1, x_2, x_3, t)\, dx_1 dx_2 dx_3 = \frac{d\mathbf{p}(t)}{dt}$$

or, componentwise,

$$\iint\limits_{S(t)} \sigma_i(x_1, x_2, x_3; n_1, n_2, n_3, t)\, ds + \iiint\limits_{V(t)} \rho(x_1, x_2, x_3, t)\left[F_i(x_1, x_2, x_3, t) - \frac{du_i}{dt}\right] dx_1 dx_2 dx_3 = 0,$$

which differs from the corresponding equations in the stationary case (cf. (2.3) and (2.4)) only by the term $F_i - \frac{du_i}{dt}$. Arguing as above and using an assertion similar to Lemma 2.1, we can prove that the stress vector on an arbitrary area element with the unit normal (n_1, n_2, n_3) is determined by the stress vectors on area elements perpendicular to the coordinate axes by the formula

$$\boldsymbol{\Sigma} = \begin{pmatrix} \sigma_1 \\ \sigma_2 \\ \sigma_3 \end{pmatrix} = \begin{bmatrix} \sigma_{11} & \sigma_{12} & \sigma_{13} \\ \sigma_{21} & \sigma_{22} & \sigma_{23} \\ \sigma_{31} & \sigma_{32} & \sigma_{33} \end{bmatrix} \begin{pmatrix} n_1 \\ n_2 \\ n_3 \end{pmatrix}.$$

This elegant matrix equality shows that we can obtain the stress vector on any area element by applying the stress tensor to the unit normal (n_1, n_2, n_3) to this area element.

Thus, the stress tensor completely determines the stressed state at a point in the case of a moving medium.

Remark 2.3. It should be emphasized that the assumption that the accelerations are bounded is essential. For example, in the case of shock waves, the change of the velocity is momentary (accelerations are infinite!) and it is impossible to satisfy the condition (2.1) on the wave where the stress vector makes a jump.

2.3. To study the question if the stress tensor obtained by the described method is unique, we consider the integral equalities $(i = 1, 2, 3)$

$$\iint\limits_{S(t)} (\sigma_{i1}n_1 + \sigma_{i2}n_2 + \sigma_{i3}n_3)ds + \iiint\limits_{V(t)} \rho\left[F_i - \frac{du_i}{dt}\right] dx_1 dx_2 dx_3 = 0$$

and take into account the dependence of the stresses on the direction of ds. Since $V(t)$, together with $S(t)$, can be arbitrarily chosen, we can assume that for any fixed t these equalities hold in any integration domain. It is an elementary fact of calculus that for sufficiently smooth functions these equalities are equivalent to the differential equations

$$\frac{\partial \sigma_{i1}}{\partial x_1} + \frac{\partial \sigma_{i2}}{\partial x_2} + \frac{\partial \sigma_{i3}}{\partial x_3} = \rho \left(F_i - \frac{du_i}{dt} \right), \quad i = 1, 2, 3. \tag{2.7}$$

Then the question of the uniqueness of the stress tensor is reduced to the question of the existence of a unique solution to (2.7) with the boundary data $\sigma_{i1}n_1 + \sigma_{i2}n_2 + \sigma_{i3}n_3$, where n_1, n_2, n_3 are the components of the unit normal to the boundary. Let us show that the solution is not unique. It suffices to construct a nonzero solution to the homogeneous equations

$$\frac{\partial \sigma_{11}}{\partial x_1} + \frac{\partial \sigma_{12}}{\partial x_2} + \frac{\partial \sigma_{13}}{\partial x_3} = 0,$$
$$\frac{\partial \sigma_{21}}{\partial x_1} + \frac{\partial \sigma_{22}}{\partial x_2} + \frac{\partial \sigma_{23}}{\partial x_3} = 0,$$

$$\frac{\partial \sigma_{31}}{\partial x_1} + \frac{\partial \sigma_{32}}{\partial x_2} + \frac{\partial \sigma_{33}}{\partial x_3} = 0$$

with the boundary conditions

$$\sigma_{11}n_1 + \sigma_{12}n_2 + \sigma_{13}n_3 = 0,$$
$$\sigma_{21}n_1 + \sigma_{22}n_2 + \sigma_{23}n_3 = 0,$$
$$\sigma_{31}n_1 + \sigma_{32}n_2 + \sigma_{33}n_3 = 0.$$

For a domain we take the cube $-1 \leqslant x_1 \leqslant 1$, $-1 \leqslant x_2 \leqslant 1$, $-1 \leqslant x_3 \leqslant 1$ and consider the boundary conditions

$$\sigma_{11} = \sigma_{21} = \sigma_{31} = 0 \text{ for } x_1 = \pm 1,$$
$$\sigma_{12} = \sigma_{22} = \sigma_{32} = 0 \text{ for } x_2 = \pm 1,$$
$$\sigma_{13} = \sigma_{23} = \sigma_{33} = 0 \text{ for } x_3 = \pm 1.$$

It is easy to verify that the problem has the nonzero solution

$$\sigma_{11} = \cos \pi x_1 \cos \pi x_2 + \cos \pi x_2,$$
$$\sigma_{12} = \sigma_{21} = \sin \pi x_1 \sin \pi x_2,$$
$$\sigma_{22} = \cos \pi x_1 \cos \pi x_2 + \cos \pi x_1,$$
$$\sigma_{33} = \sigma_{13} = \sigma_{31} = \sigma_{23} = \sigma_{32} = 0.$$

This example shows that some additional conditions should be imposed in order to determine the stresses in a unique way.

2.4. We derive this important property of the stress tensor $\|\sigma_{ij}\|$ from the moment of momentum conservation law under some assumption (cf. Postulate 2.2 below).

• The *moment of momentum* of a continuous medium contained in a region V is the integral

$$\iiint\limits_V \mathbf{x} \times (\rho \mathbf{u})\, dx_1 dx_2 dx_3 \tag{2.8}$$

of the vector product of the position vector $\mathbf{x} = (x_1, x_2, x_3)$ and the momentum $\rho\mathbf{u} = (\rho u_1, \rho u_2, \rho u_3)$ (per unit volume).

In this definition, the moment is taken relative to the origin. If we need the moment relative to some other point $\mathbf{x}_0 = (x_{10}, x_{20}, x_{30})$, the following integral should be used instead of (2.8):

$$\iiint\limits_V (\mathbf{x} - \mathbf{x}_0) \times (\rho \mathbf{u})\, dx_1 dx_2 dx_3$$

or, componentwise,

$$\iiint\limits_V [(x_2 - x_{20})u_3 - (x_3 - x_{30})u_2]\rho\, dx_1 dx_2 dx_3,$$

$$\iiint\limits_V [(x_3 - x_{30})u_1 - (x_1 - x_{10})u_3]\rho\, dx_1 dx_2 dx_3,$$

$$\iiint\limits_V [(x_1 - x_{10})u_2 - (x_2 - x_{20})u_1]\rho\, dx_1 dx_2 dx_3.$$

The role of $\mathbf{x}_0$ is often played by the center of gravity of the mass.

Postulate 2.2. *The change of the moment of momentum with time may be generated only by the moment of applied forces.*

Remark 2.4. We will follow Postulate 2.2. However, it should be noted that in many interesting cases it is necessary to take into account the change of the moment caused by an "implicit" moment of momentum which is not expressed by the integral (2.8). One of such examples is a medium consisting of small cells such that each cell contains a small spinning top. We do not take them into account in the study of the medium. But if the spinning tops are decelerated, they transfer their moment of momentum to the corresponding cells. Then we observe the appearance of the moment of momentum caused by this fact. Hence such a medium contains some sources of the moment of momentum that are not covered by the moment of external forces acting on the medium. As a rule, such sources should be taken into account if we study processes in media of complicated microstructure. One of such examples is the demagnetization caused by heating. If a magnet is suspended over some heat sources, it starts to rotate in the process of demagnetization.

Media with implicit sources of the moment of momentum have been studied in continuum mechanics recently. In the classical contexts, such sources are not taken into account. We follow the classical approach and consider only the moments of external forces, which is fixed by Postulate 2.2. However, the reader should keep in mind this restriction in order to understand the area of application of the theory developed in this book. Since we ignore implicit sources of moment distributed in the medium, our study does not cover processes with fast magnetization or demagnetization.

In the classical case, for any fixed point $\mathbf{x}_0 = (x_{10}, x_{20}, x_{30})$ and moving region $V(t)$ the equation for the moment of momentum is as follows:

$$\frac{d}{dt}\iiint\limits_{V(t)} (\mathbf{x}-\mathbf{x}_0)\times(\rho\mathbf{u})\,dx_1dx_2dx_3$$
$$= \iint\limits_{S(t)} (\mathbf{x}-\mathbf{x}_0)\times\boldsymbol{\Sigma}\,ds + \iiint\limits_{V(t)} (\mathbf{x}-\mathbf{x}_0)\times(\rho\mathbf{F})\,dx_1dx_2dx_3. \tag{2.9}$$

The double integral on the right-hand side of (2.9) expresses the moment of the stress vector $\boldsymbol{\Sigma} = \boldsymbol{\Sigma}(x_1, x_2, x_3; n_1, n_2, n_3)$, on the boundary $S(t)$ of $V(t)$, whereas the triple integral is the moment of the body force acting in $V(t)$.

We consider Eq. (2.9) for a sequence $V_1(t)$, $V_2(t), \ldots,$ $V_N(t), \ldots,$ where each region at some time $t = t^*$ is a cube $V_N(t^*)$ with edge h_N, where $h_N \to 0$ as $N \to \infty$. The cubes contract to a point (x_{10}, x_{20}, x_{30}) as $N \to \infty$. For example, for $V_N(t^*)$ we can take the cube $x_{10} \leqslant x_1 \leqslant x_{10} + h_N$, $x_{20} \leqslant x_2 \leqslant x_{20} + h_N$, $x_{30} \leqslant x_3 \leqslant x_{30} + h_N$. If the velocity, acceleration, density, body forces, and stresses are bounded smooth functions, then

$$\iiint\limits_{V_N(t^*)} (\mathbf{x}-\mathbf{x}_0)\times\mathbf{F}\,dx_1dx_2dx_3 = O(h_N^4),$$

$$\left[\frac{d}{dt}\iiint\limits_{V_N(t)} (\mathbf{x}-\mathbf{x}_0)\times(\rho\mathbf{u})\,dx_1dx_2dx_3\right]_{t=t^*} = O(h_N^4).$$

Hence (2.9) implies that

$$\iint\limits_{S_N} (\mathbf{x}-\mathbf{x}_0)\times\boldsymbol{\Sigma}\,ds = O(h_N^4)$$

at $t = t^*$, where S_N is the boundary surface of the cube $x_{i0} \leqslant x_i \leqslant x_{i0} + h_N$.

To prove the required property of the stress tensor, we analyze the integral (the moment of stresses) in more detail:

$$\iint_{S_N} (\mathbf{x} - \mathbf{x}_0) \times \mathbf{\Sigma}\, ds.$$

We first compute the moment of forces applied to the pair of faces $x_3 = x_{30}+h_N$ and $x_3 = x_{30}$. On the face $x_3 = x_{30}+h_N$, the stress vector is expressed as follows:

$$\mathbf{\Sigma} = \begin{pmatrix} \sigma_{13} \\ \sigma_{23} \\ \sigma_{33} \end{pmatrix} = \begin{pmatrix} \sigma_{13}^0 \\ \sigma_{23}^0 \\ \sigma_{33}^0 \end{pmatrix} + O(h_N), \quad \sigma_{ij}^0 = \sigma_{ij}(x_{10}, x_{20}, x_{30}, t^*),$$

and the position vector of the point with coordinates x_1, x_2, x_3 relative to the point x_{10}, x_{20}, x_{30} takes the form

$$\mathbf{x} - \mathbf{x}_0 = \begin{pmatrix} x_1 - x_{10} \\ x_2 - x_{20} \\ h_N \end{pmatrix}.$$

Integrating over the face the vector product

$$(\mathbf{x} - \mathbf{x}_0) \times \mathbf{\Sigma} = \begin{pmatrix} (x_2 - x_{20})\sigma_{33}^0 - h_N\sigma_{23}^0 \\ h_N\sigma_{13}^0 - (x_1 - x_{10})\sigma_{33}^0 \\ (x_1 - x_{10})\sigma_{23}^0 - (x_2 - x_{20})\sigma_{13}^0 \end{pmatrix} + O(h_N^2),$$

we find

$$\int_{x_{10}}^{x_{10}+h_N} \int_{x_{20}}^{x_{20}+h_N} (\mathbf{x} - \mathbf{x}_0) \times \mathbf{\Sigma}\, dx_1 dx_2 = h_N^3 \begin{pmatrix} \frac{1}{2}\sigma_{33}^0 - \sigma_{23}^0 \\ \sigma_{13}^0 - \frac{1}{2}\sigma_{33}^0 \\ \frac{1}{2}\sigma_{23}^0 - \frac{1}{2}\sigma_{13}^0 \end{pmatrix} + O(h_N^4).$$

On the face $x_3 = x_{30}$, the vector product of the stress vector $\mathbf{\Sigma} = \begin{pmatrix} -\sigma_{13}^0 \\ -\sigma_{23}^0 \\ -\sigma_{33}^0 \end{pmatrix} +$ $O(h_N)$ and the position vector $\mathbf{x} - \mathbf{x}_0 = \begin{pmatrix} x_1 - x_{10} \\ x_2 - x_{20} \\ 0 \end{pmatrix}$ yields

$$(\mathbf{x} - \mathbf{x}_0) \times \mathbf{\Sigma} = \begin{pmatrix} -(x_2 - x_{20})\sigma_{33}^0 + 0 \cdot \sigma_{23}^0 \\ -0 \cdot \sigma_{13}^0 + (x_1 - x_{10})\sigma_{33}^0 \\ -(x_1 - x_{10})\sigma_{23}^0 + (x_2 - x_{20})\sigma_{13}^0 \end{pmatrix} + O(h_N^2).$$

Integrating the last equality, we get

$$\int_{x_{10}}^{x_{10}+h_N}\int_{x_{20}}^{x_{20}+h_N} (\mathbf{x}-\mathbf{x}_0)\times\mathbf{\Sigma}\,dx_1dx_2$$

$$= h_N^3\begin{pmatrix} -\frac{1}{2}\sigma_{33}^0 + 0\cdot\sigma_{23}^0 \\ -0\cdot\sigma_{13}^0 + \frac{1}{2}\sigma_{33}^0 \\ -\frac{1}{2}\sigma_{23}^0 + \frac{1}{2}\sigma_{13}^0 \end{pmatrix} + O(h_N^4) = h_N^3\begin{pmatrix} -\frac{1}{2}\sigma_{33}^0 \\ +\frac{1}{2}\sigma_{33}^0 \\ -\frac{1}{2}\sigma_{23}^0 + \frac{1}{2}\sigma_{13}^0 \end{pmatrix} + O(h_N^4).$$

Adding the moments over the faces $x_3 = x_{30} + h_N$ and $x_3 = x_{30}$, we obtain the total moment over the pair of faces:

$$h_N^3\begin{pmatrix} -\sigma_{23}^0 \\ \sigma_{13}^0 \\ 0 \end{pmatrix} + O(h_N^4). \tag{2.10}$$

In the same way, we find the moments over the pairs $x_1 = x_{10}$, $x_1 = x_{10}+h_N$ and $x_2 = x_{20}$, $x_2 = x_{20} + h_N$:

$$h_N^3\begin{pmatrix} 0 \\ -\sigma_{31}^0 \\ \sigma_{21}^0 \end{pmatrix} + O(h_N^4), \qquad h_N^3\begin{pmatrix} \sigma_{32}^0 \\ 0 \\ -\sigma_{12}^0 \end{pmatrix} + O(h_N^4). \tag{2.11}$$

The total moment of stresses on the boundary of the cube is obtained by adding (2.10) and (2.11):

$$\iint_{S_N} (\mathbf{x}-\mathbf{x}_0)\times\mathbf{\Sigma}\,ds = h_N^3\begin{pmatrix} \sigma_{32}^0 - \sigma_{23}^0 \\ \sigma_{13}^0 - \sigma_{31}^0 \\ \sigma_{21}^0 - \sigma_{12}^0 \end{pmatrix} + O(h_N^4).$$

On the other hand, we know that the total moment is of order $O(h_N^4)$, which can happen only if

$$\sigma_{32}^0 - \sigma_{23}^0 = \sigma_{32}(x_{10}, x_{20}, x_{30}, t^*) - \sigma_{23}(x_{10}, x_{20}, x_{30}, t^*) = 0,$$
$$\sigma_{13}^0 - \sigma_{31}^0 = \sigma_{13}(x_{10}, x_{20}, x_{30}, t^*) - \sigma_{31}(x_{10}, x_{20}, x_{30}, t^*) = 0,$$
$$\sigma_{21}^0 - \sigma_{12}^0 = \sigma_{21}(x_{10}, x_{20}, x_{30}, t^*) - \sigma_{12}(x_{10}, x_{20}, x_{30}, t^*) = 0.$$

Since x_{10}, x_{20}, x_{30}, and t^* are arbitrary, the equalities

$$\sigma_{32} - \sigma_{23} = \sigma_{13} - \sigma_{31} = \sigma_{21} - \sigma_{12} = 0$$

are proved for all x, t. Thus we can make the following conclusion.

- The stress tensor $||\sigma_{ij}||$ is symmetric: $\sigma_{ik} = \sigma_{ki}$.

We again emphasize that the symmetry of the stress tensor was established under the assumption expressed in Postulate 2.2.

§ 3. Stressed State at a Point

In this section, we analyze the stressed state of a medium at a point. We formally use the conclusions of Section 2 which are based on reasonable physical arguments.

3.1. It turns out that the stress tensor is transformed by the same rule as (1.21).

Suppose that we are given the array $\|\sigma_{ik}(0,0,0)\| \equiv \|\sigma_{ik}\|$ characterizing the stressed state of the medium at some point (x_1, x_2, x_3) (without loss of generality, we assume that this point is the origin $x_1 = 0$, $x_2 = 0$, $x_3 = 0$). Making the change of coordinates

$$\mathbf{x}' = \begin{pmatrix} x_1' \\ x_2' \\ x_3' \end{pmatrix} = W \begin{pmatrix} x_1 \\ x_2 \\ x_3 \end{pmatrix} = W\mathbf{x}, \quad WW^* = I, \tag{3.1}$$

where W is some matrix, we can characterize the stressed state of the medium at the point $x_1' = 0$, $x_2' = 0$, $x_3' = 0$ by the array $\|\sigma'_{pq}(0,0,0)\| \equiv \|\sigma'_{pq}\|$. We indicate how to determine this array from $\|\sigma_{ik}\|$ and W. We consider the unit vector $\mathbf{x}^{(1)} = (x_1^{(1)}, x_2^{(1)}, x_3^{(1)})$ which is transformed into the unit vector $\mathbf{e}^{(1)}$ directed along the x_1'-axis:

$$\mathbf{e}^{(1)} \equiv \begin{pmatrix} 1 \\ 0 \\ 0 \end{pmatrix} = W \begin{pmatrix} x_1^{(1)} \\ x_2^{(1)} \\ x_3^{(1)} \end{pmatrix} = W\mathbf{x}^{(1)}. \tag{3.2}$$

From (3.2) we find

$$\mathbf{x}^{(1)} = W^*\mathbf{e}^{(1)}. \tag{3.3}$$

Similarly, we find the vectors $\mathbf{x}^{(2)} = (x_1^{(2)}, x_2^{(2)}, x_3^{(2)})$ and $\mathbf{x}^{(3)} = (x_1^{(3)}, x_2^{(3)}, x_3^{(3)})$ which go to the unit vectors $\mathbf{e}^{(2)} = (0,1,0)$ and $\mathbf{e}^{(3)} = (0,0,1)$ directed along the x_2'-axis and the x_3'-axis respectively:

$$\mathbf{x}^{(2)} = W^*\mathbf{e}^{(2)}, \quad \mathbf{x}^{(3)} = W^*\mathbf{e}^{(3)}. \tag{3.4}$$

Since we know the "old" coordinates of the vectors $\mathbf{e}^{(q)}$, $q = 1,2,3$, we can find the stress vectors $\mathbf{\Sigma}^{(q)}$, $q = 1,2,3$, on the orthogonal area elements:

$$\mathbf{\Sigma}^{(q)} = \begin{pmatrix} \sigma_1^q \\ \sigma_2^q \\ \sigma_3^q \end{pmatrix} = \begin{bmatrix} \sigma_{11} & \sigma_{12} & \sigma_{13} \\ \sigma_{21} & \sigma_{22} & \sigma_{23} \\ \sigma_{31} & \sigma_{32} & \sigma_{33} \end{bmatrix} W^*\mathbf{e}^{(q)} = \|\sigma_{ij}\| W^*\mathbf{e}^{(q)}. \tag{3.5}$$

Then we find σ'_{pq} as the projections of the vector $\mathbf{\Sigma}^{(q)}$ on the pth unit vector $W^*\mathbf{e}^{(p)}$ by the formula

$$\sigma'_{pq} = \left(\|\sigma_{ij}\| W^*\mathbf{e}^{(q)}, W^*\mathbf{e}^{(p)}\right) = \left(W\|\sigma_{ij}\| W^*\mathbf{e}^{(q)}, \mathbf{e}^{(p)}\right).$$

Hence

$$\|\sigma'_{pq}\| = W\|\sigma_{ij}\|W^*. \tag{3.6}$$

In the notation

$$W = \begin{bmatrix} w_{11} & w_{12} & w_{13} \\ w_{21} & w_{22} & w_{23} \\ w_{31} & w_{32} & w_{33} \end{bmatrix}, \quad W^* = \begin{bmatrix} w_{11} & w_{21} & w_{31} \\ w_{12} & w_{22} & w_{32} \\ w_{13} & w_{23} & w_{33} \end{bmatrix},$$

we can write (3.5) componentwise (cf. (1.21)):

$$\sigma'_{pq} = \sum_{i=1}^{3}\sum_{j=1}^{3} w_{pi}\sigma_{ij}w_{qj}. \tag{3.7}$$

We will use the rule (3.7) in the study of the stressed state of a medium at a point.

3.2. As is known from linear algebra, any symmetric matrix can be represented in the form (cf. also (1.8))

$$\|\sigma_{ij}\| = U \begin{bmatrix} s_1 & 0 & 0 \\ 0 & s_2 & 0 \\ 0 & 0 & s_3 \end{bmatrix} U^* \quad (UU^* - I)$$

where U is an orthogonal matrix. Making the change of variables (3.1) with an orthogonal matrix W, we see that the stress tensor $\|\sigma_{ij}\|$ takes the form

$$\|\sigma'_{ij}\| = W\|\sigma_{ij}\|W^* = WU \begin{bmatrix} s_1 & 0 & 0 \\ 0 & s_2 & 0 \\ 0 & 0 & s_3 \end{bmatrix} U^*W^*.$$

Setting $W = U^*$, we thereby pass to coordinates in which the stress tensor has the following simple form:

$$\|\sigma'_{ij}\| = U^*U \begin{bmatrix} s_1 & 0 & 0 \\ 0 & s_2 & 0 \\ 0 & 0 & s_3 \end{bmatrix} U^*U = \begin{bmatrix} s_1 & 0 & 0 \\ 0 & s_2 & 0 \\ 0 & 0 & s_3 \end{bmatrix}. \tag{3.8}$$

• The passage from the stress tensor $\|\sigma_{ij}\|$ to the diagonal stress tensor $\|\sigma'_{ij}\|$ (cf. (3.8)) by the above procedure is called the *reduction of the stress tensor to the principal axes.*

Consider three area elements perpendicular to the principal axes. The corresponding stress vectors on these area elements are directed along the principal axes and their values are s_1, s_2, s_3 respectively.

• The numbers s_1, s_2, s_3 in (3.8) are called the *principal stresses* and can be found as the eigenvalues of the stress tensor $\|\sigma_{ij}\|$:

$$\begin{vmatrix} \sigma_{11}-s & \sigma_{12} & \sigma_{13} \\ \sigma_{21} & \sigma_{22}-s & \sigma_{23} \\ \sigma_{31} & \sigma_{32} & \sigma_{33}-s \end{vmatrix} = 0. \tag{3.9}$$

Remark 3.1. In the literature, the principal stresses are often denoted by σ_1, σ_2, σ_3. Recall that we use these letters for the components of the stress vector Σ on some area element.

3.3. Suppose that, at some point, the stress tensor is reduced to the principal axes and, consequently, takes the form

$$\|\sigma_{ik}\| = \begin{bmatrix} s_1 & 0 & 0 \\ 0 & s_2 & 0 \\ 0 & 0 & s_3 \end{bmatrix}. \tag{3.10}$$

Consider an arbitrary area element containing this point. Let $\mathbf{n} = (n_1, n_2, n_3)$, $n_1^2 + n_2^2 + n_3^2 = 1$, be the unit normal vector to this area element. Using (3.10), we can write the stress vector on this area element as follows:

$$\begin{aligned} \boldsymbol{\Sigma} &= \begin{pmatrix} \sigma_1 \\ \sigma_2 \\ \sigma_3 \end{pmatrix} = \begin{bmatrix} \sigma_{11} & \sigma_{12} & \sigma_{13} \\ \sigma_{21} & \sigma_{22} & \sigma_{23} \\ \sigma_{31} & \sigma_{32} & \sigma_{33} \end{bmatrix} \begin{pmatrix} n_1 \\ n_2 \\ n_3 \end{pmatrix} \\ &= \begin{bmatrix} s_1 & 0 & 0 \\ 0 & s_2 & 0 \\ 0 & 0 & s_3 \end{bmatrix} \begin{pmatrix} n_1 \\ n_2 \\ n_3 \end{pmatrix} = \begin{pmatrix} s_1 n_1 \\ s_2 n_2 \\ s_3 n_3 \end{pmatrix}. \end{aligned} \tag{3.11}$$

• The length of the projection of the stress vector Σ on the normal $\mathbf{n} = (n_1, n_2, n_3)$ to the area element is called the *normal stress*. The length of the projection of the stress vector Σ on the area element is called the *shear stress*.

It is obvious that

$$\sigma_{\mathrm{n}} = (\boldsymbol{\Sigma}, \mathbf{n}) = s_1 n_1^2 + s_2 n_2^2 + s_3 n_3^2, \tag{3.12}$$

$$\sigma_{\mathrm{n}}^2 + \sigma_{\mathrm{s}}^2 = (\boldsymbol{\Sigma}, \boldsymbol{\Sigma}) = s_1^2 n_1^2 + s_2^2 n_2^2 + s_3^2 n_3^2. \tag{3.13}$$

3.4. Thus, if we know the stress tensor at some point and fix an area element, we can find the principal stresses s_1, s_2, s_3, as well as the normal stress σ_{n} and the shear stress σ_{s} relative to this area element. The question arises: For given $s_1, s_2, s_3, \sigma_{\mathrm{n}}, \sigma_{\mathrm{s}}$, is it possible to find the corresponding area element, i.e., to determine the direction of the normal to it? We indicate the conditions under which this question has a positive answer.

By (3.12) and (3.13), we have the following system with respect to n_1^2, n_2^2, n_3^2:

$$\begin{aligned} &n_1^2 + n_2^2 + n_3^2 = 1, \\ &s_1 n_1^2 + s_2 n_2^2 + s_3 n_3^2 = \sigma_n, \\ &s_1^2 n_1^2 + s_2^2 n_2^2 + s_3^2 n_3^2 = \sigma_n^2 + \sigma_s^2. \end{aligned} \tag{3.14}$$

The Vandermonde determinant

$$\begin{vmatrix} 1 & 1 & 1 \\ s_1 & s_2 & s_3 \\ s_1^2 & s_2^2 & s_3^2 \end{vmatrix}$$

does not vanish if all the principal stresses s_1, s_2, s_3 are distinct. Solving the system (3.14), we find

$$\begin{aligned} n_1^2 &= \begin{bmatrix} 1 & 1 & 1 \\ \sigma_n & s_2 & s_3 \\ \sigma_n^2 + \sigma_s^2 & s_2^2 & s_3^2 \end{bmatrix} \Big\backslash \begin{bmatrix} 1 & 1 & 1 \\ s_1 & s_2 & s_3 \\ s_1^2 & s_2^2 & s_3^2 \end{bmatrix} \\ &= \frac{s_2 s_3 (s_3 - s_2) + \sigma_n (s_2^2 - s_3^2) + (\sigma_n^2 + \sigma_s^2)(s_3 - s_2)}{(s_1 - s_2)(s_1 - s_3)(s_3 - s_2)} \\ &= \frac{s_2 s_3 - \sigma_n (s_2 + s_3) + \sigma_n^2 + \sigma_s^2}{(s_1 - s_2)(s_1 - s_3)} \\ &= \frac{\left(\sigma_n - \dfrac{s_2 + s_3}{2}\right)^2 + \sigma_s^2 - \left(\dfrac{s_2 - s_3}{2}\right)^2}{(s_1 - s_2)(s_1 - s_3)}. \end{aligned}$$

Computing n_2^2 and n_3^2 in a similar way, we obtain the system

$$\begin{aligned} n_1^2 &= \frac{\left(\sigma_n - \dfrac{s_2 + s_3}{2}\right)^2 + \sigma_s^2 - \left(\dfrac{s_2 - s_3}{2}\right)^2}{(s_1 - s_2)(s_1 - s_3)}, \\ n_2^2 &= \frac{\left(\sigma_n - \dfrac{s_3 + s_1}{2}\right)^2 + \sigma_s^2 - \left(\dfrac{s_3 - s_1}{2}\right)^2}{(s_2 - s_3)(s_2 - s_1)}, \\ n_3^2 &= \frac{\left(\sigma_n - \dfrac{s_1 + s_2}{2}\right)^2 + \sigma_s^2 - \left(\dfrac{s_1 - s_2}{2}\right)^2}{(s_3 - s_1)(s_3 - s_2)}. \end{aligned} \tag{3.15}$$

Assume that the axes are enumerated in such a way that $s_3 > s_2 > s_1$. Let us verify if formulas (3.15) yield nonnegative numbers n_1^2, n_2^2, n_3^2. Since $(s_1 -$

$s_2)(s_1 - s_3) > 0$, $(s_2 - s_3)(s_2 - s_1) < 0$, $(s_3 - s_1)(s_3 - s_2) > 0$, it is necessary and sufficient that

$$\left(\sigma_n - \frac{s_2 + s_3}{2}\right)^2 + \sigma_s^2 \geqslant \left(\frac{s_2 - s_3}{2}\right)^2,$$
$$\left(\sigma_n - \frac{s_3 + s_1}{2}\right)^2 + \sigma_s^2 \leqslant \left(\frac{s_3 - s_1}{2}\right)^2, \tag{3.16}$$
$$\left(\sigma_n - \frac{s_1 + s_2}{2}\right)^2 + \sigma_s^2 \geqslant \left(\frac{s_1 - s_2}{2}\right)^2.$$

These inequalities indicate the range of σ_n and σ_s.

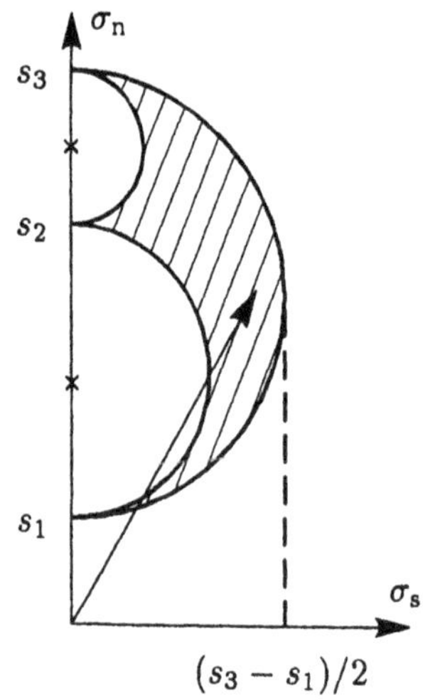

Fig. 2

3.5. In Fig. 2, the admissible domain is located between half-circles whose diameters are the segments $[s_1, s_2]$, $[s_2, s_3]$, $[s_1, s_3]$ of the σ_n-axes. This picture is called "Mohr's circles." It is seen that the maximal shear stress is equal to $(s_3 - s_1)/2$, whereas the normal stress is equal to $(s_3 + s_1)/2$. To determine an area element on which the shear stress is maximal, for given $\sigma_s = (s_3 - s_1)/2$ and $\sigma_n = (s_3 + s_1)/2$ we find n_1^2, n_2^2, n_3^2 according to formulas (3.15). We have $n_1^2 = 1/2$, $n_2^2 = 0$, $n_3^2 = 1/2$. Thus, the shear stress on area elements with the normal $\mathbf{n} = (n_1, n_2, n_3)$, $n_1 = \pm 1/\sqrt{2}$, $n_2 = 0$, $n_3 = \pm(\mp)1/\sqrt{2}$, is maximal, i.e., the shear stress is maximal on the planes bisecting the angle between the planes corresponding to the maximal and minimal stresses s_1 and s_3.

3.6. The matrix characterizing the stressed state is represented as the sum of two matrices (a multiple of the identity matrix and a matrix with zero *trace* (the sum of diagonal entries), called the *deviator* of the stress tensor):

$$\begin{bmatrix} \sigma_{11} & \sigma_{12} & \sigma_{13} \\ \sigma_{21} & \sigma_{22} & \sigma_{23} \\ \sigma_{31} & \sigma_{32} & \sigma_{33} \end{bmatrix} = \sigma \begin{bmatrix} 1 & 0 & 0 \\ 0 & 1 & 0 \\ 0 & 0 & 1 \end{bmatrix} + \begin{bmatrix} \sigma_{11}-\sigma & \sigma_{12} & \sigma_{13} \\ \sigma_{21} & \sigma_{22}-\sigma & \sigma_{23} \\ \sigma_{31} & \sigma_{32} & \sigma_{33}-\sigma \end{bmatrix}, \tag{3.17}$$

where $\sigma = (\sigma_{11} + \sigma_{22} + \sigma_{33})/3$ is called the *mean stress.* If

$$\|\sigma_{ik}\| = U \begin{bmatrix} s_1 & 0 & 0 \\ 0 & s_2 & 0 \\ 0 & 0 & s_3 \end{bmatrix} U^* \qquad (UU^* = I),$$

then

$$\left\| \sigma_{ik} - \delta_{ik} \frac{\sigma_{11} + \sigma_{22} + \sigma_{33}}{3} \right\|$$
$$= U \begin{bmatrix} s_1 - \dfrac{s_1+s_2+s_3}{3} & 0 & 0 \\ 0 & s_2 - \dfrac{s_1+s_2+s_3}{3} & 0 \\ 0 & 0 & s_3 - \dfrac{s_1+s_2+s_3}{3} \end{bmatrix} U^*. \tag{3.18}$$

We note that the shear stress remains unchanged if all the diagonal entries of $\|\sigma_{ik}\|$ simultaneously increase or decrease by the same number. This fact is obvious in view of Mohr's circles. Thus, the shear stresses can be characterized by the deviator of the stress tensor.

3.7. Consider the stress vector Σ on the area element perpendicular to some vector $\mathbf{x}$:

$$\mathbf{\Sigma} = \begin{pmatrix} \sigma_1 \\ \sigma_2 \\ \sigma_3 \end{pmatrix} = \begin{bmatrix} \sigma_{11} & \sigma_{12} & \sigma_{13} \\ \sigma_{21} & \sigma_{22} & \sigma_{23} \\ \sigma_{31} & \sigma_{32} & \sigma_{33} \end{bmatrix} \begin{pmatrix} \dfrac{x_1}{\sqrt{x_1^2+x_2^2+x_3^2}} \\ \dfrac{x_2}{\sqrt{x_1^2+x_2^2+x_3^2}} \\ \dfrac{x_3}{\sqrt{x_1^2+x_2^2+x_3^2}} \end{pmatrix}.$$

The projection of Σ on a vector $\mathbf{y}$ can be written as follows:

$$\frac{1}{\sqrt{y_1^2+y_2^2+y_3^2}}(\mathbf{\Sigma}, \mathbf{y}) = \frac{\sigma_i y_i}{\sqrt{y_1^2+y_2^2+y_3^2}} = \frac{(\sigma_{ij}x_j)y_i}{\sqrt{x_1^2+x_2^2+x_3^2}\sqrt{y_1^2+y_2^2+y_3^2}}$$
$$= \frac{\sigma_{ij}y_i x_j}{\sqrt{x_1^2+x_2^2+x_3^2}\sqrt{y_1^2+y_2^2+y_3^2}}.$$

Since $\sigma_{ij} = \sigma_{ji}$, this expression is symmetric with respect to $\mathbf{y}$ and $\mathbf{x}$.

• *The projection on a vector* $\mathbf{y}$ *of the stress vector acting on an area element perpendicular to a vector* $\mathbf{x}$ *is equal to the projection on the vector* $\mathbf{x}$ *of the stress vector on the area element perpendicular to the vector* $\mathbf{y}$.

3.8. The bilinear form $\sigma_{ij}y_i x_j$ determines the quadratic form $\sigma_{ij}x_i x_j$ which, in turn, determines the bilinear form

$$\sigma_{ij}x_i y_j = \sigma_{ij}y_i x_j = \frac{1}{2}\left[\sigma_{ij}(x_i + y_i)(x_j + y_j) - \sigma_{ij}x_i x_j - \sigma_{ij}y_i y_j\right].$$

Since the bilinear form $\sigma_{ij}y_i x_j$ is invariant under any orthogonal change of coordinates, the same is true for the quadratic form $\sigma_{ij}x_i x_j$. To distinguish the deviator is equivalent to the representation

$$\sigma_{ij}x_i x_j = \frac{\sigma_{11} + \sigma_{22} + \sigma_{33}}{3}(x_1^2 + x_2^2 + x_3^2) + \left(\sigma_{ij} - \delta_{ij}\frac{\sigma_{11} + \sigma_{22} + \sigma_{33}}{3}\right) x_i x_j$$

which is also invariant.

We note that the "deviator polynomial"

$$P = \left(\sigma_{ij} - \delta_{ij}\frac{\sigma_{11} + \sigma_{22} + \sigma_{33}}{3}\right) x_i x_j$$

is harmonic, i.e., satisfies the *Laplace equation*

$$\Delta P = \frac{\partial^2 P}{\partial x_1^2} + \frac{\partial^2 P}{\partial x_2^2} + \frac{\partial^2 P}{\partial x_3^2} = 0.$$

Let the polynomial of the second degree $Q(x_1, x_2, x_3) = \sigma_{ij}x_i x_j$ be formed by the components of the stress tensor σ_{ij}. Then

$$\Delta Q = \frac{\partial^2 Q}{\partial x_1^2} + \frac{\partial^2 Q}{\partial x_2^2} + \frac{\partial^2 Q}{\partial x_3^2} = 2(\sigma_{11} + \sigma_{22} + \sigma_{33}).$$

With the deviator of the stress tensor we can associate the quadratic form P which can be expressed in terms of Q by the formula

$$P = Q - \frac{x_1^2 + x_2^2 + x_3^2}{6}\Delta Q = Q - \frac{1}{6}(x_1^2 + x_2^2 + x_3^2)\left(\frac{\partial^2 Q}{\partial x_1^2} + \frac{\partial^2 Q}{\partial x_2^2} + \frac{\partial^2 Q}{\partial x_3^2}\right),$$

which implies that the operator sending Q to P is invariant under rotations.

With each rotation we can associate a linear transformation of σ_{ij}. Then we say that σ_{ij} is transformed by some representation of the rotation group. The deviator $\sigma_{ij} - \delta_{ij}(\sigma_{11} + \sigma_{22} + \sigma_{33})/3$ and the sum $\sigma_{11} + \sigma_{22} + \sigma_{33}$ are transformed by some representation of this group. The sum $\sigma_{11} + \sigma_{22} + \sigma_{33}$ remains unchanged under the transformation. Thus, the tensor σ_{ij} is decomposed into two independent 1- and 5-dimensional representations so that the sum $\sigma_{11} + \sigma_{22} + \sigma_{33}$ is transformed by the 1-dimensional representation and the deviator is transformed according to the 5-dimensional representation. We note that the 5-dimensional representation cannot be decomposed into representations of smaller dimension, i.e., it is *irreducible.*

In order to characterize the shear stresses, it is sometimes useful to use orthogonal invariants. For such invariants we can take the eigenvalues $\varkappa_1$, $\varkappa_2$, $\varkappa_3$ of the deviator, i.e., solutions of the equation

$$\det \begin{bmatrix} \sigma_{11}-\sigma-\varkappa & \sigma_{12} & \sigma_{13} \\ \sigma_{21} & \sigma_{22}-\sigma-\varkappa & \sigma_{23} \\ \sigma_{31} & \sigma_{32} & \sigma_{33}-\sigma-\varkappa \end{bmatrix} = 0,$$

where $\sigma = (\sigma_{11}+\sigma_{22}+\sigma_{33})/3$, or symmetric functions of these solutions. Every symmetric function can be expressed in terms of the basic functions $\varkappa_1+\varkappa_2+\varkappa_3$, $\varkappa_1\varkappa_2+\varkappa_2\varkappa_3+\varkappa_3\varkappa_1$, $\varkappa_1\varkappa_2\varkappa_3$. It is easy to prove that $\varkappa_i = s_i - \dfrac{s_1+s_2+s_3}{3}$. Hence $\varkappa_1+\varkappa_2+\varkappa_3 = 0$ and there are only two independent invariants of the deviator: $\varkappa_1\varkappa_2+\varkappa_2\varkappa_3+\varkappa_3\varkappa_1$ and $\varkappa_1\varkappa_2\varkappa_3$.

As a rule, the intensity of the shear stresses is measured by either the expression

$$\sqrt{\frac{2}{3}(\varkappa_1^2+\varkappa_2^2+\varkappa_3^2)} = \sqrt{\frac{(s_2-s_3)^2+(s_3-s_1)^2+(s_1-s_2)^2}{6}}$$

or the maximal shear stress

$$\max\left|\frac{s_i-s_j}{2}\right| = \max\left|\frac{\varkappa_i-\varkappa_j}{2}\right|.$$

§ 4. Isotropic Elastic Medium, Equation of State, and the Murnaghan Formulas

In Section 1, we used the formal assertion that the stressed state of a continuous medium changes under deformations. This assertion was not formalized there because we had no means to describe stresses. Now, such a formalism is developed and we can describe connections between stresses and deformations.

4.1. Some additional difficulties arise in the case of an elastic medium because we should take into account the dependence of stresses on the thermal state of the medium, for example, on the temperature. The simplest example of an elastic medium is an ideal gas. In this case, the stress tensor is diagonal and is proportional to the identity matrix in any orthogonal coordinates:

$$\|\sigma_{ij}\| = \begin{bmatrix} \sigma & 0 & 0 \\ 0 & \sigma & 0 \\ 0 & 0 & \sigma \end{bmatrix} = \begin{bmatrix} -p & 0 & 0 \\ 0 & -p & 0 \\ 0 & 0 & -p \end{bmatrix} = -pI, \tag{4.1}$$

where p, called the *pressure*, depends on the density ρ and the temperature T of the medium.

As is known, the work dA produced under expansion (compression) of a gas element of mass Δm is expressed by the formula

$$dA = p(\rho, T)d(\Delta V) = \Delta m\, p(\rho, T)\, d\frac{1}{\rho}, \tag{4.2}$$

where $\Delta V = \dfrac{1}{\rho}\Delta m$.

In thermodynamics, the following assertion is postulated and can be justified on the basis of phenomenological arguments or statistical physics:

Postulate 4.1. *There exists a function $S = S(\rho, T)$ that remains unchanged during reversible adiabatic processes, i.e., when there is no heat transfer and there is no heat sources connected with the change of energy caused, for example, by chemical reactions or by the work of viscous stresses.*

We note that if a gas element receives some amount of heat dQ, then the change of $S(\rho, T)$ is proportional to dQ:

$$\Delta m\, dS = \frac{1}{\tau} dQ, \tag{4.3}$$

where $1/\tau$ depends only on the density ρ and the temperature T.

• The function $S = S(\rho, T)$ is called the *entropy*, and Postulate 4.1 is often referred to as the *entropy principle*.

The *first principle of thermodynamics* expressing the conservation of energy in thermal processes asserts that there exists a function $E = E(\rho, T)$, called the *specific internal energy* (per unit mass). If the volume of a gas element changes due to the heat influx dQ, then

$$\Delta m\, dE(\rho, T) = -dA + dQ, \tag{4.5}$$

i.e., in view of (4.2) and (4.3),

$$dE = -pd\left(\frac{1}{\rho}\right) + \tau dS. \tag{4.6}$$

Here, the work dA is taken with the sign "$-$" because it is done by a gas element upon the surrounding medium.

By the second principle of thermodynamics, admitting various but equivalent formulations, the entropy $S = S(\rho, T)$ can be replaced with any function $f(S) = f[S(\rho, T)]$ such that in the equality

$$dE = -p\, d\left(\frac{1}{\rho}\right) + \frac{\tau}{f'(S)} df(S) \tag{4.7}$$

the coefficient $\tau/f'(S)$ depends only on the temperature T. Such a choice can be justified on the basis of any formulation of the second principle of thermodynamics.[1)]

Therefore, we can regard $f(S)$ as the measure of entropy and $\tau/f'(S)$ as the measure of temperature. Replacing $f(S)$ with S and $\tau/f'(S)$ with T, we can write (4.7) in the form

$$dE = -p\,d\left(\frac{1}{\rho}\right) + TdS. \tag{4.8}$$

Let $E = E(\rho, S)$. Then we obtain the *thermodynamical identity*

$$dE(\rho, S) = \frac{p(\rho, S)}{\rho^2}d\rho + TdS, \tag{4.9}$$

from which we can find the dependence of the pressure p and temperature T on ρ and S:

$$p = p(\rho, S) = \rho^2 E_\rho(\rho, S), \quad T = T(\rho, S) = E_S(\rho, S). \tag{4.10}$$

Instead of the density ρ, the specific volume $V = 1/\rho$ is often used. Then (4.8) is written in the form

$$dE(V, S) = -p(V, S)dV + T(V, S)dS,$$

which implies (cf. (4.10))

$$p(V, S) = -E_V(V, S), \quad T(V, S) = E_S(V, S). \tag{4.11}$$

A gas with the equation of state $E = E(\rho, S)$ is the simplest example of an elastic medium. In this case, E depends only on the entropy S and the density ρ of the medium (ρ/ρ_0 is the seal of the medium under deformation and ρ_0 is the density before deformation). In particular, the change of energy E caused by a reversible deformation, i.e., $S = \text{const}$, is completely determined by ρ_0 and the value of the density ρ after deformation.

4.2. To describe an isotropic elastic medium in the general case, we accept the following postulate.

Postulate 4.2. *The specific energy E (per unit mass) and the entropy S are functions of parameters describing the medium such that* (cf. (4.5)):

$$\Delta m\, dE = dA + dQ, \quad dQ = \Delta m\, T\, dS, \tag{4.12}$$

where dA is the work done during a deformation of a mass element Δm and dQ is the amount of heat received by this element.

[1)] In Section 22, we will discuss the second principle of thermodynamics for gases and formulate it in the form of the well-posedness condition for the corresponding differential equations.

Here, dA is taken with the sign "+" because, unlike the case of gas, the work dA is done over the element but not by this element itself.

Below, we will introduce the notion of a state of a medium and choose parameters describing the state.

Let us consider an elastic medium, for example, a metal of density ρ_0 in a nonstressed state at some standard temperature T_0 (for example, the room temperature). We consider a cube whose edges are parallel to the coordinate axes and heat it without deformation. For this purpose, we place the cube in a box made of a rigid material so that no expansion of the cube is possible. Then a certain pressure appears on the walls of the cube in view of internal stresses. Since the cube contracts under cooling, we "paste together" the cube and the box. So no deformations of the cube occurs. If we heat the cube, the entropy and energy change:

$$dQ = \Delta m\, T\, dS, \quad dE\, \Delta m = dQ.$$

Since no deformations occur, it is natural to suppose that E depends only on the entropy S, i.e., $E = E(S)$. Then the temperature of the cube is determined by the entropy: $T = E_S(S)$. Moreover, the amount of heat depends only on the entropy S (or the temperature $T(S)$):

$$Q(S) = \Delta m \int_{S_0}^{S} T(S)\, dS = \Delta m[E(S) - E(S_0)], \tag{4.13}$$

where S_0 is the entropy of the medium in the nonstressed state at the standard temperature T_0. Sometimes (for example, in the case of an ideal gas), it is more convenient to consider, instead of the state at the standard temperature, the state of the medium with a certain standard pressure p_0 (for example, the atmospheric pressure).

- The *standard state* of a medium with entropy S is obtained from the nonstressed state of the medium at the standard temperature T_0 (respectively, the entropy S_0) by heating (or cooling) without deformation. For any S the density of the medium in the standard state is ρ_0.

Postulate 4.3. *The state of an elastic medium with the entropy S can be transformed into the standard state with the density ρ_0 by a reversible adiabatic deformation (i.e., S is a constant).*

Under this assumption, any state with the entropy S of an elastic medium is associated by means of a reversible deformation with the standard state with the entropy S and the density ρ_0 which, in turn, is associated by means of heating or cooling without deformation with the nonstressed state of the medium at the standard temperature T_0 (the entropy S_0).

• A medium is said to be *elastic* if the stressed state depends only on the entropy S and the linear transformation C connecting the standard state with the entropy S with the stressed state under consideration (cf. Section 1)

4.3. It is reasonable to regard the entries c_{ij} of the matrix C as parameters characterizing the state of an elastic medium and write the equation of state in the form

$$E = E(c_{11}, c_{12}, c_{13}, c_{21}, c_{22}, c_{23}, c_{31}, c_{32}, c_{33}, S). \tag{4.14}$$

Later we will specify the character of possible dependence of E on C and S using the fact that the matrix C describes not only deformations but also rotations of the medium. Since the energy E remains unchanged under rotations, it is natural to assume that, in the case of the identity matrix C, the energy $E(c_{11}, c_{12}, \dots, c_{33}, S) = E(1,0,0,0,1,0,0,0,1,S)$ coincides with the energy $E(S)$ relative to the standard state with entropy S. (Recall that the energy $E(S)$ is defined in terms of the amount of heat necessary for transformation of the unstressed state with the standard temperature T_0 (the entropy S_0) into the standard state with the entropy S.)

4.4. In an isotropic medium, the number of independent variables in the equation of state (4.14) can be decreased. As was shown in Section 1, each matrix C with positive determinant[2)] can be represented in the form

$$C = UKV, \quad UU^* = I, \quad VV^* = I, \quad K = \begin{bmatrix} k_1 & 0 & 0 \\ 0 & k_2 & 0 \\ 0 & 0 & k_3 \end{bmatrix}, \tag{4.15}$$

where $\det U = \det V = 1$. Hence the linear transformation C consists of the rotation V, the extension K along the coordinate axes, and the rotation U. The internal energy E changes only under the extension K and remains unchanged under rotations. The rotation V transforms some three orthogonal axes into the coordinate axes along which the extension K is performed. The rotation U transforms the coordinate axes into some other position. Thus, the transformation $C = UKV$ realizes an extension along some three orthogonal axes and a rotation.

In an isotropic medium, where all the directions are equal in rights, the change of energy E caused by deformations depends on the extension coefficients k_1, k_2, k_3 but not on the direction of the axes along which the extension is performed. Thus, if $C = UKV$ transforms the standard state of a medium with the entropy S into a state for which the stress tensor coincides with the stress

---

2) Transformations C with $\det C < 0$ are not suitable for describing deformations of elastic media because, in this case, there is no one-parameter continuous sequence of matrices $C(t)$ or, more exactly, a curve $C(t)$, $\det C(t) \neq 0$, in the space of matrices joining C with the identity matrix. If a matrix $C(t)$ is singular, at least one of k_i's vanishes.

tensor of the element Δm of the elastic medium, then with the stressed state we associate the specific internal energy E depending only on k_1, k_2, k_3, S, i.e., $E = E(k_1, k_2, k_3, S)$.

We emphasize that for parameters describing the stressed state we take the extension coefficients k_1, k_2, k_3, and the rotation matrix U but not the matrix V because, the standard state of an isotropic medium with the entropy S remains unchanged under any rotation V Indeed, a stressed stated is associated with two matrices

$$C^{(1)} = U \begin{bmatrix} k_1 & 0 & 0 \\ 0 & k_2 & 0 \\ 0 & 0 & k_3 \end{bmatrix} V^{(1)}, \quad C^{(2)} = U \begin{bmatrix} k_1 & 0 & 0 \\ 0 & k_2 & 0 \\ 0 & 0 & k_3 \end{bmatrix} V^{(2)},$$

which differ only by the right orthogonal factor $V^{(1)}$ or $V^{(2)}$, and are the same in an isotropic medium. That is why for parameters describing the stressed state we take only the orthogonal matrix U and the extension coefficients k_1, k_2, k_3.

• *In an isotropic elastic medium, the equation of state* (4.14) *takes the form*

$$E = E(k_1, k_2, k_3, S). \tag{4.16}$$

Since the coordinate axes are equal in rights, the function E is symmetric with respect to k_1, k_2, k_3, i.e., it remains the same under permutations of variables. By the entropy principle and the second principle of thermodynamics, we can assert the following.

• *If an element of a medium (of mass Δm) in the nonstressed state heats without deformation (i.e., k_1, k_2, k_3 are constant) then*

$$dQ = \Delta m\, T\, dS = \Delta m\, dE(k_1, k_2, k_3, S)\big|_{k_i = \text{const}}. \tag{4.17}$$

Note that, as a rule, the stressed state changes under heating even if k_1, k_2, k_3 are constant).

In other words, the derivative $E_S(k_1, k_2, k_3, S)$ can be identified with the temperature:

$$E_S(k_1, k_2, k_3, S) = T. \tag{4.18}$$

If the derivatives E_{k_1}, E_{k_2}, E_{k_3} are known, we can find the stresses in the deformed medium in the same way as we can find the pressure $p = \rho^2 E_\rho(\rho, S)$ from the derivative $E_\rho(\rho, S)$ in the case of an ideal gas.

4.5. Consider a medium in the standard state with the entropy S that occupies some region in the space (x_1, x_2, x_3) at a time $t = t_0$.

We make elementary experiments on adiabatic extension (contraction) along three mutually perpendicular axes with coefficients k_1, k_2, k_3. The case $k_i > 1$

corresponds to extension and the case $0 < k_i < 1$ corresponds to contraction. We temporarily assume that the extension axes coincide with the coordinate axes and the displacements of material particles are expressed as follows: $x_1 = k_1 x_{10}$, $x_2 = k_2 x_{20}$, $x_3 = k_3 x_{30}$, $k_i = k_i(t)$, $k_i(t_0) = 1$, $t_0 \leqslant t \leqslant t_1$. It is obvious that the density of the medium changes by the rule $\rho = \dfrac{\rho_0}{k_1 k_2 k_3}$ (ρ_0 is the density of the medium in the standard state). We also assume that the adiabaticity condition holds, i.e. the entropy S is constant.

Consider a cube $0 \leqslant x_{10} \leqslant h$, $0 \leqslant x_{20} \leqslant h$, $0 \leqslant x_{30} \leqslant h$ containing the mass $\rho_0 h^3$. We compute the work under deformation of this cube. During the time-interval (t_0, t), the energy E changes:

$$\begin{aligned}
&\rho_0 h^3 E[k_1(t), k_2(t), k_3(t), S] - \rho_0 h^3 E[k_1(t_0), k_2(t_0), k_3(t_0), S] \\
&\quad = \rho_0 h^3 E[k_1(t), k_2(t), k_3(t), S] - \rho_0 h^3 E(1, 1, 1, S).
\end{aligned}$$

If the extension coefficients change from k_i to $k_i + \delta k_i$, the work is expressed as the increment in energy of the cube:

$$\rho_0 h^3 \delta E = \rho_0 h^3 (E_{k_1} \delta k_1 + E_{k_2} \delta k_2 + E_{k_3} \delta k_3). \tag{4.19}$$

On the other hand, this work can be found from the stresses and displacements relative to the faces of the cube.

We consider *homogeneous deformations* of the cube, i.e., at each time t the matrix

$$C = \|c_{ij}\| = \left\| \frac{\partial x_i(x_{10}, x_{20}, x_{30}, t)}{\partial x_{j0}} \right\| = \begin{bmatrix} k_1(t) & 0 & 0 \\ 0 & k_2(t) & 0 \\ 0 & 0 & k_3(t) \end{bmatrix}$$

is independent of x_1, x_2, x_3 and the components σ_{ij} of the stress tensor are the same at all points of the cube ($\sigma_{ij} = \sigma_{ij}(t)$). Recall that the condition that σ_{ij} depend only on c_{ij} and S was postulated as the basic characteristic property of an elastic medium.

The faces of the parallelepiped $x_1 = k_1 h$, $x_1 = 0$ with the normals $(1, 0, 0)$, $(-1, 0, 0)$ are acted by the stress vectors with components

$$\begin{pmatrix} \sigma_{11} \\ \sigma_{21} \\ \sigma_{31} \end{pmatrix}, \quad -\begin{pmatrix} \sigma_{11} \\ \sigma_{21} \\ \sigma_{31} \end{pmatrix}.$$

These faces are obtained from the faces $x_{10} = h$ and $x_{10} = 0$ of the original cube by deformation. The faces $x_2 = k_2 h$, $x_2 = 0$ and $x_3 = k_3 h$, $x_3 = 0$ are acted by the stress vectors

$$\begin{pmatrix} \sigma_{12} \\ \sigma_{22} \\ \sigma_{32} \end{pmatrix}, \quad -\begin{pmatrix} \sigma_{12} \\ \sigma_{22} \\ \sigma_{32} \end{pmatrix}, \quad \begin{pmatrix} \sigma_{13} \\ \sigma_{23} \\ \sigma_{33} \end{pmatrix}, \quad -\begin{pmatrix} \sigma_{13} \\ \sigma_{23} \\ \sigma_{33} \end{pmatrix}.$$

On the faces $x_1 = k_1 h$ and $x_1 = 0$, we distinguish small area elements

$$\begin{cases} x_1 = k_1 h, \\ \overline{x}_2 \leqslant x_2 \leqslant \overline{x}_2 + \Delta x_2, \\ \overline{x}_3 \leqslant x_3 \leqslant \overline{x}_3 + \Delta x_3, \end{cases} \qquad \begin{cases} x_1 = 0, \\ \overline{x}_2 \leqslant x_2 \leqslant \overline{x}_2 + \Delta x_2, \\ \overline{x}_3 \leqslant x_3 \leqslant \overline{x}_3 + \Delta x_3 \end{cases}$$

and compute the forces acting on these elements:

$$\begin{pmatrix} \sigma_{11} \\ \sigma_{21} \\ \sigma_{31} \end{pmatrix} \Delta x_2 \Delta x_3, \qquad -\begin{pmatrix} \sigma_{11} \\ \sigma_{21} \\ \sigma_{31} \end{pmatrix} \Delta x_2 \Delta x_3.$$

Under the variations δk_1, δk_2, δk_3, the displacements of points of these area elements are as follows:

$$\delta x_1 = \delta k_1 h, \quad \delta x_1 = 0,$$
$$\delta x_2 = \frac{\delta k_2}{k_2}\overline{x}_2, \quad \delta x_2 = \frac{\delta k_2}{k_2}\overline{x}_2,$$
$$\delta x_3 = \frac{\delta k_3}{k_3}\overline{x}_3, \quad \delta x_3 = \frac{\delta k_3}{k_3}\overline{x}_3.$$

Then the work done by the stresses on these area elements is as follows:

$$\sigma_{11}\Delta x_2 \Delta x_3 \delta k_1 h + \sigma_{21}\Delta x_2 \Delta x_3 \frac{\delta k_2}{k_2}\overline{x}_2 + \sigma_{31}\Delta x_2 \Delta x_3 \frac{\delta k_3}{k_3}\overline{x}_3,$$
$$-\sigma_{11}\Delta x_2 \Delta x_3 0 - \sigma_{21}\Delta x_2 \Delta x_3 \frac{\delta k_2}{k_2}\overline{x}_2 - \sigma_{31}\Delta x_2 \Delta x_3 \frac{\delta k_3}{k_3}\overline{x}_3.$$

Hence the total work on both elements is equal to $\sigma_{11}\delta k_1 h \Delta x_2 \Delta x_3$. Adding the works on all area elements on thc faces $x_1 = k_1 h$, $x_1 = 0$, i.e., integrating over x_2 and x_3, we find

$$\int_0^{k_3 h}\int_0^{k_2 h} \sigma_{11}\delta k_1 h \, dx_2 dx_3 = \sigma_{11}\delta k_1 h (k_2 h)(k_3 h) = h^3 \sigma_{11} k_2 k_3 \delta k_1.$$

Similarly, the total work on the pair of faces $x_2 = k_2 h$ and $x_2 = 0$ is equal to $h^3 \sigma_{22} k_3 k_1 \delta k_2$ and the total work on. the faces $x_3 = k_3 h$ and $x_3 = 0$ is equal to $h^3 \sigma_{33} k_1 k_2 \delta k_3$. The total work done by the stresses under the variation δk_1, δk_2, δk_3 is expressed as follows:

$$h^3(\sigma_{11} k_2 k_3 \delta k_1 + \sigma_{22} k_3 k_1 \delta k_2 + \sigma_{33} k_1 k_2 \delta k_3). \tag{4.20}$$

This work should be equal to increment in the energy of the cube (4.19). Indeed, since the process is adiabatic, an increment in the energy can happen only due to the work. Since $\delta S = 0$. the last term in the formula $\delta E = E_{k_1}\delta k_1 + E_{k_2}\delta k_2 + E_{k_3}\delta k_3 + E_S \delta S$ can be omitted. From (4.19) and (4.20) we conclude that

$$\sigma_{11} k_2 k_3 \delta k_1 + \sigma_{22} k_3 k_1 \delta k_2 + \sigma_{33} k_1 k_2 \delta k_3 = \rho_0 E_{k_1}\delta k_1 + \rho_0 E_{k_2}\delta k_2 + \rho_0 E_{k_3}\delta k_3. \tag{4.21}$$

Since the variations δk_i are arbitrary, from (4.21) we can express some components of the stress tensor in terms of the derivatives of $E(k_1, k_2, k_3, S)$:

$$\sigma_{11} = \frac{\rho_0}{k_2 k_3} E_{k_1}, \quad \sigma_{22} = \frac{\rho_0}{k_3 k_1} E_{k_2}, \quad \sigma_{33} = \frac{\rho_0}{k_1 k_2} E_{k_3}. \tag{4.22}$$

Unfortunately, we have representations only for three components of the stress tensor. But what about the remaining components? It turns out that they vanish:

$$\sigma_{12} = \sigma_{21} = 0, \quad \sigma_{13} = \sigma_{31} = 0, \quad \sigma_{23} = \sigma_{32} = 0. \tag{4.23}$$

4.6. To prove the equalities $\sigma_{12} = \sigma_{21} = 0$, we try to find the energy needed for an adiabatic deformation of the cube $0 \leqslant x_{10} \leqslant h$, $0 \leqslant x_{20} \leqslant h$, $0 \leqslant x_{30} \leqslant h$ that is realized by the transformation

$$\begin{pmatrix} x_1 \\ x_2 \\ x_3 \end{pmatrix} = C \begin{pmatrix} x_{10} \\ x_{30} \\ x_{30} \end{pmatrix} = \begin{bmatrix} k_1 \cos\alpha & -k_1 \sin\alpha & 0 \\ k_2 \sin\alpha & k_2 \cos\alpha & 0 \\ 0 & 0 & k_3 \end{bmatrix} \begin{pmatrix} x_{10} \\ x_{30} \\ x_{30} \end{pmatrix}$$
$$= \begin{bmatrix} k_1 & 0 & 0 \\ 0 & k_2 & 0 \\ 0 & 0 & k_3 \end{bmatrix} \begin{bmatrix} \cos\alpha & -\sin\alpha & 0 \\ \sin\alpha & \cos\alpha & 0 \\ 0 & 0 & 1 \end{bmatrix} \begin{pmatrix} x_{10} \\ x_{30} \\ x_{30} \end{pmatrix}.$$

This transformation consists of the rotation by an angle α about the x_{30}-axis and the extension along the coordinate axes with coefficients k_1, k_2, k_3. Under such a rotation, the energy remains unchanged. The amount of energy expended for the extension (contraction) is equal to

$$\rho_0 h^3 \big[E(k_1, k_2, k_3, S) - E(1, 1, 1, S)\big].$$

The matrix C can be represented as the product

$$\begin{bmatrix} k_1 \cos\alpha & -k_1 \sin\alpha & 0 \\ k_2 \sin\alpha & k_2 \cos\alpha & 0 \\ 0 & 0 & k_3 \end{bmatrix} = \begin{bmatrix} \cos\alpha & -\dfrac{k_1}{k_2} \sin\alpha & 0 \\ \dfrac{k_2}{k_1} \sin\alpha & \cos\alpha & 0 \\ 0 & 0 & 1 \end{bmatrix} \begin{bmatrix} k_1 & 0 & 0 \\ 0 & k_2 & 0 \\ 0 & 0 & k_3 \end{bmatrix}.$$

Therefore, if the medium in the standard state with the entropy S is subject to the adiabatic extension along the coordinate axes with coefficients k_1, k_2, k_3, then the deformation is described by the matrix

$$\begin{bmatrix} \cos\alpha & -\dfrac{k_1}{k_2} \sin\alpha & 0 \\ \dfrac{k_2}{k_1} \sin\alpha & \cos\alpha & 0 \\ 0 & 0 & 1 \end{bmatrix} \tag{4.24}$$

no work is produced. We consider only such variations with an infinitesimal parameter α. Then the matrix (4.24) admits the approximation

$$\begin{bmatrix} 1 & -\frac{k_1}{k_2}\alpha & 0 \\ \frac{k_2}{k_1}\alpha & 1 & 0 \\ 0 & 0 & 1 \end{bmatrix} = \begin{bmatrix} \cos\alpha & -\frac{k_1}{k_2}\sin\alpha & 0 \\ \frac{k_2}{k_1}\sin\alpha & \cos\alpha & 0 \\ 0 & 0 & 1 \end{bmatrix} + O(\alpha^2). \tag{4.25}$$

Consider the parallelepiped $0 \leqslant x_1 \leqslant k_1 h$, $0 \leqslant x_2 \leqslant k_2 h$, $0 \leqslant x_3 \leqslant k_3 h$, obtained from the cube $0 \leqslant x_{i0} \leqslant h$ by the extension and apply to it the transformation with the infinitesimal parameter α:

$$x_1' = x_1 - \alpha\frac{k_1}{k_2}x_2, \quad x_2' = \alpha\frac{k_2}{k_1}x_1 + x_2, \quad x_3' = x_3.$$

Let us estimate the work of the stresses on the faces of the parallelepiped.

Points of the faces $x_3 = 0$ and $x_3 = k_3 h$ with the same coordinates x_2 and x_1 have the same displacements $\delta x_1 = -\alpha\frac{k_1}{k_2}x_2$, $\delta x_2 = \alpha\frac{k_2}{k_1}x_1$, $\delta x_3 = 0$, whereas the applied stresses are equal but point in the opposite directions,

$$\begin{pmatrix} \sigma_{13} \\ \sigma_{23} \\ \sigma_{33} \end{pmatrix} \text{ on } x_3 = k_3 h, \qquad \begin{pmatrix} -\sigma_{13} \\ -\sigma_{23} \\ -\sigma_{33} \end{pmatrix} \text{ on } x_3 = 0.$$

Therefore, the total work done by these stresses vanishes.

The situation is different on the pairs of faces $x_1 = k_1 h$, $x_1 = 0$ and $x_2 = k_2 h$, $x_2 = 0$. The points with coordinates x_2, x_3 on the faces $x_1 = k_1 h$ and $x_1 = 0$ have displacements

$$\begin{aligned} &\delta x_1 = -\alpha\frac{k_1}{k_2}x_2, \quad \delta x_1 = -\alpha\frac{k_1}{k_2}x_2, \\ &\delta x_2 = \alpha\frac{k_2}{k_1}k_1 h, \quad \delta x_2 = 0, \\ &\delta x_3 = 0, \quad \delta x_3 = 0. \end{aligned}$$

The stress vectors

$$\begin{pmatrix} \sigma_{11} \\ \sigma_{21} \\ \sigma_{31} \end{pmatrix}, \qquad \begin{pmatrix} -\sigma_{11} \\ -\sigma_{21} \\ -\sigma_{31} \end{pmatrix},$$

entropy S is subject to the transformation $C = UKV$, it is natural to assume that the stresses are independent of the rotation V. Indeed, we obtain the same stressed state from the standard state if use the transformation $C = UKV$ or $C = UKU^*$. The last transformation is more convenient, and we prefer to use it in the further consideration.

The transformation UKU^* consists of the rotation U^* sending some three orthogonal axes to the coordinate axes, the extension K along these axes, and the rotation U which is inverse to the first rotation U^*.

Thus, the transformation UKU^* is the extension with coefficients k_1, k_2, k_3 along the orthogonal axes obtained by rotation U of the coordinate axes. Taking the extension axes for the axes of some auxiliary coordinate system x_1', x_2', x_3', we obtain the array

$$\begin{bmatrix} \rho k_1 E_{k_1} & 0 & 0 \\ 0 & \rho k_2 E_{k_2} & 0 \\ 0 & 0 & \rho k_3 E_{k_3} \end{bmatrix}.$$

We can pass from the auxiliary coordinates x_1', x_2', x_3' to the original coordinates x_1, x_2, x_3 by the formulas

$$\begin{pmatrix} x_1 \\ x_2 \\ x_3 \end{pmatrix} = U \begin{pmatrix} x_1' \\ x_2' \\ x_3' \end{pmatrix}.$$

Using the rule of transformation of the stress tensor under rotations of coordinate axes (cf. Section 3) we make the following conclusion.

- *The stressed state is described in the coordinates x_1, x_2, x_3 by the stress tensor*

$$\begin{bmatrix} \sigma_{11} & \sigma_{12} & \sigma_{13} \\ \sigma_{21} & \sigma_{22} & \sigma_{23} \\ \sigma_{31} & \sigma_{32} & \sigma_{33} \end{bmatrix} = U \begin{bmatrix} \rho k_1 E_{k_1} & 0 & 0 \\ 0 & \rho k_2 E_{k_2} & 0 \\ 0 & 0 & \rho k_3 E_{k_3} \end{bmatrix} U^*. \tag{4.30}$$

This stressed state is obtained from the standard state by the transformation $C = UKV$.

For example, if E depends only on the entropy S and the density $\rho = \dfrac{\rho_0}{k_1 k_2 k_3}$, then

$$k_1 E_{k_1}\big|_{S=\text{const}} = k_1 \rho_{k_1} E_\rho\big|_{S=\text{const}} = -k_1 \frac{\rho_0}{k_1^2 k_2 k_3} E_\rho\big|_{S=\text{const}} = -\rho E_\rho,$$

$$k_1 E_{k_1} = k_2 E_{k_2} = k_3 E_{k_3} = -\rho E_\rho$$

and, consequently,

$$s_1 = s_2 = s_3 = \rho k_1 E_{k_1} = \rho k_2 E_{k_2} = \rho k_3 E_{k_3} = -\rho^2 E_\rho.$$

where

$$\begin{aligned} s_1 &= \frac{\rho_0}{k_2 k_3} E_{k_1}(k_1, k_2, k_3, S) = \frac{\rho_0}{k_1 k_2 k_3} k_1 E_{k_1} = \rho k_1 E_{k_1}(k_1, k_2, k_3, S), \\ s_2 &= \rho k_2 E_{k_2}(k_1, k_2, k_3, S), \\ s_3 &= \rho k_3 E_{k_3}(k_1, k_2, k_3, S). \end{aligned} \tag{4.27}$$

The temperature of the medium is determined by the formula

$$T = E_S(k_1, k_2, k_3, S). \tag{4.28}$$

Expressing dE from formulas (4.27) and (4.28), we obtain the *thermodynamical identity*

$$\begin{aligned} dE(k_1, k_2, k_3, S) &= E_{k_1} dk_1 + E_{k_2} dk_2 + E_{k_3} dk_3 + E_S dS \\ &= \frac{s_1}{\rho} \frac{dk_1}{k_1} + \frac{s_2}{\rho} \frac{dk_2}{k_2} + \frac{s_3}{\rho} \frac{dk_3}{k_3} + T dS \end{aligned} \tag{4.29}$$

which is similar to the thermodynamical identity (4.9) for an ideal gas.

Formulas (4.26)–(4.29) for the stresses in an elastic medium are called the *Murnaghan formulas*. They were first obtained by the American mathematician Murnaghan in 1937. Murnaghan also suggested a rule of computing stresses in the case of a nondiagonal matrix C. Now, we derive this rule.

If $k_1 = k_2 = k_3 = 1$, the stresses are found from the Murnaghan formulas:

$$\begin{bmatrix} \sigma_{11} & \sigma_{12} & \sigma_{13} \\ \sigma_{21} & \sigma_{22} & \sigma_{23} \\ \sigma_{31} & \sigma_{32} & \sigma_{33} \end{bmatrix} = \begin{bmatrix} \sigma & 0 & 0 \\ 0 & \sigma & 0 \\ 0 & 0 & \sigma \end{bmatrix},$$

$$\sigma = s_1 = \rho E_{k_1}(1,1,1,S) = s_2 = \rho E_{k_2}(1,1,1,S) = s_3 = \rho E_{k_3}(1,1,1,S).$$

(The principal stresses coincide because the medium is isotropic and, consequently, $E(k_1, k_2, k_3, S)$ is a symmetric function of k_1, k_2, k_3.)

The stress vector Σ on an area element with the unit normal $\mathbf{n} = (n_1, n_2, n_3)$ is found by the formula

$$\mathbf{\Sigma} = \begin{bmatrix} \sigma & 0 & 0 \\ 0 & \sigma & 0 \\ 0 & 0 & \sigma \end{bmatrix} \begin{pmatrix} n_1 \\ n_2 \\ n_3 \end{pmatrix} = \sigma \begin{pmatrix} n_1 \\ n_2 \\ n_3 \end{pmatrix}.$$

Note that the stress vector is always orthogonal to the area element on which it acts, and the modulus of this stress is independent of the direction.

• The stress tensor is said to be *spherical* if it is proportional to the identity matrix.

By definition, the stress tensor corresponding to the standard state of the medium with the entropy S is spherical. The state characterized by a spherical stress tensor has no distinguished directions. The state remains unchanged under any rotation V. Therefore, if a medium in the standard state with the entropy S is subject to the transformation $C = UKV$, it is natural to assume that the stresses are independent of the rotation V. Indeed, we obtain the same stressed state from the standard state if use the transformation $C = UKV$ or $C = UKU^*$. The last transformation is more convenient, and we prefer to use it in the further consideration.

The transformation UKU^* consists of the rotation U^* sending some three orthogonal axes to the coordinate axes, the extension K along these axes, and the rotation U which is inverse to the first rotation U^*.

Thus, the transformation UKU^* is the extension with coefficients k_1, k_2, k_3 along the orthogonal axes obtained by rotation U of the coordinate axes. Taking the extension axes for the axes of some auxiliary coordinate system x_1', x_2', x_3', we obtain the array

$$\begin{bmatrix} \rho k_1 E_{k_1} & 0 & 0 \\ 0 & \rho k_2 E_{k_2} & 0 \\ 0 & 0 & \rho k_3 E_{k_3} \end{bmatrix}.$$

We can pass from the auxiliary coordinates x_1', x_2', x_3' to the original coordinates x_1, x_2, x_3 by the formulas

$$\begin{pmatrix} x_1 \\ x_2 \\ x_3 \end{pmatrix} = U \begin{pmatrix} x_1' \\ x_2' \\ x_3' \end{pmatrix}.$$

Using the rule of transformation of the stress tensor under rotations of coordinate axes (cf. Section 3) we make the following conclusion.

- *The stressed state is described in the coordinates x_1, x_2, x_3 by the stress tensor*

$$\begin{bmatrix} \sigma_{11} & \sigma_{12} & \sigma_{13} \\ \sigma_{21} & \sigma_{22} & \sigma_{23} \\ \sigma_{31} & \sigma_{32} & \sigma_{33} \end{bmatrix} = U \begin{bmatrix} \rho k_1 E_{k_1} & 0 & 0 \\ 0 & \rho k_2 E_{k_2} & 0 \\ 0 & 0 & \rho k_3 E_{k_3} \end{bmatrix} U^*. \tag{4.30}$$

This stressed state is obtained from the standard state by the transformation $C = UKV$.

For example, if E depends only on the entropy S and the density $\rho = \dfrac{\rho_0}{k_1 k_2 k_3}$, then

$$k_1 E_{k_1}\big|_{S=\text{const}} = k_1 \rho_{k_1} E_\rho\big|_{S=\text{const}} = -k_1 \frac{\rho_o}{k_1^2 k_2 k_3} E_\rho\big|_{S=\text{const}} = -\rho E_\rho,$$
$$k_1 E_{k_1} = k_2 E_{k_2} = k_3 E_{k_3} = -\rho E_\rho$$

and, consequently,

$$s_1 = s_2 = s_3 = \rho k_1 E_{k_1} = \rho k_2 E_{k_2} = \rho k_3 E_{k_3} = -\rho^2 E_\rho.$$

For any orthogonal matrix U we have

$$U\begin{bmatrix}-\rho^2 E_\rho & 0 & 0\\ 0 & -\rho^2 E_\rho & 0\\ 0 & 0 & -\rho^2 E_\rho\end{bmatrix}U^* = \begin{bmatrix}-\rho^2 E_\rho & 0 & 0\\ 0 & -\rho^2 E_\rho & 0\\ 0 & 0 & -\rho^2 E_\rho\end{bmatrix}.$$

In this case, the stress tensor is spherical and is independent of the choice of orthogonal coordinates. The value of the normal stress taken with the opposite sign is called the *pressure* (cf. (4.1)) and is denoted by p, i.e., $p = \rho^2 E_\rho(\rho, S)$.

Thus, we have shown that the Murnaghan formulas lead to the known expressions in the case of gases or fluids.

§ 5. Viscous Stresses in an Isotropic Medium

In Section 1, we postulated that viscous stresses are functions of the strain rate tensor $||u_{ik}||$

$$u_{ik} = \frac{1}{2}\left(\frac{\partial u_i}{\partial x_k} + \frac{\partial u_k}{\partial x_i}\right). \tag{5.1}$$

5.1. By the isotropy condition, we have to impose very strong conditions on an admissible form of the dependence of the stress tensor on the strain rate tensor. Namely,

$$\sigma_{ik} = f_{ik}(u_{11}, u_{12}, u_{13}, u_{21}, u_{22}, u_{23}, u_{31}, u_{32}, u_{33}), \tag{5.2}$$

where f_{ik} is independent of the choice of orthogonal coordinates. In other words, for any orthogonal matrix W and σ'_{ik} and u'_{ik} such that

$$||\sigma'_{ik}|| = W||\sigma_{ik}||W^*, \quad ||u'_{ik}|| = W||u_{ik}||W^*,$$

the dependence (5.2) preserves the form:

$$\sigma'_{ik} = f_{ik}(u'_{11}, u'_{12}, u'_{13}, u'_{21}, u'_{22}, u'_{23}, u'_{31}, u'_{32}, u'_{33}). \tag{5.3}$$

• We say that formula (5.2) determines a *tensor-valued function* $||\sigma_{ik}||$ *of tensor variable* $||u_{ik}||$.

Our goal is to describe such functions.

Let W be an orthogonal matrix reducing $||u_{ik}||$ to the diagonal form:

$$||u_{ik}|| = W\begin{bmatrix}\lambda_1 & 0 & 0\\ 0 & \lambda_2 & 0\\ 0 & 0 & \lambda_3\end{bmatrix}W^* \qquad (WW^* = I,\ \det W = +1). \tag{5.4}$$

We choose coordinates with respect to which the tensor $||u_{ik}||$ is diagonal:

$$||u_{ik}|| = \begin{bmatrix} \lambda_1 & 0 & 0 \\ 0 & \lambda_2 & 0 \\ 0 & 0 & \lambda_3 \end{bmatrix} \tag{5.5}$$

and assume that the tensor $||\sigma_{ik}||$ is represented in these coordinates as follows:

$$||\sigma_{ik}|| = \begin{bmatrix} s_1 & \tau_3 & \tau_2 \\ \tau_3 & s_2 & \tau_1 \\ \tau_2 & \tau_1 & s_3 \end{bmatrix}. \tag{5.6}$$

Since σ_{ik} are found from u_{ik} by formula (5.2), we have

$$s_1 = f_1(\lambda_1, \lambda_2, \lambda_3), \quad s_2 = f_2(\lambda_1, \lambda_2, \lambda_3), \quad s_3 = f_3(\lambda_1, \lambda_2, \lambda_3),$$
$$\tau_1 = f_4(\lambda_1, \lambda_2, \lambda_3), \quad \tau_2 = f_5(\lambda_1, \lambda_2, \lambda_3), \quad \tau_3 = f_6(\lambda_1, \lambda_2, \lambda_3).$$

Making the change of coordinates with $W = \begin{bmatrix} \theta_1 & 0 & 0 \\ 0 & \theta_2 & 0 \\ 0 & 0 & \theta_3 \end{bmatrix}$, where $\theta_k = \pm 1$, from the representations (5.5) and (5.6) of $||u_{ik}||$ and $||\sigma_{ik}||$ respectively we find

$$\begin{bmatrix} \theta_1 & 0 & 0 \\ 0 & \theta_2 & 0 \\ 0 & 0 & \theta_3 \end{bmatrix} \begin{bmatrix} \lambda_1 & 0 & 0 \\ 0 & \lambda_2 & 0 \\ 0 & 0 & \lambda_3 \end{bmatrix} \begin{bmatrix} \theta_1 & 0 & 0 \\ 0 & \theta_2 & 0 \\ 0 & 0 & \theta_3 \end{bmatrix} = \begin{bmatrix} \theta_1^2\lambda_1 & 0 & 0 \\ 0 & \theta_2^2\lambda_2 & 0 \\ 0 & 0 & \theta_3^2\lambda_3 \end{bmatrix}$$
$$= \begin{bmatrix} \lambda_1 & 0 & 0 \\ 0 & \lambda_2 & 0 \\ 0 & 0 & \lambda_3 \end{bmatrix},$$

$$\begin{bmatrix} \theta_1 & 0 & 0 \\ 0 & \theta_2 & 0 \\ 0 & 0 & \theta_3 \end{bmatrix} \begin{bmatrix} s_1 & \tau_3 & \tau_2 \\ \tau_3 & s_2 & \tau_1 \\ \tau_2 & \tau_1 & s_3 \end{bmatrix} \begin{bmatrix} \theta_1 & 0 & 0 \\ 0 & \theta_2 & 0 \\ 0 & 0 & \theta_3 \end{bmatrix} = \begin{bmatrix} s_1\theta_1^2 & \tau_3\theta_1\theta_2 & \tau_2\theta_1\theta_3 \\ \tau_3\theta_1\theta_2 & s_2\theta_2^2 & \tau_1\theta_2\theta_3 \\ \tau_2\theta_1\theta_3 & \tau_1\theta_2\theta_3 & s_3\theta_3^2 \end{bmatrix}.$$

We see that the tensor (5.5) remains unchanged, whereas the tensor (5.6) goes to the tensor

$$||\sigma'_{ik}|| = \begin{bmatrix} s_1\theta_1^2 & \tau_3\theta_1\theta_2 & \tau_2\theta_1\theta_3 \\ \tau_3\theta_1\theta_2 & s_2\theta_2^2 & \tau_1\theta_2\theta_3 \\ \tau_2\theta_1\theta_3 & \tau_1\theta_2\theta_3 & s_3\theta_3^2 \end{bmatrix} = \begin{bmatrix} s_1 & \tau_3\theta_1\theta_2 & \tau_2\theta_1\theta_3 \\ \tau_3\theta_1\theta_2 & s_2 & \tau_1\theta_2\theta_3 \\ \tau_2\theta_1\theta_3 & \tau_1\theta_2\theta_3 & s_3 \end{bmatrix}.$$

Since formula (5.2) should be invariant under any orthogonal change of coordinates, we have

$$s_1 = f_1(\lambda_1, \lambda_2, \lambda_3), \quad s_2 = f_2(\lambda_1, \lambda_2, \lambda_3), \quad s_3 = f_3(\lambda_1, \lambda_2, \lambda_3), \tag{5.7}$$

$$\tau_1\theta_2\theta_3 = f_4(\lambda_1, \lambda_2, \lambda_3), \quad \tau_2\theta_1\theta_3 = f_5(\lambda_1, \lambda_2, \lambda_3),$$
$$\tau_3\theta_1\theta_2 = f_6(\lambda_1, \lambda_2, \lambda_3). \tag{5.8}$$

Hence the products $\tau_1\theta_2\theta_3$, $\tau_2\theta_1\theta_3$, $\tau_3\theta_1\theta_2$ should be independent of the choice of the triple θ_1, θ_2, θ_3. In particular, comparing $(\theta_1' = +1,\ \theta_2' = +1,\ \theta_3' = +1)$ with $(\theta_1'' = -1,\ \theta_2'' = +1,\ \theta_3'' = -1)$, we see that $\tau_1\theta_2'\theta_3' = \tau_1\theta_2''\theta_3''$, i.e., $\tau_1 = -\tau_1$ and, consequently, $\tau_1 = 0$. Similarly, $\tau_2 = 0$ and $\tau_3 = 0$.

• *If the tensor $\|u_{ik}\|$ is reduced by an orthogonal transformation to the diagonal form* (5.5), *then the tensor-valued function $\|\sigma_{ik}\|$ of the form* (5.2) *of tensor variable $\|u_{ik}\|$ takes the diagonal form*

$$\|\sigma_{ik}\| = \begin{bmatrix} s_1 & 0 & 0 \\ 0 & s_2 & 0 \\ 0 & 0 & s_3 \end{bmatrix}.$$

If

$$\|u_{ik}\| = W \begin{bmatrix} \lambda_1 & 0 & 0 \\ 0 & \lambda_2 & 0 \\ 0 & 0 & \lambda_3 \end{bmatrix} W^*,$$

then

$$\|\sigma_{ik}\| = W \begin{bmatrix} s_1 & 0 & 0 \\ 0 & s_2 & 0 \\ 0 & 0 & s_3 \end{bmatrix} W^*.$$

It remains to find admissible functions f_1, f_2, f_3 in (5.7). Since the coordinate axes are equal in rights by the isotropy condition, there exists a function f such that

$$\begin{aligned} s_1 &= f(\lambda_1, \lambda_2, \lambda_3) = f(\lambda_1, \lambda_3, \lambda_2), \\ s_2 &= f(\lambda_2, \lambda_1, \lambda_3) = f(\lambda_2, \lambda_3, \lambda_1), \\ s_3 &= f(\lambda_3, \lambda_1, \lambda_2) = f(\lambda_3, \lambda_2, \lambda_1). \end{aligned}$$

Hence $f(x, y, z) = f(x, z, y)$, i.e., for a fixed x the function f is symmetric with respect to y and z. Hence f depends only on the basic symmetric functions $y + z$, yz, i.e., $f(x, y, z) = \tilde{f}(x, yz, y + z)$. We set

$$F(x, x + y + z, xyz) = \tilde{f}(x, yz, y + z) = f(x, y, z) = f(x, z, y).$$

Then

$$\begin{aligned} F(x, x + y + z, xyz) &= f(x, y, z) = f(x, z, y), \\ F(y, x + y + z, xyz) &= f(y, z, x) = f(y, x, z), \\ F(z, x + y + z, xyz) &= f(z, x, y) = f(z, y, x). \end{aligned}$$

There exists a quadratic polynomial $P(\xi) = a_0 + a_1\xi + a_2\xi^2$ with coefficients a_0, a_1, a_2 depending on x, y, z such that

$$\begin{aligned} P(x) &= F(x, x+y+z, xyz), \\ P(y) &= F(y, x+y+z, xyz), \\ P(z) &= F(z, x+y+z, xyz). \end{aligned}$$

The coefficients a_0, a_1, a_2 of the polynomial P are found from the linear system

$$\begin{aligned} a_0 + a_1 x + a_2 x^2 &= F(x, x+y+z, xyz), \\ a_0 + a_1 y + a_2 y^2 &= F(y, x+y+z, xyz), \\ a_0 + a_1 z + a_2 z^2 &= F(z, x+y+z, xyz) \end{aligned}$$

and are expressed as follows:

$$a_0 = \frac{1}{\Delta} \begin{vmatrix} F(x, x+y+z, xyz) & x & x^2 \\ F(y, x+y+z, xyz) & y & y^2 \\ F(z, x+y+z, xyz) & z & z^2 \end{vmatrix},$$

$$a_1 = \frac{1}{\Delta} \begin{vmatrix} 1 & F(x, x+y+z, xyz) & x^2 \\ 1 & F(y, x+y+z, xyz) & y^2 \\ 1 & F(z, x+y+z, xyz) & z^2 \end{vmatrix},$$

$$a_2 = \frac{1}{\Delta} \begin{vmatrix} 1 & x & F(x, x+y+z, xyz) \\ 1 & y & F(y, x+y+z, xyz) \\ 1 & z & F(z, x+y+z, xyz) \end{vmatrix},$$

where

$$\Delta = \begin{vmatrix} 1 & x & x^2 \\ 1 & y & y^2 \\ 1 & z & z^2 \end{vmatrix}.$$

It is obvious that a_0, a_1, a_2 remain unchanged under any permutations of the variables x, y, z, i.e., they are symmetric functions and can be expressed in terms of the basic symmetric functions $x+y+z$, $xy+yz+zx$, xyz.

Thus,

$$\begin{aligned} s_1 &= a_0(J_1, J_2, J_3) + \lambda_1 a_1(J_1, J_2, J_3) + \lambda_1^2 a_2(J_1, J_2, J_3), \\ s_2 &= a_0(J_1, J_2, J_3) + \lambda_2 a_1(J_1, J_2, J_3) + \lambda_2^2 a_2(J_1, J_2, J_3), \\ s_3 &= a_0(J_1, J_2, J_3) + \lambda_3 a_1(J_1, J_2, J_3) + \lambda_3^2 a_2(J_1, J_2, J_3), \end{aligned} \tag{5.9}$$

where

$$\begin{aligned} J_1 &= \lambda_1 + \lambda_2 + \lambda_3, \\ J_2 &= \lambda_2\lambda_3 + \lambda_3\lambda_1 + \lambda_1\lambda_2, \\ J_3 &= \lambda_1\lambda_2\lambda_3. \end{aligned} \tag{5.10}$$

By these equalities, we have

$$\begin{aligned} \|\sigma_{ik}\| &= W \begin{bmatrix} s_1 & 0 & 0 \\ 0 & s_2 & 0 \\ 0 & 0 & s_3 \end{bmatrix} W^* \\ &= a_0 W \begin{bmatrix} 1 & 0 & 0 \\ 0 & 1 & 0 \\ 0 & 0 & 1 \end{bmatrix} W^* + a_1 W \begin{bmatrix} \lambda_1 & 0 & 0 \\ 0 & \lambda_2 & 0 \\ 0 & 0 & \lambda_3 \end{bmatrix} W^* \\ &\quad + a_2 W \begin{bmatrix} \lambda_1^2 & 0 & 0 \\ 0 & \lambda_2^2 & 0 \\ 0 & 0 & \lambda_3^2 \end{bmatrix} W^* \\ &= a_0 I + a_1 \|u_{ik}\| + a_2 \|u_{ik}\|^2, \end{aligned} \tag{5.11}$$

where a_0, a_1, a_2 depend only on J_1, J_2, J_3 which coincide with the orthogonal invariants of $\|u_{ik}\|$:

$$\begin{aligned} J_1 &= \lambda_1 + \lambda_2 + \lambda_3 = u_{11} + u_{22} + u_{33}, \\ J_2 &= \lambda_2\lambda_3 + \lambda_3\lambda_1 + \lambda_1\lambda_2 = \begin{vmatrix} u_{11} & u_{12} \\ u_{21} & u_{22} \end{vmatrix} + \begin{vmatrix} u_{22} & u_{23} \\ u_{32} & u_{33} \end{vmatrix} + \begin{vmatrix} u_{33} & u_{31} \\ u_{13} & u_{11} \end{vmatrix}, \\ J_3 &= \lambda_1\lambda_2\lambda_3 = \begin{vmatrix} u_{11} & u_{12} & u_{13} \\ u_{21} & u_{22} & u_{23} \\ u_{31} & u_{32} & u_{33} \end{vmatrix}. \end{aligned} \tag{5.12}$$

Thus, we have found the general form of a tensor-valued function $\|\sigma_{ik}\|$ of tensor variable $\|u_{ik}\|$ under the assumption that the matrices $\|\sigma_{ik}\|$ and $\|u_{ik}\|$ are symmetric and that the dependence preserves the form under any orthogonal change of the coordinate system.

5.2. It is of interest to compare the general form of tensor-valued functions of tensor variable with *matrix functions of matrix variable* studied in linear algebra. Let a symmetric matrix $\|u_{ik}\|$ admit the canonical representation (5.4). A matrix-valued function of matrix variable is usually defined as

$$f(\|u_{ik}\|) = W \begin{bmatrix} f(\lambda_1) & 0 & 0 \\ 0 & f(\lambda_2) & 0 \\ 0 & 0 & f(\lambda_3) \end{bmatrix} W^*.$$

Using the general form of an invariant tensor-valued function of tensor variable, we can write

$$||\sigma_{ik}|| = W \begin{bmatrix} \tilde{f}(\lambda_1,\lambda_2+\lambda_3,\lambda_2\lambda_3) & 0 & 0 \\ 0 & \tilde{f}(\lambda_2,\lambda_1+\lambda_3,\lambda_1\lambda_3) & 0 \\ 0 & 0 & \tilde{f}(\lambda_3,\lambda_1+\lambda_2,\lambda_1\lambda_2) \end{bmatrix} W^*$$

we can use a more general dependence of the eigenvalues of $||\sigma_{ik}||$ on the eigenvalues λ_1, λ_2, λ_3 of $||u_{ik}||$.

To derive the general form of a tensor-valued function $||\sigma_{ik}||$ of tensor variable $||u_{ik}||$, we used only the fact that all orthogonal coordinate systems are, in a sense, equal. Applying this result to the viscous stress tensor $||\sigma_{ik}||$ expressed in terms of the strain rate tensor $||u_{ik}||$, we conclude that only representations of the form $\sigma_{ik} = a_0 + a_1 u_{ik} + a_2 u_{ij} u_{jk}$ are admissible, where the coefficients a_0, a_1, a_2 are expressed in terms of the invariants J_1, J_2, J_3 of the strain rate tensor. These coefficients also depend on properties of the medium in a neighborhood of the point under consideration. They are different for different materials and can depend on the density and temperature: $a_0 = a_0(\rho, T, J_1, J_2, J_3)$, $a_1 = a_1(\rho, T, J_1, J_2, J_3)$, $a_2 = a_2(\rho, T, J_1, J_2, J_3)$. The dependence can be more complicated. For example, the viscous forces described by σ_{ik} may also depend on the matrix $||c_{ik}||$ characterizing the deformation.

Taking into account arguments coming from thermodynamics, we should impose certain conditions on the functions $a_k = a_k(\rho, T, J_1, J_2, J_3)$. Namely, the work done to overcome the internal friction forces under deformation should be strictly positive. This work is expended for heating the medium.

5.3. We develop the mathematical tools necessary for a mathematical statement of the above postulate and obtain necessary conditions on the coefficients a_0, a_1, and a_2. We also introduce a very important notion of the work of internal surface stresses and motivate it.

As was mentioned in Section 1, displacements of a material point of a medium during a small time-interval produce displacement and rotation of the neighborhood of the point, as well as a deformation (change of shape) described by the linear transformation with the orthogonal matrix

$$W \begin{bmatrix} 1+\lambda_1\Delta t & 0 & 0 \\ 0 & 1+\lambda_2\Delta t & 0 \\ 0 & 0 & 1+\lambda_3\Delta t \end{bmatrix} W^* \tag{5.13}$$

which can be interpreted as the extension with coefficients $1+\lambda_k\Delta t$ along three orthogonal axes whose directed vectors form the columns of the matrix W. As was shown, in this case, the viscous stress tensor takes the form

$$||\sigma_{ik}|| = W \begin{bmatrix} s_1 & 0 & 0 \\ 0 & s_2 & 0 \\ 0 & 0 & s_3 \end{bmatrix} W^*. \tag{5.14}$$

Therefore, by rotation, we reach a situation where the matrices describing deformation and stresses are diagonal:

$$\begin{bmatrix} 1+\lambda_1\Delta t & 0 & 0 \\ 0 & 1+\lambda_2\Delta t & 0 \\ 0 & 0 & 1+\lambda_3\Delta t \end{bmatrix}, \qquad \begin{bmatrix} s_1 & 0 & 0 \\ 0 & s_2 & 0 \\ 0 & 0 & s_3 \end{bmatrix}$$

We distinguish a small cube whose edges have length h and are parallel to the coordinate axes. The deformed cube is a parallelepiped with edges $(1+\lambda_1\Delta t)h$, $(1+\lambda_2\Delta t)h$, $(1+\lambda_3\Delta t)h$. The forces counteracting separation of faces of the parallelepiped are perpendicular to the faces and are equal to s_1h^2, s_2h^2, s_3h^2. The work done by these forces, $s_1h^2\lambda_1\Delta t\, h + s_2h^2\lambda_2\Delta t\, h + s_3h^2\lambda_3\Delta t\, h = (s_1\lambda_1 + s_2\lambda_2 + s_3\lambda_3)h^3\Delta t$, is proportional to the volume h^3 of the cube and the time-interval Δt. Therefore, it is natural to regard $s_1\lambda_1 + s_2\lambda_2 + s_3\lambda_3$ as the specific power of the work done by friction forces under deformation (unit volume per unit time). It turns out that $s_1\lambda_1 + s_2\lambda_2 + s_3\lambda_3$ can be computed without reducing matrices to the canonical form. Indeed, consider the product of the matrices

$$\begin{bmatrix} \sigma_{11} & \sigma_{12} & \sigma_{13} \\ \sigma_{21} & \sigma_{22} & \sigma_{23} \\ \sigma_{31} & \sigma_{32} & \sigma_{33} \end{bmatrix} \begin{bmatrix} u_{11} & u_{12} & u_{13} \\ u_{21} & u_{22} & u_{23} \\ u_{31} & u_{32} & u_{33} \end{bmatrix}$$

$$= W \begin{bmatrix} s_1 & 0 & 0 \\ 0 & s_2 & 0 \\ 0 & 0 & s_3 \end{bmatrix} W^* W \begin{bmatrix} \lambda_1 & 0 & 0 \\ 0 & \lambda_2 & 0 \\ 0 & 0 & \lambda_3 \end{bmatrix} W^* = W \begin{bmatrix} s_1\lambda_1 & 0 & 0 \\ 0 & s_2\lambda_2 & 0 \\ 0 & 0 & s_3\lambda_3 \end{bmatrix} W^*$$

and verify that $s_1\lambda_1 + s_2\lambda_2 + s_3\lambda_3$ is the trace of this product. Using the rule of multiplication of matrices, we find that the trace is equal to $\sigma_{ik}u_{ki}$. The equality $s_1\lambda_1 + s_2\lambda_2 + s_3\lambda_3 \equiv s_i\lambda_i = \sigma_{ki}u_{ik}$ is proved. Thus, we have established the following assertion.

- *Under a deformation of a volume element $dx_1dx_2dx_3$ the work done by viscous forces in time dt is equal to $\sigma_{ik}u_{ki}dx_1dx_2dx_3$.*

We assume that $\|\sigma_{ik}\|$ and $\|u_{ki}\|$ can be simultaneously reduced to the diagonal form by the same rotation matrix W. However, this assumption is not essential. We will use the formula $\sigma_{ik}u_{ki}h^3\Delta t$ in order to compute the work done by arbitrary stresses σ_{ik} (not only the stresses generated by of the internal friction) in the time-interval Δt under deformation of a small cube whose edges have length h and are parallel to the coordinate axes when the medium is deformed with the strain rate tensor $\|u_{ik}\|$. For an orthogonal matrix W we again write

$$\|u_{ik}\| = W \begin{bmatrix} \lambda_1 & 0 & 0 \\ 0 & \lambda_2 & 0 \\ 0 & 0 & \lambda_3 \end{bmatrix} W^*, \qquad (5.15)$$

but, in this case, $\|\sigma_{ik}\|$ is not necessarily reducible to the diagonal form. Generally speaking, we have

$$\|\sigma_{ik}\| = W \begin{bmatrix} s_1 & \tau_3 & \tau_2 \\ \tau_3 & s_2 & \tau_1 \\ \tau_2 & \tau_1 & s_3 \end{bmatrix} W^*. \tag{5.16}$$

During the time-interval Δt, the cube becomes the parallelepiped with edges $(1+\lambda_1\Delta t)h$, $(1+\lambda_2\Delta t)h$, $(1+\lambda_3\Delta t)h$. The stresses caused by these deformations

$$\begin{pmatrix} s_1 \\ \tau_3 \\ \tau_2 \end{pmatrix}, \quad \begin{pmatrix} \tau_3 \\ s_2 \\ \tau_1 \end{pmatrix}, \quad \begin{pmatrix} \tau_2 \\ \tau_1 \\ s_3 \end{pmatrix},$$

do the work

$$s_1h^2\lambda_1\Delta t\, h + s_2h^2\lambda_2\Delta t\, h + s_3h^2\lambda_3\Delta t\, h = (s_1\lambda_1 + s_2\lambda_2 + s_3\lambda_3)h^3\Delta t.$$

This formula means that the sum of the works done by τ_j vanishes. Further we argue as in Section 4 at the beginning of the proof of the Murnaghan formulas. By (5.14) and (5.15),

$$\|u_{ik}\|\,\|\sigma_{kj}\| = W \begin{bmatrix} \lambda_1 & 0 & 0 \\ 0 & \lambda_2 & 0 \\ 0 & 0 & \lambda_3 \end{bmatrix} W^*W \begin{bmatrix} s_1 & \tau_3 & \tau_2 \\ \tau_3 & s_2 & \tau_1 \\ \tau_2 & \tau_1 & s_3 \end{bmatrix} W^*$$

$$= W \begin{bmatrix} \lambda_1 s_1 & \lambda_1\tau_3 & \lambda_1\tau_2 \\ \lambda_2\tau_3 & \lambda_2 s_2 & \lambda_2\tau_1 \\ \lambda_3\tau_2 & \lambda_3\tau_1 & \lambda_3 s_3 \end{bmatrix} W^*,$$

$$u_{ik}\sigma_{ki} = \operatorname{tr}(\|u_{ik}\|\,\|\sigma_{kj}\|) = \operatorname{tr} \begin{bmatrix} \lambda_1 s_1 & \lambda_1\tau_3 & \lambda_1\tau_2 \\ \lambda_2\tau_3 & \lambda_2 s_2 & \lambda_2\tau_1 \\ \lambda_3\tau_2 & \lambda_3\tau_1 & \lambda_3 s_3 \end{bmatrix} = \lambda_1 s_1 + \lambda_2 s_2 + \lambda_3 s_3.$$

Hence we conclude that, in this more general case, $u_{ik}\sigma_{ki}h^3\Delta t$ is the work done by σ_{ij} under deformation of a small volume. We note that we do not use the possibility to reduce $\|\sigma_{ik}\|$ and $\|u_{ik}\|$ to the diagonal form simultaneously. The above arguments justify the following definition.

- The integral

$$\iiint \delta t\, \sigma_{ki}u_{ik}\, dx_1dx_2dx_3 = \delta t \iiint \sigma_{ki}\frac{1}{2}\left(\frac{\partial u_i}{\partial x_k} + \frac{\partial u_k}{\partial x_i}\right) dx_1dx_2dx_3$$

$$= \iiint \sigma_{ki}\frac{1}{2}\left(\frac{\partial(u_i\delta t)}{\partial x_k} + \frac{\partial(u_k\delta t)}{\partial x_i}\right) dx_1dx_2dx_3$$

$$= \iiint \sigma_{ki}\frac{1}{2}\left[\frac{\partial(\delta x_i)}{\partial x_k} + \frac{\partial(\delta x_k)}{\partial x_i}\right] dx_1dx_2dx_3,$$

is called the *work done by the internal surface forces (stresses)* in producing deformation of this volume caused by the displacements $\delta x_1 = u_1 \delta t$, $\delta x_2 = u_2 \delta t$, $\delta x_3 = u_3 \delta t$ of material points. This work is produced by the corresponding decrease of the kinetic energy and is expanded for increasing the internal energy.

The work done by the internal surface forces is often expressed as the same integral but with the opposite sign "−"

$$A = -\iiint \delta t\, \sigma_{ki} u_{ik}\, dx_1 dx_2 dx_3.$$

In Section 15 below, the sum $\sigma_{ik} u_{ki} = \sigma_{ik} \dfrac{\partial u_k}{\partial x_i}$ will appear as a term on the right-hand side of differential equations describing the energy balance in a continuous medium. This term is one of sources of increments in the internal energy.

5.4. Here, we formulate some postulates.

Postulate 5.1. *The work done by the internal friction forces under deformations of a medium is nonnegative and can be formulated as the inequality*

$$\sigma_{ik} u_{ki} = s_i \lambda_i \equiv s_1 \lambda_1 + s_2 \lambda_2 + s_3 \lambda_3 \geqslant 0. \tag{5.17}$$

Since $s_i = a_0 + a_1 \lambda_i + a_2 \lambda_i^2$, Postulate 5.1 implies the inequality

$$a_0(\lambda_1 + \lambda_2 + \lambda_3) + a_1(\lambda_1^2 + \lambda_2^2 + \lambda_3^2) + a_2(\lambda_1^3 + \lambda_2^3 + \lambda_3^3) \geqslant 0. \tag{5.18}$$

The sums $\lambda_1 + \lambda_2 + \lambda_3$, $\lambda_1^2 + \lambda_2^2 + \lambda_3^2$, $\lambda_1^3 + \lambda_2^3 + \lambda_3^3$ are expressed in terms of the invariants $\lambda_1 + \lambda_2 + \lambda_3 = J_1$, $\lambda_1^2 + \lambda_2^2 + \lambda_3^2 = J_1^2 - 2J_2$, $\lambda_1^3 + \lambda_2^3 + \lambda_3^3 = J_1^3 - 3J_1 J_2 + 3J_3$.

Thus, according to the dissipativity postulate, we assume that

$$\begin{aligned} &J_1 a_0(\rho, T, J_1, J_2, J_3) + (J_1^2 - 2J_2) a_1(\rho, T, J_1, J_2, J_3) \\ &\quad + (J_1^3 - 3J_1 J_2 + 3J_3) a_2(\rho, T, J_1, J_2, J_3) \geqslant 0. \end{aligned} \tag{5.19}$$

Usually, one requires only the condition that the components of the viscous stress tensor are linear functions of the components of the strain rate tensor which is equivalent to the equalities $a_0 = mJ_1 = m(u_{11} + u_{22} + u_{33}) = mu_{jj}$, $a_1 = n$, $a_2 = 0$ (m and n are independent of u_{ik}). Moreover, $\sigma_{ik} = m\delta_{ik} u_{jj} + nu_{ik}$. By the equalities $J_1 = \lambda_1 + \lambda_2 + \lambda_3$ and $J_2 = \lambda_1\lambda_2 + \lambda_2\lambda_3 + \lambda_3\lambda_1$, the dissipativity postulate

$$\begin{aligned} \sigma_{ik} u_{ki} &= J_1 a_0 + (J_1^2 - 2J_2) a_1 + (J_1^3 - 3J_1 J_2 + 3J_3) a_2 \\ &= J_1^2 m + (J_1^2 - 2J_2) n = (m+n) J_1^2 - 2nJ_2 \geqslant 0 \end{aligned}$$

is reduced to the condition that the following quadratic form is positive semidefinite:

$$\begin{aligned}(m+n)J_1^2 - 2nJ_2 &= (m+n)(\lambda_1+\lambda_2+\lambda_3)^2 - 2n(\lambda_1\lambda_2+\lambda_2\lambda_3+\lambda_3\lambda_1)\\ &= \frac{3m+n}{3}(\lambda_1+\lambda_2+\lambda_3)^2 + n\sum_{i=1}^{3}\left(\lambda_i - \frac{\lambda_1+\lambda_2+\lambda_3}{3}\right)^2\end{aligned}$$

Now, it is obvious that the inequalities $n \geqslant 0$ and $3m+n \geqslant 0$ are sufficient and necessary for dissipativity.

In the notation $n = 2\eta$, $3m+n = 3\zeta$ $(m = \zeta - (2/3)\eta)$ which is often used in physics and mechanics, we can write the equality $\sigma_{ik} = m\delta_{ik}u_{jj} + nu_{ik}$ in the form

$$\sigma_{ik} = \eta\left(\frac{\partial u_i}{\partial x_k} + \frac{\partial u_k}{\partial x_i} - \frac{2}{3}\delta_{ik}\frac{\partial u_l}{\partial x_l}\right) + \zeta\delta_{ik}\frac{\partial u_l}{\partial x_l}. \qquad (5.20)$$

The condition that $\sigma_{ik}u_{ki}$ is positive semidefinite means that $\eta \geqslant 0$ and $\zeta \geqslant 0$.

• The coefficient η is called the *shear viscosity coefficient* and ζ is referred to as the *bulk viscosity coefficient.*

In one-dimensional problems, $u_1 = u_1(x_1, t)$, $u_2 = 0$, $u_3 = 0$, the viscous stresses are given by the formulas

$$\sigma_{11} = \eta\left(2\frac{\partial u_1}{\partial x_1} - \frac{2}{3}\frac{\partial u_1}{\partial x_1}\right) + \zeta\frac{\partial u_1}{\partial x_1} = \left(\frac{4}{3}\eta + \zeta\right)\frac{\partial u_1}{\partial x_1}, \qquad (5.21)$$

$$\sigma_{22} = \sigma_{33} = \left(\zeta - \frac{2}{3}\eta\right)\frac{\partial u_1}{\partial x_1}, \qquad \sigma_{ij} = 0 \text{ if } i \neq j. \qquad (5.22)$$

To simplify the expression (5.19) for σ_{11}, we introduce the viscosity coefficient $\mu = \dfrac{4}{3}\eta + \zeta$. Then $\sigma_{11} = \mu\dfrac{\partial u_1}{\partial x_1}$.

Thus we have discussed possible expressions for the viscous stress tensor in an isotropic medium. Further, we will consider only the linear representation with respect to the velocity gradient and assume that the viscosity coefficients η and ζ depend on the density and temperature of the medium. To fix an idea, we do not touch more complicated cases where the viscous stress tensor can also depend on deformations.

In the further consideration, we will assume that the stress of a medium can be represented as the sums of the "viscous" stress described by the formulas derived in this section and the "elastic" stress described by the Murnaghan formulas (cf. Section 4). This conjecture is often used in applications. In particular, owing to this conjecture, we were able to simplify the analysis of the dependence of stresses on deformations and strain rate.

§ 6. Strain Tensors and Distortion Tensor

In this section, we continue to study deformations (cf. Section 1). Now, we can characterize the stressed state of an elastic medium after deformation (cf. Section 4), which allows us to introduce a special array describing deformations. This array also gives use all the necessary starting information on the change of size and shape of elements of the medium if we want to compute the stress tensor in the case of an isotropic medium with the equation of state $E = E(k_1, k_2, k_3, S)$.

6.1. We know that the stress tensor at any point of an isotropic medium is determined by the Murnaghan formulas (cf. (4.26) and (4.26))

$$\begin{aligned} \|\sigma_{ij}\| &= U \begin{bmatrix} \frac{\rho_0}{k_2 k_3} E_{k_1} & 0 & 0 \\ 0 & \frac{\rho_0}{k_3 k_1} E_{k_2} & 0 \\ 0 & 0 & \frac{\rho_0}{k_1 k_2} E_{k_3} \end{bmatrix} U^* \\ &= \rho U \begin{bmatrix} k_1 E_{k_1} & 0 & 0 \\ 0 & k_2 E_{k_2} & 0 \\ 0 & 0 & k_3 E_{k_3} \end{bmatrix} U^*, \end{aligned} \tag{6.1}$$

where $\rho = \dfrac{\rho_0}{k_1 k_2 k_3}$, if the stressed state of the medium in a neighborhood of a fixed point can be obtained from the standard state by adiabatic deformation (with constant entropy S) described by a linear transformation

$$\begin{gathered} C = \|c_{ij}\| = U \begin{bmatrix} k_1 & 0 & 0 \\ 0 & k_2 & 0 \\ 0 & 0 & k_3 \end{bmatrix} V = UKV, \\ UU^* = I = VV^*, \quad \det U = \det V = +1. \end{gathered} \tag{6.2}$$

Comparing (6.1) with (6.2), we see that, unlike the stress tensor (6.1), the decomposition of C contains the matrix V. The question arises: Is it possible to consider, instead of C, some other characteristics containing information only on the extension coefficients k_1, k_2, k_3 and the rotation matrix U in (6.1). It turns out that for such a characteristic we can take the so-called metric deformation tensor.

- By the *metric deformation tensor* we mean the matrix

$$G = \|g_{ij}\| = U \begin{bmatrix} 1/k_1^2 & 0 & 0 \\ 0 & 1/k_2^2 & 0 \\ 0 & 0 & 1/k_3^2 \end{bmatrix} U^* = UK^{-2}U^*. \tag{6.3}$$

We note that the symmetric matrix $\|g_{ij}\|$ $(g_{ij} = g_{ji})$ can be found from the matrix C without reducing C to the canonical form: $\|g_{ij}\| = (C^{-1})^* C^{-1}$. Indeed, $C = UKV$, $C^{-1} = V^* K^{-1} U^*$, $(C^{-1})^* = UK^{-1}V$, $(C^{-1})^* C^{-1} = UK^{-1}VV^*K^{-1}U^* = UK^{-2}U^* = G$.

To motivate the term "metric deformation tensor" we consider an infinitesimal vector $dx_0 = (dx_{10}, dx_{20}, dx_{30})$ joining two neighboring points x_0 and $x_0 + dx_0$ of the medium in the initial (before deformation) state and analyze the change of its length under the transformation $x = x(x_0)$ or, componentwise,

$$x_1 = x_1(x_{10}, x_{20}, x_{30}),\ x_2 = x_2(x_{10}, x_{20}, x_{30}),\ x_3 = x_3(x_{10}, x_{20}, x_{30}).$$

We know that dx is obtained from dx_0 by the linear transformation $dx = C dx_0$, where $C = \|c_{ij}\| = \left\| \dfrac{\partial x_i}{\partial x_{j0}} \right\|$. We consider the set of vectors dx that are located on the sphere with radius dr before the deformation. Since

$$dr^2 = (dx_0, dx_0) = (C^{-1}dx, C^{-1}dx) = ((C^{-1})^* C^{-1} dx, dx) = (G dx, dx),$$

this set is an ellipsoid of the form

$$(G dx, dx) = g_{ik} dx_i dx_k \left(\equiv \sum_{i,k} g_{ik} dx_i dx_k \right) = dr^2.$$

It is clear that $dr = \sqrt{g_{ik}(x_1, x_2, x_3) dx_i dx_k}$ is the length of this vector before deformation. If l is a curve in the deformed medium defined by the equation $x_i = x_i(\lambda)$, $0 \leqslant \lambda \leqslant 1$, then the integral

$$\int_l \sqrt{g_{ij}(x_1, x_2, x_3) dx_i dx_j} = \int_0^1 \sqrt{g_{ij}(x_1(\lambda), x_2(\lambda), x_3(\lambda)) \frac{dx_i}{d\lambda} \frac{dx_j}{d\lambda}}\, d\lambda$$

represents the length of the curve before deformation. Thus, in the deformed medium, $\|g_{ij}\|$ plays the role of a metric and it coincides with the Euclidean metric $dr = \sqrt{dx_{10}^2 + dx_{20}^2 + dx_{30}^2}$ before deformation. That is why $\|g_{ij}\|$ is called the metric deformation tensor.

It is easy to show that the metric deformation tensor is a second-order orthogonal tensor in the sense of the definition in the Section 1. Indeed, the rule (1.21) follows from the equalities $dx' = W dx$, $dx = W^* dx'$, $dr^2 = (G dx, dx) = (G W^* dx', W^* dx') = (W G W^* dx', dx') = (G' dx', dx')$.

6.2. In the literature, some other symmetric matrices, called *strain tensors*, are often used in order to describe deformations. They are expressed in terms of $K = \begin{bmatrix} k_1 & 0 & 0 \\ 0 & k_2 & 0 \\ 0 & 0 & k_3 \end{bmatrix}$ and U.

We list the most known strain tensors. The verification of the fact that all these tensors are orthogonal second-order tensors, i.e., satisfy (1.21), is left to the reader.

The Cauchy strain tensor:

$$(CC^*)^{1/2} - I = UKU^* - I = U \begin{bmatrix} k_1 - 1 & 0 & 0 \\ 0 & k_2 - 1 & 0 \\ 0 & 0 & k_3 - 1 \end{bmatrix} U^*.$$

The Green strain tensor:

$$\frac{1}{2}(CC^* - I) = U \begin{bmatrix} \frac{1}{2}(k_1^2 - 1) & 0 & 0 \\ 0 & \frac{1}{2}(k_2^2 - 1) & 0 \\ 0 & 0 & \frac{1}{2}(k_3^2 - 1) \end{bmatrix} U^*.$$

The Henky strain tensor:

$$U \begin{bmatrix} \ln k_1 & 0 & 0 \\ 0 & \ln k_2 & 0 \\ 0 & 0 & \ln k_3 \end{bmatrix} U^*.$$

The Almansi strain tensor:

$$\frac{1}{2}(I - (C^*)^{-1}C^{-1}) = U \begin{bmatrix} \frac{1}{2}(1 - 1/k_1^2) & 0 & 0 \\ 0 & \frac{1}{2}(1 - 1/k_2^2) & 0 \\ 0 & 0 & \frac{1}{2}(1 - 1/k_3^2) \end{bmatrix} U^*$$

$$= \frac{1}{2}(I - G) = \left\| \frac{1}{2}(\delta_{ij} - g_{ij}) \right\| = ||\varepsilon_{ij}||.$$

The metric deformation tensor $||g_{ij}||$ is similar to the Almansi strain tensor $||\varepsilon_{ij}||$ and differs from it only by the constant tensor term $|| - \delta_{ij}||$ and the coefficient $-1/2$. By the presence of the term $-\delta_{ij}$, the Almansi strain tensor ε_{ij} is close to the zero tensor, whereas the metric deformation tensor $||g_{ij}||$ is close to the identity tensor δ_{ij} in the case of small deformations.

We will mainly need the metric deformation tensor $||g_{ij}||$ but, sometimes, it will be convenient to use the Almansi strain tensor $||\varepsilon_{ij}||$.

6.3. If we need a more detailed description of the linear transformation $C = UKV$, including an information on both rotations U, V and extension K, we can use the inverse matrix

$$A = \|a_{ij}\| = C^{-1} = V^* K^{-1} U = V^* \begin{bmatrix} 1/k_1 & 0 & 0 \\ 0 & 1/k_2 & 0 \\ 0 & 0 & 1/k_3 \end{bmatrix} U. \tag{6.4}$$

The matrix A is the Jacobi matrix of the inverse transformation to C which transforms the stressed state into the standard state:

$$A = \|a_{ij}\| = \left\| \frac{\partial x_{i0}}{\partial x_j} \right\|. \tag{6.5}$$

• The matrix A defined by (6.4) or (6.5) is called the *distortion tensor.*

Let us show that A is a second-order orthogonal tensor in the sense of the definition in Section 1. For this purpose, we consider the change of the matrix A under a simultaneous orthogonal transformation W of the Cartesian coordinates before and after deformation of the medium: $\mathbf{x}_0' = W\mathbf{x}_0$, $\mathbf{x}' = W\mathbf{x}$. If the vector $d\mathbf{x}_0$ goes to the vector $d\mathbf{x}$, then $d\mathbf{x}_0 = Ad\mathbf{x}$. It is obvious that $d\mathbf{x}_0 = W^* d\mathbf{x}_0'$, $d\mathbf{x} = W^* d\mathbf{x}'$, $W^* d\mathbf{x}_0' = AW^* d\mathbf{x}'$, $d\mathbf{x}_0' = WAW^* d\mathbf{x}' = A' d\mathbf{x}'$. Hence $A' = WAW^*$, which expresses the rule of transformation of A under rotations of the Cartesian coordinates. This rule is the same as (1.21).

Since $G = \|g_{ij}\| = (C^{-1})^* C^{-1}$ and $C^{-1} = A$, we have $G = \|g_{ij}\| = A^* A$ or, componentwise,

$$\begin{aligned} g_{ij} &= a_{ki} a_{kj} \equiv a_{1i} a_{1j} + a_{2i} a_{2j} + a_{3i} a_{3j} \\ &= \frac{\partial x_{k0}}{\partial x_i} \frac{\partial x_{k0}}{\partial x_j} \equiv \frac{\partial x_{10}}{\partial x_i} \frac{\partial x_{10}}{\partial x_j} + \frac{\partial x_{20}}{\partial x_i} \frac{\partial x_{20}}{\partial x_j} + \frac{\partial x_{30}}{\partial x_i} \frac{\partial x_{30}}{\partial x_j}. \end{aligned}$$

We note that the density ρ of the deformed medium can be determined from the density ρ_0 of the medium before deformation and the extension coefficients k_1, k_2, k_3 by the formula

$$\rho = \frac{\rho_0}{k_1 k_2 k_3} = \rho_0 (\det K)^{-1}. \tag{6.6}$$

Using the relations

$$\begin{aligned} &\det C = \det U \det K \det V = 1 \cdot \det K \cdot 1 = \det K, \\ &\det A = \det C^{-1} = (\det K)^{-1}, \\ &\det G = \det A^* A = \det A^* \det A = (\det A)^2 = (\det K)^{-2}, \end{aligned}$$

we find

$$\rho = \frac{\rho_0}{\det C} = \rho_0 \det A = \rho_0 \sqrt{\det G}. \tag{6.7}$$

These formulas will be repeatedly used in the further consideration.

6.4. Sometimes, we need to preserve an information about the extension K and the first rotation V but ignore the final rotation U in the decomposition $C =$

UKV. For this purpose, we will use strain tensors, called *strain tensors in the Lagrangian description*, unlike the above-listed strain tensors which are referred to as *strain tensors in the Eulerian description*. The terms *Lagrangian* and *Eulerian* descriptions are initiated by the notion of the Lagrangian and Eulerian coordinates. We recall that the *Lagrangian coordinates* remain unchanged during the motion of the medium. The simplest typical example of the Lagrangian coordinates are the coordinates x_{10}, x_{20}, x_{30} of particles at an initial time $t = t_0$. The Eulerian coordinates are connected with points of the space where the motion of a medium is considered. The Cartesian coordinates x_1, x_2, x_3 serve as an example of the Eulerian coordinates.

In order to describe the motion in the form

$$x_1 = x_1(x_{10}, x_{20}, x_{30}, t), \; x_2 = x_2(x_{10}, x_{20}, x_{30}, t), \; x_3 = x_3(x_{10}, x_{20}, x_{30}, t)$$

we should indicate a correspondence between the Eulerian coordinates and the Lagrangian coordinates at distinct time moments.

Table 1

List of Strain Tensors

Strain tensor	Strain tensor in the Eulerian description	Strain tensor in the Lagrangian description
The Cauchy strain tensor	$UKU^* - I$	$V^*KV - I$
The Green tensor	$\frac{1}{2}(UK^2U^* - I)$	$\frac{1}{2}(V^*K^2V - I)$
The Henky strain tensor	$U\begin{bmatrix} \ln k_1 & 0 & 0 \\ 0 & \ln k_2 & 0 \\ 0 & 0 & \ln k_3 \end{bmatrix}U^*$	$V^*\begin{bmatrix} \ln k_1 & 0 & 0 \\ 0 & \ln k_2 & 0 \\ 0 & 0 & \ln k_3 \end{bmatrix}V$
The Almansi strain tensor	$\frac{1}{2}(I - UK^{-2}U^*) = \|\varepsilon_{ij}\|$	$\frac{1}{2}(I - V^*K^{-2}V)$
The metric deformation tensor	$UK^{-2}U^* = A^*A = G$	$V^*K^{-2}V = AA^* = \|\gamma_{ij}\|$

By the representation $\left\| \dfrac{\partial x_i}{\partial x_{j0}} \right\| = C = UKV$, we can indicate the directions along which the extension K is performed in the nondeformed space x_{10}, x_{20}, x_{30} (with the Lagrangian coordinates) using the orthogonal matrix V, as well as the in the space x_1, x_2, x_3 (with the Eulerian coordinates) using the orthogonal matrix U. Thus, V and U determine the position of the principal axes along which the medium is deformed in the Lagrangian coordinates and the Eulerian coordinates respectively. This explains why the strain tensor constructed from K and U is called the strain tensor in the Eulerian description, whereas the tensor constructed from K and V is referred to the strain tensor in the Lagrangian description. The strain tensors in the Lagrangian and Eulerian descriptions are given in Table 1.

If we need to use a strain tensor in the Lagrangian description, we will take, as a rule, the metric deformation tensor whose components are expressed in terms of the distortion tensor by the formula

$$\gamma_{ij} = a_{ki}a_{kj} \equiv a_{1i}a_{1j} + a_{2i}a_{2j} + a_{3i}a_{3j}. \tag{6.8}$$

§ 7. Murnaghan Formulas in the Tensor Form and Small Deformations

We recall (cf. Sections 4 and 6) that if, at some point, a deformation of an isotropic elastic medium is described by the metric deformation tensor

$$G = \|g_{ij}\| = U \begin{bmatrix} 1/k_1^2 & 0 & 0 \\ 0 & 1/k_2^2 & 0 \\ 0 & 0 & 1/k_3^2 \end{bmatrix} U^*, \quad UU^* = I, \quad \det U = +1, \tag{7.1}$$

then the stress tensor $\|\sigma_{ij}\|$ is expressed by the Murnaghan formulas in the form

$$\|\sigma_{ij}\| = \rho U \begin{bmatrix} k_1 E_{k_1} & 0 & 0 \\ 0 & k_2 E_{k_2} & 0 \\ 0 & 0 & k_3 E_{k_3} \end{bmatrix} U^*. \tag{7.2}$$

If $E = E(k_1, k_2, k_3, S)$ is the equation of state, the stress tensor $\|\sigma_{ij}\|$ is uniquely determined from the metric deformation tensor $\|g_{ij}\|$. We write the corresponding formulas in a more elegant form, which, in addition, allows us to perform calculations without reducing the strain tensor to the canonical form.

7.1. Since $1/k_1^2$, $1/k_2^2$, $1/k_3^2$ are the eigenvalues of the matrix G, they can be uniquely determined up to an order by the invariants of G:

$$\begin{aligned}
\mathcal{I}_1 &= \frac{1}{k_1^2} + \frac{1}{k_2^2} + \frac{1}{k_3^2} = g_{11} + g_{22} + g_{33}, \\
\mathcal{I}_2 &= \frac{1}{k_2^2k_3^2} + \frac{1}{k_3^2k_1^2} + \frac{1}{k_1^2k_2^2} = \begin{vmatrix} g_{22} & g_{23} \\ g_{32} & g_{33} \end{vmatrix} + \begin{vmatrix} g_{33} & g_{31} \\ g_{13} & g_{11} \end{vmatrix} + \begin{vmatrix} g_{11} & g_{12} \\ g_{21} & g_{22} \end{vmatrix}, \\
\mathcal{I}_3 &= \frac{1}{k_1^2k_2^2k_3^2} = \begin{vmatrix} g_{11} & g_{12} & g_{13} \\ g_{21} & g_{22} & g_{23} \\ g_{31} & g_{32} & g_{33} \end{vmatrix} = \det G.
\end{aligned} \tag{7.3}$$

Under our assumptions, the specific internal energy $E = E(k_1, k_2, k_3, S)$ is a symmetric function of k_1, k_2, k_3. Therefore, E depends on $\mathcal{I}_1$, $\mathcal{I}_2$, $\mathcal{I}_3$ and can be expressed in terms of g_{ik}. Therefore, we can write $E = E(\mathcal{I}_1, \mathcal{I}_2, \mathcal{I}_3, S)$. Introduce the notation

$$g_1 = 1/k_1^2, \quad g_2 = 1/k_2^2, \quad g_3 = 1/k_3^2 \tag{7.4}$$

and note that

$$k_1 E_{k_1} = -2g_1 E_{g_1} = -2g_1 \left[\frac{\partial E}{\partial \mathcal{I}_1}\frac{\partial \mathcal{I}_1}{\partial g_1} + \frac{\partial E}{\partial \mathcal{I}_2}\frac{\partial \mathcal{I}_2}{\partial g_1} + \frac{\partial E}{\partial \mathcal{I}_3}\frac{\partial \mathcal{I}_3}{\partial g_1} \right].$$

Since

$$\begin{aligned} \frac{\partial \mathcal{I}_1}{\partial g_1} &= \frac{\partial}{\partial g_1}(g_1 + g_2 + g_3) = 1, \\ \frac{\partial \mathcal{I}_2}{\partial g_1} &= \frac{\partial}{\partial g_1}(g_2 g_3 + g_3 g_1 + g_1 g_2) = g_2 + g_3 = \mathcal{I}_1 - g_1, \\ \frac{\partial \mathcal{I}_3}{\partial g_1} &= \frac{\partial}{\partial g_1}(g_1 g_2 g_3) = g_2 g_3 = \frac{\mathcal{I}_3}{g_1}, \end{aligned}$$

we find

$$k_1 E_{k_1} = -2g_1 \left(\frac{\partial E}{\partial \mathcal{I}_1} + \mathcal{I}_1 \frac{\partial E}{\partial \mathcal{I}_2} - g_1 \frac{\partial E}{\partial \mathcal{I}_2} + g_1^{-1} \mathcal{I}_3 \frac{\partial E}{\partial \mathcal{I}_3} \right). \tag{7.5}$$

Similarly,

$$k_2 E_{k_2} = -2g_2 \left(\frac{\partial E}{\partial \mathcal{I}_1} + \mathcal{I}_1 \frac{\partial E}{\partial \mathcal{I}_2} - g_2 \frac{\partial E}{\partial \mathcal{I}_2} + g_2^{-1} \mathcal{I}_3 \frac{\partial E}{\partial \mathcal{I}_3} \right), \tag{7.6}$$

$$k_3 E_{k_3} = -2g_3 \left(\frac{\partial E}{\partial \mathcal{I}_1} + \mathcal{I}_1 \frac{\partial E}{\partial \mathcal{I}_2} - g_3 \frac{\partial E}{\partial \mathcal{I}_2} + g_3^{-1} \mathcal{I}_3 \frac{\partial E}{\partial \mathcal{I}_3} \right). \tag{7.7}$$

By (7.2) and (7.5)–(7.7), we can transform the stress tensor $\|\sigma_{ij}\|$ as follows:

$$\begin{aligned} \begin{bmatrix} \sigma_{11} & \sigma_{12} & \sigma_{13} \\ \sigma_{21} & \sigma_{22} & \sigma_{23} \\ \sigma_{31} & \sigma_{32} & \sigma_{33} \end{bmatrix} &= \rho U \begin{bmatrix} k_1 E_{k_1} & 0 & 0 \\ 0 & k_2 E_{k_2} & 0 \\ 0 & 0 & k_3 E_{k_3} \end{bmatrix} U^* \\ = \rho U \Bigg[-2 \left(\frac{\partial E}{\partial \mathcal{I}_1} + \mathcal{I}_1 \frac{\partial E}{\partial \mathcal{I}_2} \right) & \begin{bmatrix} g_1 & 0 & 0 \\ 0 & g_2 & 0 \\ 0 & 0 & g_3 \end{bmatrix} + 2 \frac{\partial E}{\partial \mathcal{I}_2} \begin{bmatrix} g_1^2 & 0 & 0 \\ 0 & g_2^2 & 0 \\ 0 & 0 & g_3^2 \end{bmatrix} \\ - 2\mathcal{I}_3 \frac{\partial E}{\partial \mathcal{I}_3} & \begin{bmatrix} 1 & 0 & 0 \\ 0 & 1 & 0 \\ 0 & 0 & 1 \end{bmatrix} \Bigg] U^* \end{aligned}$$

$$= -2\rho\left(\frac{\partial E}{\partial \mathcal{I}_1} + \mathcal{I}_1\frac{\partial E}{\partial \mathcal{I}_2}\right) U \begin{bmatrix} g_1 & 0 & 0 \\ 0 & g_2 & 0 \\ 0 & 0 & g_3 \end{bmatrix} U^* + 2\rho\frac{\partial E}{\partial \mathcal{I}_2} U \begin{bmatrix} g_1^2 & 0 & 0 \\ 0 & g_2^2 & 0 \\ 0 & 0 & g_3^2 \end{bmatrix} U^*$$

$$- 2\rho\mathcal{I}_3\frac{\partial E}{\partial \mathcal{I}_3}\begin{bmatrix} 1 & 0 & 0 \\ 0 & 1 & 0 \\ 0 & 0 & 1 \end{bmatrix}$$

$$= -2\rho\left(\frac{\partial E}{\partial \mathcal{I}_1} + \mathcal{I}_1\frac{\partial E}{\partial \mathcal{I}_2}\right) G + 2\rho\frac{\partial E}{\partial \mathcal{I}_2} G^2 - 2\rho\mathcal{I}_3\frac{\partial E}{\partial \mathcal{I}_3} I$$

$$= -2\rho G\left\{\left(\frac{\partial E}{\partial \mathcal{I}_1} + \mathcal{I}_1\frac{\partial E}{\partial \mathcal{I}_2}\right) I - \frac{\partial E}{\partial \mathcal{I}_2} G + \mathcal{I}_3\frac{\partial E}{\partial \mathcal{I}_3} G^{-1}\right\}. \tag{7.8}$$

It turns out that the matrix in the curl brackets on the right-hand side of (7.8) can be written in the following elegant form:

$$\left(\frac{\partial E}{\partial \mathcal{I}_1} + \mathcal{I}_1\frac{\partial E}{\partial \mathcal{I}_2}\right) I - \frac{\partial E}{\partial \mathcal{I}_2} G + \mathcal{I}_3\frac{\partial E}{\partial \mathcal{I}_3} G^{-1} = \left\|\frac{\partial E}{\partial g_{ij}}\right\|. \tag{7.9}$$

If (7.9) is true, we have

$$\|\sigma_{ik}\| = -2\rho G\left\|\frac{\partial E}{\partial g_{jk}}\right\| = -2\rho\|g_{ij}\|\left\|\frac{\partial E}{\partial g_{jk}}\right\| \tag{7.10}$$

or, shortly,

$$\sigma_{ik} = -2\rho g_{ij}\frac{\partial E}{\partial g_{jk}}. \tag{7.11}$$

To prove (7.9), we compute the derivatives of the invariants $\mathcal{I}_1$, $\mathcal{I}_2$, $\mathcal{I}_3$ (2) with respect to g_{ij}. Since $g_{ik} = g_{ki}$, we have

$$\left\|\frac{\partial \mathcal{I}_1}{\partial g_{ik}}\right\| = \begin{bmatrix} 1 & 0 & 0 \\ 0 & 1 & 0 \\ 0 & 0 & 1 \end{bmatrix} = I, \tag{7.12}$$

$$\left\|\frac{\partial \mathcal{I}_2}{\partial g_{ik}}\right\| = \begin{bmatrix} g_{22}+g_{33} & -g_{21} & -g_{31} \\ -g_{12} & g_{33}+g_{11} & -g_{32} \\ -g_{13} & -g_{23} & g_{11}+g_{22} \end{bmatrix}$$

$$= (g_{11}+g_{22}+g_{33})I - G = \mathcal{I}_1 I - G, \tag{7.13}$$

$$\left\|\frac{\partial \mathcal{I}_3}{\partial g_{ik}}\right\| = \begin{bmatrix} \text{matrix} \\ \text{of algebraic} \\ \text{complements} \\ \text{of elements} \end{bmatrix} = \mathcal{I}_3 G^{-1}. \tag{7.14}$$

We note that formula (7.14) written in the form

$$\left\|\frac{\partial \det G}{\partial g_{ik}}\right\| = (\det G)G^{-1} \tag{7.15}$$

will be used in Section 8.

Using the relation

$$\frac{\partial E}{\partial g_{ik}} = \frac{\partial E}{\partial \mathcal{I}_1}\frac{\partial \mathcal{I}_1}{\partial g_{ik}} + \frac{\partial E}{\partial \mathcal{I}_2}\frac{\partial \mathcal{I}_2}{\partial g_{ik}} + \frac{\partial E}{\partial \mathcal{I}_3}\frac{\partial \mathcal{I}_3}{\partial g_{ik}}$$

and substituting the expressions (7.12), (7.13), (7.14) for $\dfrac{\partial \mathcal{I}_1}{\partial g_{ik}}, \dfrac{\partial \mathcal{I}_2}{\partial g_{ik}}, \dfrac{\partial \mathcal{I}_3}{\partial g_{ik}}$ respectively, we obtain (7.9).

We write formula (7.11) componentwise:

$$\begin{aligned}
\sigma_{11} &= -2\rho\left(g_{11}\frac{\partial E}{\partial g_{11}} + g_{12}\frac{\partial E}{\partial g_{21}} + g_{13}\frac{\partial E}{\partial g_{31}}\right),\\
\sigma_{12} &= -2\rho\left(g_{11}\frac{\partial E}{\partial g_{12}} + g_{12}\frac{\partial E}{\partial g_{22}} + g_{13}\frac{\partial E}{\partial g_{32}}\right),\\
\sigma_{21} &= -2\rho\left(g_{21}\frac{\partial E}{\partial g_{11}} + g_{22}\frac{\partial E}{\partial g_{21}} + g_{23}\frac{\partial E}{\partial g_{31}}\right).
\end{aligned} \tag{7.16}$$

It seems that the symmetry of the stress tensor, $\sigma_{ik} = \sigma_{ki}$, does not follow from (7.16). In particular, it is not clear if σ_{12} is equal to σ_{21}. But, as we know, the symmetry property of $||\sigma_{ij}||$ holds in view of the (7.2). Moreover, one can establish this fact using the property that E is not an arbitrary function of g_{ik} but depends on the invariants $\mathcal{I}_1$, $\mathcal{I}_2$, $\mathcal{I}_3$.

• *We derived formula* (7.11) *using the Murnaghan formulas in the matrix form* (*cf.* *Section* 4). *Formulas* (7.11) *or* (7.16) *are also referred to as the Murnaghan formulas* (*in the tensor form*).

In Section 8, we give another proof of the Murnaghan formulas.

7.2. In the case of small deformations, we assume that the equation of state is given as the dependence of E on the invariants of the Almansi strain tensor $||\varepsilon_{ij}||$ and the entropy S:

$$E = E\left(\varepsilon_{11} + \varepsilon_{22} + \varepsilon_{33}, \begin{vmatrix}\varepsilon_{22} & \varepsilon_{23}\\ \varepsilon_{32} & \varepsilon_{33}\end{vmatrix} + \begin{vmatrix}\varepsilon_{33} & \varepsilon_{31}\\ \varepsilon_{13} & \varepsilon_{11}\end{vmatrix} + \begin{vmatrix}\varepsilon_{11} & \varepsilon_{12}\\ \varepsilon_{21} & \varepsilon_{22}\end{vmatrix}, \begin{vmatrix}\varepsilon_{11} & \varepsilon_{12} & \varepsilon_{13}\\ \varepsilon_{21} & \varepsilon_{22} & \varepsilon_{23}\\ \varepsilon_{31} & \varepsilon_{32} & \varepsilon_{33}\end{vmatrix}, S\right). \tag{7.17}$$

Since $g_{ij} = \delta_{ij} - 2\varepsilon_{ij}$, the Murnaghan formulas can be written in the form

$$\sigma_{ij} = \rho(\delta_{ik} - 2\varepsilon_{ik})\frac{\partial E}{\partial \varepsilon_{kj}}. \tag{7.18}$$

We expand E with respect to ε_{ij} and S in a neighborhood of a point $\varepsilon_{ij} = 0$, $S = S_0$. In linear elasticity, only first- and second-order terms of the expansion of E are taken into account. We have

$$E = E_0 + E_S(S - S_0) + \frac{1}{2}E_{SS}(S - S_0)^2 + E_{S1}(S - S_0)(\varepsilon_{11} + \varepsilon_{22} + \varepsilon_{33})$$
$$+ E_1(\varepsilon_{11} + \varepsilon_{22} + \varepsilon_{33}) + \frac{1}{2}E_{11}(\varepsilon_{11} + \varepsilon_{22} + \varepsilon_{33})^2$$
$$+ \frac{1}{2}E_{22}\left(\begin{vmatrix}\varepsilon_{22} & \varepsilon_{23}\\ \varepsilon_{32} & \varepsilon_{33}\end{vmatrix} + \begin{vmatrix}\varepsilon_{33} & \varepsilon_{31}\\ \varepsilon_{13} & \varepsilon_{11}\end{vmatrix} + \begin{vmatrix}\varepsilon_{11} & \varepsilon_{12}\\ \varepsilon_{21} & \varepsilon_{22}\end{vmatrix}\right), \tag{7.19}$$

where E_0 is the initial energy (without loss of generality, we further assume that $E_0 = 0$ since the energy is uniquely determined up to an additive constant), the subscript "S" means the differentiation with respect to S and the subscripts "1" and "2" of E mean the differentiation with respect to the first and second arguments of E of the form (7.17). The derivatives are taken at $\varepsilon_{ij} = 0$, $S = S_0$. Assuming that the medium is in the nonstressed state (i.e., $\sigma_{ij} = 0$) at the standard temperature T_0) and using (7.18), we compute

$$E_S\Big|_{\substack{S=S_0\\ \varepsilon_{ij}=0}} = T_0, \quad \frac{\partial E}{\partial \varepsilon_{ij}}\Big|_{\substack{S=S_0\\ \varepsilon_{ij}=0}} = 0.$$

The last equality holds only if $E_1 = 0$. Hence we can write (7.19) as follows:

$$E = T_0(S - S_0) + \frac{1}{2}E_{SS}(S - S_0)^2 + E_{S1}(S - S_0)(\varepsilon_{11} + \varepsilon_{22} + \varepsilon_{33})$$
$$+ \left(\frac{E_{11}}{2} + \frac{E_{22}}{4}\right)(\varepsilon_{11} + \varepsilon_{22} + \varepsilon_{33})^2 - \frac{E_{22}}{4}[(\varepsilon_{11}\varepsilon_{11} + \varepsilon_{12}\varepsilon_{21} + \varepsilon_{13}\varepsilon_{31})$$
$$+ (\varepsilon_{21}\varepsilon_{12} + \varepsilon_{22}\varepsilon_{22} + \varepsilon_{23}\varepsilon_{32}) + (\varepsilon_{31}\varepsilon_{13} + \varepsilon_{32}\varepsilon_{23} + \varepsilon_{33}\varepsilon_{33})]$$

or, introducing parameters ω, π, λ, μ,

$$E = T_0(S - S_0) + \frac{\omega}{2}(S - S_0)^2 - \frac{\pi}{\rho_0}(S - S_0)(\varepsilon_{11} + \varepsilon_{22} + \varepsilon_{33})$$
$$+ \frac{\lambda}{2\rho_0}(\varepsilon_{11} + \varepsilon_{22} + \varepsilon_{33})^2 + \frac{\mu}{\rho_0}\varepsilon_{ij}\varepsilon_{ji}. \tag{7.20}$$

- The coefficients λ and μ in (7.20) are called the *Lamé constants.*

Thus, to write out the equation of state, it is necessary to know the Lamé coefficients λ, μ and parameters ω and π which are known if the thermal capacity q and the coefficient of thermal expansion β are given (cf. below).

From (7.20) we obtain the approximate formula for the temperature

$$T = E_S = T_0 - \frac{\pi}{\rho_0}(\varepsilon_{11} + \varepsilon_{22} + \varepsilon_{33}) + \omega(S - S_0) \tag{7.21}$$

and the stress tensor

$$\sigma_{ij} = \rho_0 \frac{\partial E}{\partial \varepsilon_{ij}} = \left[-\pi(S - S_0) + \lambda(\varepsilon_{11} + \varepsilon_{22} + \varepsilon_{33})\right]\delta_{ij} + 2\mu\varepsilon_{ij}. \tag{7.22}$$

To derive (7.22), we transformed the second-order terms in the Murnaghan formulas

$$\sigma_{ij} = \rho(\delta_{ik} - 2\varepsilon_{ik})\frac{\partial E}{\partial \varepsilon_{ki}} = \rho_0 \frac{\partial E}{\partial \varepsilon_{ij}} + \{(\rho - \rho_0)\frac{\partial E}{\partial \varepsilon_{ij}} - 2\rho\varepsilon_{ik}\frac{\partial E}{\partial \varepsilon_{kj}}\}$$

and used the relation

$$\rho = \rho_0\sqrt{\det(\delta_{ij} - 2\varepsilon_{ij})} = \rho_0[1 - (\varepsilon_{11} + \varepsilon_{22} + \varepsilon_{33})] + \dots,$$

where the dots denote second-order terms with respect to ε_{ij}.

7.3. Thus, we have derived the following formulas of linear elasticity:

$$\begin{aligned}
&E = T_0(S - S_0) + \frac{\omega}{2}(S - S_0)^2 - \frac{\pi}{\rho_0}(S - S_0)(\varepsilon_{11} + \varepsilon_{22} + \varepsilon_{33}) \\
&\quad + \frac{\lambda}{2\rho_0}(\varepsilon_{11} + \varepsilon_{22} + \varepsilon_{33})^2 + \frac{\mu}{\rho_0}\varepsilon_{ij}\varepsilon_{ji}, \\
&\sigma_{11} = -\pi(S - S_0) + \lambda(\varepsilon_{11} + \varepsilon_{22} + \varepsilon_{33}) + 2\mu\varepsilon_{11}, \\
&\sigma_{22} = -\pi(S - S_0) + \lambda(\varepsilon_{11} + \varepsilon_{22} + \varepsilon_{33}) + 2\mu\varepsilon_{22}, \\
&\sigma_{33} = -\pi(S - S_0) + \lambda(\varepsilon_{11} + \varepsilon_{22} + \varepsilon_{33}) + 2\mu\varepsilon_{33}, \\
&\sigma_{ij} = 2\mu\varepsilon_{ij} \quad \text{if } i \neq j, \\
&T - T_0 = \omega(S - S_0) - \frac{\pi}{\rho_0}(\varepsilon_{11} + \varepsilon_{22} + \varepsilon_{33}), \\
&1 - \frac{\rho}{\rho_0} = \varepsilon_{11} + \varepsilon_{22} + \varepsilon_{33}.
\end{aligned} \tag{7.23}$$

In the nonstressed state ($\sigma_{ij} = 0$, $\frac{\omega}{T_0}(S - S_0) \lll |S - S_0|$), from (7.23) we obtain the following approximations:

$$\begin{aligned}
&\pi(S - S_0) = (\lambda + 2\mu/3)\left(1 - \frac{\rho}{\rho_0}\right), \\
&T - T_0 = \omega(S - S_0) - \frac{\pi}{\rho_0}\left(1 - \frac{\rho}{\rho_0}\right), \\
&E = T_0(S - S_0)
\end{aligned} \tag{7.24}$$

or, expressing $S - S_0$,

$$\begin{aligned}
&1 - \frac{\rho}{\rho_0} = \frac{\rho_0\pi}{\rho_0\omega(\lambda + 2\mu/3) - \pi^2}(T - T_0) = \beta(T - T_0), \\
&E = \frac{\rho_0 T_0(\lambda + 2\mu/3)}{\rho_0\omega(\lambda + 2\mu/3) - \pi^2}(T - T_0) = q(T - T_0),
\end{aligned} \tag{7.25}$$

where

$$\beta = \frac{\rho_0\pi}{\rho_0\omega(\lambda + 2\mu/3) - \pi^2}, \quad q = \frac{\rho_0 T_0(\lambda + 2\mu/3)}{\rho_0\omega(\lambda + 2\mu/3) - \pi^2}$$

are the *coefficient of thermal volume expansion* and the *thermal capacity*.

In turn, if $(\lambda + 2\mu/3)$, T_0, β, q are known, we can compute ω and π by the formulas

$$\omega = \frac{T_0}{q}\left(1 + \frac{(\lambda + 2\mu/3)\beta^2 T_0}{\rho_0 q}\right), \quad \pi = \frac{(\lambda + 2\mu/3)\beta T_0}{q}.$$

7.4. The following formulas expressing the linear dependence of stresses on deformations are called *Hook's law*:

$$\sigma_{ij} = \left[-\pi(S - S_0) + \lambda(\varepsilon_{11} + \varepsilon_{22} + \varepsilon_{33})\right]\delta_{ij} + 2\mu\varepsilon_{ij}. \tag{7.26}$$

By (7.21), we have

$$S - S_0 = \frac{1}{\omega}(T - T_0) + \frac{\pi}{\rho_0 \omega}(\varepsilon_{11} + \varepsilon_{22} + \varepsilon_{33}). \tag{7.27}$$

Using (7.27), we can write the Hook law in the form

$$\sigma_{ij} = \left[-\frac{\pi}{\omega}(T - T_0) + \left(\lambda - \frac{\pi^2}{\rho_0 \omega}\right)(\varepsilon_{11} + \varepsilon_{22} + \varepsilon_{33})\right]\delta_{ij} + 2\mu\varepsilon_{ij}. \tag{7.28}$$

As a rule, formulas (7.26) and (7.29) are referred to as the "Hook law" only for isentropic and isothermal processes:

isentropic process: $\sigma_{ij} = \lambda(\varepsilon_{11} + \varepsilon_{22} + \varepsilon_{33})\delta_{ij} + 2\mu\varepsilon_{ij}$,

isothermal process: $\sigma_{ij} = \left(\lambda - \dfrac{\pi^2}{\rho_0 \omega}\right)(\varepsilon_{11} + \varepsilon_{22} + \varepsilon_{33})\delta_{ij} + 2\mu\varepsilon_{ij}$.

§ 8. Moving Medium. Murnaghan Formulas

Describing the motion of a continuous medium by the equalities

$$x_1 = x_1(x_{10}, x_{20}, x_{30}, t), \ x_2 = x_2(x_{10}, x_{20}, x_{30}, t), \ x_3 = x_3(x_{10}, x_{20}, x_{30}, t),$$

we thereby describe the trajectories of material particles with coordinates x_{10}, x_{20}, x_{30} at a time $t = t_0$ The components of the velocity of the particles are the partial derivatives

$$u_i = \frac{\partial}{\partial t} x_i(x_{10}, x_{20}, x_{30}, t)\Big|_{x_{j0}=\text{const}}$$

in some fixed Lagrangian coordinates (for example, the initial coordinates of the particles). In the space equipped with the Eulerian coordinates x_1, x_2, x_3, these trajectories are the integral curves of the equations $\dfrac{dx_i}{dt} = u_i$. Along each trajectory, the initial coordinates x_{i0} are constant:

$$\frac{dx_{i0}}{dt} = \frac{\partial x_{i0}}{\partial t} + \frac{\partial x_{i0}}{\partial x_1}\frac{dx_1}{dt} + \frac{\partial x_{i0}}{\partial x_2}\frac{dx_2}{dt} + \frac{\partial x_{i0}}{\partial x_3}\frac{dx_3}{dt} = \frac{\partial x_{i0}}{\partial t} + u_k\frac{\partial x_{i0}}{\partial x_k} = 0. \quad (8.1)$$

As usual, we denote by $\frac{d}{dt}$ the so-called *substantional derivative*

$$\frac{d}{dt} = \frac{\partial}{\partial t} + u_k\frac{\partial}{\partial x_k} \equiv \frac{\partial}{\partial t} + u_1\frac{\partial}{\partial x_1} + u_2\frac{\partial}{\partial x_2} + u_3\frac{\partial}{\partial x_3}, \quad (8.2)$$

i.e., the time-derivative along the trajectory of a fixed material particle.

8.1. We differentiate Eq. (8.1) with respect to x_j:

$$\frac{\partial}{\partial t}\left(\frac{\partial x_{i0}}{\partial x_j}\right) + u_k\frac{\partial}{\partial x_k}\left(\frac{\partial x_{i0}}{\partial x_j}\right) + \frac{\partial u_k}{\partial x_j}\left(\frac{\partial x_{i0}}{\partial x_k}\right) = 0. \quad (8.3)$$

Recalling the definition of the distortion tensor $A = \|a_{ij}\| = \left\|\frac{\partial x_{i0}}{\partial x_j}\right\|$ (cf. Section 6), we write (8.3) in the form

$$\frac{d}{dt}A + AW \equiv \frac{\partial}{\partial t}A + u_k\frac{\partial}{\partial x_k}A + AW = 0, \quad (8.4)$$

where $W = \left\|\frac{\partial u_i}{\partial x_k}\right\|$, or, componentwise,

$$\frac{\partial a_{ij}}{\partial t} + u_k\frac{\partial a_{ij}}{\partial x_k} + a_{ik}\frac{\partial u_k}{\partial x_j} = 0. \quad (8.5)$$

8.2. Since the metric deformation tensor $G = \|g_{ij}\|$ is expressed in terms of the distortion tensor $A = \|a_{ij}\|$ by the formula $G = A^*A$, we can write a similar equation for G if we know an equation describing the behavior of A with time. Passing to the transposed matrices in (8.4), we find

$$\frac{dA^*}{dt} + W^*A^* = 0. \quad (8.5)$$

Hence

$$0 = A^*\left(\frac{dA}{dt} + AW\right) + \left(\frac{dA^*}{dt} + W^*A^*\right)A$$
$$= \frac{d}{dt}(A^*A) + (A^*A)W + W^*(A^*A) = \frac{d}{dt}G + GW + W^*G.$$

Thus,

$$\frac{d}{dt}G + GW + W^*G = 0 \quad (8.7)$$

or, componentwise,

$$\frac{\partial g_{ij}}{\partial t} + u_r \frac{\partial g_{ij}}{\partial x_r} + g_{ik}\frac{\partial u_k}{\partial x_j} + \frac{\partial u_k}{\partial x_i} g_{kj} = 0. \tag{8.8}$$

This equation shows how the field of the metric deformation tensor changes with time.

8.3. We can write an equation for the matrix $C = A^{-1}$ using the relation $dC = -A^{-1}(dA)A^{-1}$ which follows from the equalities:

$$0 = d(CA) = CdA + (dC)A = A^{-1}dA + dCA.$$

Indeed, multiplying Eq. (2) from the left and right by A^{-1}, we obtain the equation

$$\frac{dC}{dt} - WC = 0 \tag{8.9}$$

or, componentwise,

$$\frac{\partial c_{ij}}{\partial t} + u_k \frac{\partial c_{ij}}{\partial x_k} - \frac{\partial u_i}{\partial x_k} c_{kj} = 0. \tag{8.10}$$

Remark 8.1. Sometimes, instead of Eq. (8.4) for $\frac{d}{dt}A$ and Eq. (8.7) for $\frac{d}{dt}G$, similar equations for variations of A and G are used. Let a material particle with the coordinates x_1, x_2, x_3 before variation take infinitesimal displacements $(\delta x_1, \delta x_2, \delta x_3)$. Then

$$\delta a_{ij} + a_{ik}\frac{\partial \delta x_k}{\partial x_j} = 0, \tag{8.11}$$

$$\delta g_{ij} + g_{ik}\frac{\partial \delta x_k}{\partial x_j} + \frac{\partial \delta x_k}{\partial x_i} g_{kj} = 0. \tag{8.12}$$

Equations (8.11) and (8.12) are reduced to Eqs. (8.5) and (8.8) respectively if we set

$$\delta x_i = u_i dt, \quad \delta a_{ij} = \frac{da_{ij}}{dt}dt, \quad \delta g_{ij} = \frac{dg_{ij}}{dt}dt.$$

8.4. By the mass conservation law, the density ρ of the deformed medium can be found from the distortion tensor A (cf. Section 6) by the formula $\rho = \rho_0 \det A$. Then the variation of the density ρ takes the form

$$\delta\rho = \rho_0 \delta \begin{vmatrix} a_{11} & a_{12} & a_{13} \\ a_{21} & a_{22} & a_{23} \\ a_{31} & a_{32} & a_{33} \end{vmatrix} = \rho_0 \delta a_{ij} M_{ji},$$

where M_{ji} are the algebraic complements of a_{ij}. As is known, M_{ji} differ from the entries c_{ij} of the inverse matrix A^{-1} only by the factor $\det A$, and we have $M_{ji} = c_{ji} \det A = \frac{\rho}{\rho_0} c_{ji}$. Therefore,

$$\delta\rho = \rho c_{ji} \delta a_{ij} = -\rho c_{ji} a_{ik} \frac{\partial(\delta x_k)}{\partial x_j} = -\rho \operatorname{tr}\left[CA\left\|\frac{\partial(\delta x_k)}{\partial x_j}\right\|\right]$$
$$= -\rho \operatorname{tr}\left[A^{-1}A\left\|\frac{\partial(\delta x_k)}{\partial x_j}\right\|\right] = -\rho \operatorname{tr}\left\|\frac{\partial(\delta x_k)}{\partial x_j}\right\| = -\rho \frac{\partial(\delta x_i)}{\partial x_i},$$

where $\operatorname{tr} H$ denotes the trace of a matrix H. Finally, we have

$$\delta\rho = -\rho \frac{\partial(\delta x_i)}{\partial x_i}. \tag{8.13}$$

Similarly, we can compute the time-derivative of the density ρ if the deformation is described by the velocity field of the particles:

$$\frac{d\rho}{dt} \equiv \frac{\partial \rho}{\partial t} + u_i \frac{\partial \rho}{\partial x_i} = -\rho \frac{\partial u_i}{\partial x_i}.$$

- The equation

$$\frac{\partial \rho}{\partial t} + u_i \frac{\partial \rho}{\partial x_i} + \rho \frac{\partial u_i}{\partial x_i} \equiv \frac{\partial \rho}{\partial t} + \frac{\partial(\rho u_i)}{\partial x_i} = 0. \tag{8.14}$$

is called the *equation of continuity*.

Remark 8.2. The equation of continuity expresses the mass conservation law and can be derived from the integral form of this law. However, for our purposes it is important to emphasize that this equation is a consequence of the equality $\rho = \rho_0 \det A$ and Eq. (8.5) describing the change of the distortion tensor $||a_{ij}||$ with time.

Problem 8.1. Prove that the following equations immediately follow from Eqs. (8.10) and (8.14):

$$\frac{\partial(\rho c_{ij})}{\partial t} + \frac{\partial(u_k \rho c_{ij})}{\partial x_k} - \rho c_{kj} \frac{\partial u_i}{\partial x_k} = 0. \tag{8.15}$$

This fact will be very important in the further consideration.

8.5. We describe another approach to obtaining the equation of continuity. It is based on the equality $\rho = \rho_0\sqrt{\det \|g_{ij}\|} = \rho_0 \det G$, which yields

$$\begin{aligned}\frac{\partial\rho}{\partial t} + u_r\frac{\partial\rho}{\partial x_r} &\equiv \frac{d\rho}{dt} = \frac{\partial\rho}{\partial g_{ij}}\frac{dg_{ij}}{dt} \equiv \frac{\partial\rho}{\partial g_{ij}}\left(\frac{\partial g_{ij}}{\partial t} + u_r\frac{\partial g_{ij}}{\partial x_r}\right)\\ &= -\frac{\partial\rho}{\partial g_{ij}}\left(g_{ik}\frac{\partial u_k}{\partial x_j} + \frac{\partial u_k}{\partial x_i}g_{kj}\right) = -\frac{\partial\rho}{\partial g_{ij}}g_{ik}\frac{\partial u_k}{\partial x_j} - \frac{\partial\rho}{\partial g_{ij}}\frac{\partial u_k}{\partial x_i}g_{kj}\\ &= -\frac{\partial\rho}{\partial g_{ji}}g_{ik}\frac{\partial u_k}{\partial x_j} - \frac{\partial\rho}{\partial g_{ij}}g_{jk}\frac{\partial u_k}{\partial x_i} = -2\frac{\partial\rho}{\partial g_{ij}}g_{jk}\frac{\partial u_k}{\partial x_i}.\end{aligned}$$

Here, we used the symmetry property $g_{jk} = g_{kj}$ and its consequence $\dfrac{\partial\rho}{\partial g_{ij}} = \dfrac{\partial\rho}{\partial g_{ji}}$. Using the expressions for the derivatives of $\det G$ with respect to g_{ij} (cf. Section 7), we easily find

$$\left\|\frac{\partial\rho}{\partial g_{ij}}\right\| = \frac{\rho_0}{2\sqrt{\det G}}\left\|\frac{\partial \det G}{\partial g_{ij}}\right\| = \frac{\rho_0}{2\sqrt{\det G}}\det G G^{-1} = \frac{\rho}{2}G^{-1},$$

$$\left\|\frac{\partial\rho}{\partial g_{ij}}g_{ik}\right\| = \frac{\rho}{2}G^{-1}G = \frac{\rho}{2}I, \qquad \left\|\frac{\partial\rho}{\partial g_{ij}}g_{jk}\frac{\partial u_k}{\partial x_r}\right\| = \frac{\rho}{2}\left\|\frac{\partial u_i}{\partial x_r}\right\|$$

and, consequently,

$$-2\frac{\partial\rho}{\partial g_{ij}}g_{jk}\frac{\partial u_k}{\partial x_i} = -\rho\,\mathrm{tr}\left\|\frac{\partial u_i}{\partial x_r}\right\| = -\rho\frac{\partial u_i}{\partial x_i}. \tag{8.16}$$

Thus, we again obtain the equation of continuity (8.14) but, in this case, from the equality

$$\frac{\partial\rho}{\partial t} + u_r\frac{\partial\rho}{\partial x_r} + \rho\frac{\partial u_i}{\partial x_i} = \frac{\partial\rho}{\partial g_{ij}}\left(\frac{\partial g_{ij}}{\partial t} + u_r\frac{\partial g_{ij}}{\partial x_r} + g_{ik}\frac{\partial u_k}{\partial x_j} + \frac{\partial u_k}{\partial x_i}g_{kj}\right)$$

whose right-hand side vanishes because of the properties of g_{ij}.

We use the representations of δg_{ij}, δa_{ij}, $\delta\rho$ in order to give another proof of the Murnaghan formulas (cf. Section 7)

$$\sigma_{ij} = -2\rho g_{ik}\frac{\partial E}{\partial g_{kj}}. \tag{8.17}$$

By definition (cf. Section 5), the work done by the internal surface forces under deformation of some volume element is the integral over this volume

$$\begin{aligned}&\iiint \sigma_{ki}\cdot\frac{1}{2}\left[\frac{\partial(\delta x_i)}{\partial x_k} + \frac{\partial(\delta x_k)}{\partial x_i}\right]dx_1dx_2dx_3\\ &\quad = \iiint \frac{\sigma_{ki}}{\rho}\cdot\frac{1}{2}\left[\frac{\partial(\delta x_i)}{\partial x_k} + \frac{\partial(\delta x_k)}{\partial x_i}\right]\rho\, dx_1dx_2dx_3\\ &\quad = \iiint \frac{\sigma_{ki}}{\rho}\cdot\frac{1}{2}\left[\frac{\partial(\delta x_i)}{\partial x_k} + \frac{\partial(\delta x_k)}{\partial x_i}\right]dm.\end{aligned}$$

Introducing the mass differential $dm = \rho dx_1 dx_2 dx_3$, we can say that the integral is taken over a deformable mass element of the medium so that the material particles with coordinates (x_1, x_2, x_3) receive displacements $(\delta x_1, \delta x_2, \delta x_3)$. The change of the internal energy of this element $\iiint \delta E dm$ is equal to the sum of the work done by the internal forces and the energy equal to an increment in heat $\iiint \delta q dm$. Since the integral is taken over an arbitrary mass element, we have

$$\delta E(g_{11}, g_{12}, \ldots, g_{33}, S) = \frac{\sigma_{ki}}{\rho} \cdot \frac{1}{2}\left[\frac{\partial(\delta x_i)}{\partial x_k} + \frac{\partial(\delta x_k)}{\partial x_i}\right] + \delta q. \tag{8.18}$$

Here, we assume that the function $E = E(k_1, k_2, k_3, S)$ determining the equation of state for an isotropic medium is expressed, as in Section 7, in terms of the metric deformation tensor g_{ij}.

Recall that the change of heat δq is expressed in terms of dS by the equality

$$\delta q = T\delta S = E_S \delta S$$

(cf. Section 4). Therefore,

$$\delta E = \frac{\partial E}{\partial g_{ij}}\delta g_{ij} + \frac{\partial E}{\partial S}\delta S = \frac{\sigma_{ki}}{\rho} \cdot \frac{1}{2}\left[\frac{\partial(\delta x_i)}{\partial x_k} + \frac{\partial(\delta x_k)}{\partial x_i}\right] + E_S \delta S$$

and, consequently,

$$\frac{\partial E}{\partial g_{ij}}\delta g_{ij} = \frac{\sigma_{ki}}{\rho} \cdot \frac{1}{2}\left[\frac{\partial(\delta x_i)}{\partial x_k} + \frac{\partial(\delta x_k)}{\partial x_i}\right]. \tag{8.19}$$

Using the symmetry property $\sigma_{ki} = \sigma_{ik}$ of the stress tensor, we can transform the right-hand side of (8.19) as follows:

$$\frac{\sigma_{ki}}{\rho} \cdot \frac{1}{2}\left[\frac{\partial(\delta x_i)}{\partial x_k} + \frac{\partial(\delta x_k)}{\partial x_i}\right] = \frac{1}{2\rho}\sigma_{ki}\frac{\partial(\delta x_i)}{\partial x_k} + \frac{1}{2\rho}\sigma_{ki}\frac{\partial(\delta x_k)}{\partial x_i}$$
$$= \frac{1}{2\rho}\sigma_{ki}\frac{\partial(\delta x_i)}{\partial x_k} + \frac{1}{2\rho}\sigma_{ik}\frac{\partial(\delta x_k)}{\partial x_i} \equiv \frac{1}{2\rho}\sigma_{ki}\frac{\partial(\delta x_i)}{\partial x_k} + \frac{1}{2\rho}\sigma_{ki}\frac{\partial(\delta x_i)}{\partial x_k} = \frac{\sigma_{ki}}{\rho}\frac{\partial(\delta x_i)}{\partial x_k}.$$

Using the expression (8.12) for the variation δg_{ij} we transform the left-hand side of (8.19) as follows:

$$\frac{\partial E}{\partial g_{ij}}\delta g_{ij} = -\frac{\partial E}{\partial g_{ij}}g_{ik}\frac{\partial(\delta x_k)}{\partial x_j} - \frac{\partial E}{\partial g_{ij}}g_{kj}\frac{\partial(\delta x_k)}{\partial x_i}$$
$$= -\frac{\partial E}{\partial g_{ij}}g_{ik}\frac{\partial(\delta x_k)}{\partial x_j} - \frac{\partial E}{\partial g_{ji}}g_{jk}\frac{\partial(\delta x_k)}{\partial x_i} \equiv -\frac{\partial E}{\partial g_{ij}}g_{ik}\frac{\partial(\delta x_k)}{\partial x_j} - \frac{\partial E}{\partial g_{ij}}g_{ik}\frac{\partial(\delta x_k)}{\partial x_j}$$
$$= -2\frac{\partial E}{\partial g_{ij}}g_{ik}\frac{\partial(\delta x_k)}{\partial x_j} = -2\frac{\partial E}{\partial g_{ji}}g_{jk}\frac{\partial(\delta x_k)}{\partial x_i} = -2g_{kj}\frac{\partial E}{\partial g_{ji}}\frac{\partial(\delta x_k)}{\partial x_i}.$$

Thus, (8.19) can be written as follows:

$$\frac{\sigma_{ki}}{\rho}\frac{\partial(\delta x_k)}{\partial x_i} = -2g_{kj}\frac{\partial E}{\partial g_{ji}}\frac{\partial(\delta x_k)}{\partial x_i}. \tag{8.20}$$

Here, we assume that the components σ_{ki} depend only on the metric deformation tensor $\|g_{ij}\|$ and the entropy S but not on variations δx_k. Thus, we obtain the equality

$$\frac{\sigma_{ki}}{\rho} = -2g_{kj}\frac{\partial E}{\partial g_{ji}}$$

which is the same as the Murnaghan formulas (cf. Section 7).

8.6. If the medium under consideration is not isotropic (for example, a crystalline medium), the energy expended for deformation of a volume element expressing by the linear transformation $C = UKV$ depends not only on the extension $K = \begin{bmatrix} k_1 & 0 & 0 \\ 0 & k_2 & 0 \\ 0 & 0 & k_3 \end{bmatrix}$ and the entropy S but also on the directions along which the extension K is performed. More exactly, E depends on k_1, k_2, k_3, S, and the location of the extension axes relative to the crystallographic axes. If the coordinates are connected with the directions of these axes in a nondeformed medium, then one should assume that E is a function of S, K, V. In this case, the internal energy E is independent of the final rotation U.

Sometimes, it is more convenient to assume that E is expressed in terms of a_{ij} and S, i.e., $E = E(a_{11}, a_{12}, \ldots, a_{33}, S)$. Then

$$\delta E = E_S \delta S + \frac{\partial E}{\partial a_{im}}\delta a_{im} = E_S\delta S - a_{ik}\frac{\partial E}{\partial a_{im}}\frac{\partial(\delta x_k)}{\partial x_m}.$$

Comparing the last formula with the formula $\delta E = E_S\delta S + \dfrac{\sigma_{km}}{\rho}\dfrac{\partial(\delta x_k)}{\partial x_m}$, we obtain the Murnaghan formulas in the form

$$\sigma_{km} = -\rho a_{ik} E_{a_{im}}. \tag{8.21}$$

Sometimes, it is convenient to assume that E depends on c_{ij} and S, i.e., $E = E(c_{11}, c_{12}, \ldots, c_{33}, S)$. In this case, the Murnaghan formulas can be obtained in the form (the detailed proof is left to the reader)

$$\sigma_{km} = \rho c_{k\alpha} E_{c_{m\alpha}}. \tag{8.22}$$

The tensor $\|\sigma_{ij}\|$ defined by (8.21), (8.22) is symmetric only if the dependence of E on a_{ij} or c_{ij} satisfies some additional conditions, for example, if E depends only on g_{ij} or which are expressed in terms of a_{ij} or c_{ij} (E can also depend on S). We note that the symmetry of the stress tensor is essentially used in the proof of the Murnaghan formulas.

Chapter II

Effective Elastic Deformation

§ 9. Relaxation of Stresses

In Chapter I, we described properties of the stressed state of a medium and indicated two reasons for appearing or changing stresses. The first reason is the reaction of the medium to the change of shape of its elements. Such stresses are said to be *elastic* (cf. Sections 4 and 8). The second reason is the internal friction between neighboring elements of the medium. Such stresses expressed through the strain rate (cf. Section 5) are said to be *viscous*.

Viscous and elastic stresses change if the strain rate tensor or the metric deformation tensor changes. In actual media, it can happen that stresses change even if the medium is in the rest state and, consequently, the strain rate tensor vanishes, whereas the metric deformation tensor is independent of time. The simplest example is the process of equalizing temperatures where some elements of the medium become cool, whereas other elements are heated, which leads to a change of stresses depending on the thermal state of the medium. As a rule, the resulting stressed state is not in equilibrium, and the medium can begin to move even if it is in the rest state before deformation. In what follows, we will consider a motionless medium.

9.1. We give formulation of the relaxation law. Stresses in a motionless medium can change in view of chemical reactions between parts of the medium, phase transitions inside its crystal grains, and other reasons. One of the most important reasons is displacements of microscopic defects, called *dislocations*, in a crystal lattice. To study such processes, it is necessary to study their kinetics. But, in many practical situations, it suffices to use a rather rough approach suggested by Maxwell when he developed ideas due to Poisson. In an isotropic Maxwell medium without any macroscopic displacements and any heat influx, the stressed state changes in such a way that the shear stress decreases. We know (cf. Section 3) that the value of the shear stress on area elements is determined by the orientation of these elements relative to the principal axes of the stress tensor and by the differences $s_1 - s_2$, $s_2 - s_3$, $s_3 - s_1$ between the eigenvalues s_i of the stress tensor $\|\sigma_{ij}\|$. Hence the decay of the shear stresses can be described as the decay of the differences $s_i - s_j$. The decay rate is characterized by some parameter τ, called the *relaxation time of shear stresses*. The change of s_1, s_2, s_3 in a Maxwell medium without macroscopic deformations is described by the differential equations

$$\begin{aligned} \frac{ds_1}{dt} &= -\frac{1}{\tau}\left(s_1 - \frac{s_1+s_2+s_3}{3}\right), \\ \frac{ds_2}{dt} &= -\frac{1}{\tau}\left(s_2 - \frac{s_1+s_2+s_3}{3}\right), \\ \frac{ds_3}{dt} &= -\frac{1}{\tau}\left(s_3 - \frac{s_1+s_2+s_3}{3}\right). \end{aligned} \tag{9.1}$$

These equations present only one of possible mathematical formulations of relaxation processes. In this model, the sum $s_1 + s_2 + s_3$ is constant:

$$\frac{ds}{dt} = -\frac{1}{\tau}(s_1 - s) - \frac{1}{\tau}(s_2 - s) - \frac{1}{\tau}(s_3 - s) = 0, \tag{9.2}$$

where $s = (s_1 + s_2 + s_3)/3$, and the principal axes are unchanged. The directed unit vectors of the principal axes form the columns of the orthogonal matrix U in the representation of the stress tensor

$$\|\sigma_{ij}\| = \begin{bmatrix} \sigma_{11} & \sigma_{12} & \sigma_{13} \\ \sigma_{21} & \sigma_{22} & \sigma_{23} \\ \sigma_{31} & \sigma_{32} & \sigma_{33} \end{bmatrix} = U \begin{bmatrix} s_1 & 0 & 0 \\ 0 & s_2 & 0 \\ 0 & 0 & s_3 \end{bmatrix} U^*. \tag{9.3}$$

Since U is constant, the law of relaxation of shear stresses admits the following equivalent formulation:

$$\frac{d}{dt}\begin{bmatrix} \sigma_{11} & \sigma_{12} & \sigma_{13} \\ \sigma_{21} & \sigma_{22} & \sigma_{23} \\ \sigma_{31} & \sigma_{32} & \sigma_{33} \end{bmatrix} = -\frac{1}{\tau}\begin{bmatrix} \sigma_{11} - \dfrac{\sigma_{kk}}{3} & \sigma_{12} & \sigma_{13} \\ \sigma_{21} & \sigma_{22} - \dfrac{\sigma_{kk}}{3} & \sigma_{23} \\ \sigma_{31} & \sigma_{32} & \sigma_{33} - \dfrac{\sigma_{kk}}{3} \end{bmatrix}$$

or, shortly,

$$\frac{d\sigma_{ij}}{dt} = -\frac{1}{\tau}\left(\sigma_{ij} - \frac{1}{3}\delta_{ij}\sigma_{kk}\right). \tag{9.4}$$

where $\sigma_{kk} = \sigma_{11} + \sigma_{22} + \sigma_{33}$.

The equivalence of the formulations (9.1) and (9.4) of the relaxation law follows from the chain of equalities

$$\frac{d}{dt}\begin{bmatrix} \sigma_{11} & \sigma_{12} & \sigma_{13} \\ \sigma_{21} & \sigma_{22} & \sigma_{23} \\ \sigma_{31} & \sigma_{32} & \sigma_{33} \end{bmatrix} = \frac{d}{dt}\left(U\begin{bmatrix} s_1 & 0 & 0 \\ 0 & s_2 & 0 \\ 0 & 0 & s_3 \end{bmatrix}U^*\right)$$

$$= U\begin{bmatrix} \frac{ds_1}{dt} & 0 & 0 \\ 0 & \frac{ds_2}{dt} & 0 \\ 0 & 0 & \frac{ds_3}{dt} \end{bmatrix}U^* = U\begin{bmatrix} -\frac{1}{\tau}(s_1 - s) & 0 & 0 \\ 0 & -\frac{1}{\tau}(s_2 - s) & 0 \\ 0 & 0 & -\frac{1}{\tau}(s_3 - s) \end{bmatrix}U^*$$

$$= -\frac{1}{\tau}U\begin{bmatrix} s_1 & 0 & 0 \\ 0 & s_2 & 0 \\ 0 & 0 & s_3 \end{bmatrix}U^* + \frac{1}{\tau}\frac{s_1 + s_2 + s_3}{3}I$$

$$= -\frac{1}{\tau}\begin{bmatrix} \sigma_{11} & \sigma_{12} & \sigma_{13} \\ \sigma_{21} & \sigma_{22} & \sigma_{23} \\ \sigma_{31} & \sigma_{32} & \sigma_{33} \end{bmatrix} + \frac{\sigma_{11} + \sigma_{22} + \sigma_{33}}{3\tau}I, \tag{9.5}$$

where $s = (s_1 + s_2 + s_3)/3$.

In terms of the deviator of the stress tensor, the relaxation law takes the form

$$\frac{d}{dt}\left[\sigma_{ij} - \frac{1}{3}\delta_{ij}\sigma_{kk}\right] = -\frac{1}{\tau}\left(\sigma_{ij} - \frac{1}{3}\delta_{ij}\sigma_{kk}\right). \tag{9.6}$$

By (9.6), it is clear that

$$\left(\sigma_{ij} - \frac{1}{3}\delta_{ij}\sigma_{kk}\right)\Big|_{t=t_1} = \left(\sigma_{ij} - \frac{1}{3}\delta_{ij}\sigma_{kk}\right)\Big|_{t=t_0} e^{-\int_{t_0}^{t_1}\frac{dt}{\tau}}, \tag{9.7}$$

i.e., each component of the deviator exponentially decay and the decay is the faster, the smaller is τ. This motivates the choice of the term the "relaxation time of shear stresses."

9.2. For our purposes, it is convenient to use some other mathematical formalization of the relaxation law for shear stresses in a Maxwell medium. Before describing this formalization, we introduce the notion of the "effective elastic strain tensor." To fix the main idea, we consider the simplest model of linear elasticity which was already considered in Section 7.

In linear elasticity, the stress tensor σ_{ij} and the strain tensor ε_{ij} (the Almansi strain tensor in the case of small deformations) are connected with the temperature T by the equality

$$\sigma_{ij} = \left(-\frac{\pi}{\omega}(T - T_0) + \left(\lambda - \frac{\pi^2}{\rho_0 \omega} \right) (\varepsilon_{11} + \varepsilon_{22} + \varepsilon_{33}) \right) \delta_{ij} + 2\mu \varepsilon_{ij}. \tag{9.8}$$

Therefore, the deviator of the stress tensor $||\sigma_{ij}||$ is proportional to the deviator of the strain tensor $||\varepsilon_{ij}||$:

$$\varepsilon_{ij} - \delta_{ij} \frac{\varepsilon_{11} + \varepsilon_{22} + \varepsilon_{33}}{3} = \frac{1}{2\mu} \left(\sigma_{ij} - \delta_{ij} \frac{\sigma_{11} + \sigma_{22} + \sigma_{33}}{3} \right). \tag{9.9}$$

To determine ε_{ij}, one can measure the stresses σ_{ij} and the temperature T and, after that, compute ε_{ij} from the equations connected deformations, temperature, and stresses.

• The tensor ε_{ij} defined by the above procedure is called the *effective elastic strain tensor*.

We recall that the proof of the Murnaghan formulas that connect stresses and deformations (cf. Sections 4 and 8) was based on the arguments concerned with the energy conservation law and the work of the internal surface stress caused by some deformations of the medium. If the deformations occur so fast that the stresses have not time to equalize, the arguments of Sections 4 and 8 remain valid in this case. But we should take into account that the change of stresses caused by relaxation during large time-intervals can lead to a change of the effective elastic strain tensor because it is computed from them.

Such a change of effective deformations happens if there are no actual deformations of the medium and displacements of the material particles.

We consider the relaxation of stresses assuming that the behavior of the deviator of the stress tensor is described by Eqs. (9.6). Using the equality (cf. (9.9))

$$\varepsilon_{ij}^{\text{eff}} - \delta_{ij} \varepsilon_{kk}^{\text{eff}} = \frac{1}{2\mu} \left(\sigma_{ij} - \frac{1}{3} \delta_{ij} \sigma_{kk} \right), \tag{9.10}$$

where

$$\varepsilon_{kk}^{\text{eff}} = \varepsilon_{11}^{\text{eff}} + \varepsilon_{22}^{\text{eff}} + \varepsilon_{33}^{\text{eff}},$$

we write (9.6) in terms of components of the effective elastic strain tensor $\varepsilon_{ij}^{\text{eff}}$:

$$\frac{d}{dt} \left(\varepsilon_{ij}^{\text{eff}} - \frac{1}{3} \delta_{ij} \varepsilon_{kk}^{\text{eff}} \right) = -\frac{1}{\tau} \left(\varepsilon_{ij}^{\text{eff}} - \frac{1}{3} \delta_{ij} \varepsilon_{kk}^{\text{eff}} \right). \tag{9.11}$$

Here, the superscript "eff" recalls that we deal with effective elastic deformations. We see that the equations for the deviator of the effective elastic strain tensor

$||\varepsilon_{ij}^{\text{eff}}||$ turn out to be the same as those for the deviator of the stress tensor $||\sigma_{ij}||$ (cf. (9.6) and (9.11)).

• *For our purpose, it is convenient to postulate the relaxation law in the form of the equations for the effective elastic strain tensor. In the case of small deformations, this formulation is the same as the mathematical formalization presented in* 9.1.

9.3. We define the metric effective elastic deformation tensor g_{ij}^{eff} without the assumption that deformations and stresses are necessarily small. Suppose that, during time-intervals $\Delta t \ll \tau$ that are significantly less than the relaxation time τ, the medium under consideration is elastic and is characterized by the equation of state $E = E(k_1, k_2, k_3, S)$. As we know, if the standard state of such a medium (obtained from the nonstressed state with the standard temperature by heating up to a given value of the entropy S) is subjected to an adiabatic deformation in a sufficiently short time-interval (less than τ), then we can compute the stresses in the following way. For the metric deformation tensor

$$||g_{ij}|| = G = U \begin{bmatrix} 1/k_1^2 & 0 & 0 \\ 0 & 1/k_2^2 & 0 \\ 0 & 0 & 1/k_3^2 \end{bmatrix} U^* \qquad (UU^* = I, \quad \det U = +1) \tag{9.12}$$

the stress tensor $||\sigma_{ij}||$ and the temperature T are expressed by the formula

$$||\sigma_{ij}|| = U \begin{bmatrix} \dfrac{\rho_0 E_{k_1}}{k_2 k_3} & 0 & 0 \\ 0 & \dfrac{\rho_0 E_{k_2}}{k_3 k_1} & 0 \\ 0 & 0 & \dfrac{\rho_0 E_{k_3}}{k_1 k_2} \end{bmatrix} U^*, \tag{9.13}$$
$$T = E_S(k_1, k_2, k_3, S).$$

If σ_{ij} and T are known, we can compute the orthogonal matrix U whose columns are the eigenvectors of $||\sigma_{ij}||$. Taking the eigenvalues s_1, s_2, s_3 of the matrix $||\sigma_{ij}||$ and the temperature T for the right-hand sides of the equations

$$\begin{aligned} &\frac{\rho_0}{k_2 k_3} E_{k_1}(k_1, k_2, k_3, S) = s_1, \\ &\frac{\rho_0}{k_3 k_1} E_{k_2}(k_1, k_2, k_3, S) = s_2, \\ &\frac{\rho_0}{k_1 k_2} E_{k_3}(k_1, k_2, k_3, S) = s_3, \\ &E_S(k_1, k_2, k_3, S) = T, \end{aligned} \tag{9.14}$$

we find k_1, k_2, k_3, and S from the system (9.14). Since the orthogonal rotation matrix U and the extension coefficients k_i are known, we thereby completely determine the metric deformation tensor $||g_{ij}||$ by formula (9.12). Thus, we can compute $||g_{ij}||$ starting with the measured stresses and temperature and using the elasticity laws.

• The tensor computed by the above procedure is called the *metric effective elastic deformation tensor* and is denoted by $||g_{ij}^{\text{eff}}||$.

The above procedure of computing g_{ij}^{eff} for given σ_{ij} and T is equivalent to the solution of the equations

$$-2\rho g_{ik}\frac{\partial E}{\partial g_{kj}} = \sigma_{ij} \quad \left(\rho = \rho_0\sqrt{\det ||g_{ij}||}\right), \tag{9.15}$$

$$E_S = T \tag{9.16}$$

with respect to g_{ij} and T, where the right-hand sides σ_{ij} and T are known. Indeed, it suffices to recall Section 7 where we proved the equivalence of different forms of the Murnaghan formulas.

Remark 9.1. In a medium with relaxation of shear stresses, the metric effective elastic deformation tensor $||g_{ij}^{\text{eff}}||$ is, as a rule, different from the metric deformation tensor $||g_{ij}||$ introduced in Section 6 by the formulas

$$||g_{ij}|| = [C^{-1}]^*[C^{-1}], \quad C = \left\|\frac{\partial x_i}{\partial x_{j0}}\right\|, \tag{9.17}$$

where

$$x_1 = x_1(x_{10}, x_{20}, x_{30}),$$
$$x_2 = x_2(x_{10}, x_{20}, x_{30}),$$
$$x_3 = x_3(x_{10}, x_{20}, x_{30}),$$

describing the actual displacements of material particles. To emphasize this difference, we often call the tensor $||g_{ij}||$ the *metric total deformation tensor*.

9.4. In Maxwell media with relaxation of shear stresses, the density ρ of the medium computed through the metric effective elastic deformation tensor $\rho = \rho_0\sqrt{\det ||g_{ij}^{\text{eff}}||}$ should coincide with the actual density of the medium which is found from the metric total deformation tensor by the formula $\rho = \rho_0\sqrt{\det ||g_{ij}||}$. If the density ρ_0 of the medium in the nonstressed state remains unchanged under the process of relaxation of shear stresses, then

$$\det ||g_{ij}|| = \det ||g_{ij}^{\text{eff}}||. \tag{9.18}$$

• A medium is said to be *solid* if the density ρ_0 of the medium in the nonstressed state remains constant under the process of relaxation of shear stresses.

Almost all metallic bodies (iron, copper, aluminum, etc.) are solid. Nonsolid media also exit in the nature, for example, porous bodies, soils, etc. In the case of a porous medium with relaxation of shear stresses, the density ρ_0, as a rule, increases.

In our further considerations, we will deal only with solid media with constant ρ_0.

In the case of solid media, we can compute the metric effective elastic deformation tensor $||g_{ij}^{\text{eff}}||$ if we know the density ρ, the temperature T, and the deviator of the stress tensor (not the entire stress tensor). Indeed, reducing the deviator of the stress tensor to the principal axes,

$$||\sigma_{ij}|| - \frac{\sigma_{kk}}{3} I = U \begin{bmatrix} s_1 - s & 0 & 0 \\ 0 & s_2 - s & 0 \\ 0 & 0 & s_3 - s \end{bmatrix} U^*,$$

where $s = (s_1 + s_2 + s_3)/3$, we find the rotation matrix U as above and the differences of the principal stresses by the formulas

$$s_i - s_j = \left(s_i - \frac{s_1 + s_2 + s_3}{3}\right) - \left(s_j - \frac{s_1 + s_2 + s_3}{3}\right).$$

Then k_1, k_2, k_3, S are found from the system

$$\begin{aligned} k_1 k_2 k_3 &= \frac{\rho_0}{\rho}, \\ k_1 E_{k_1} - k_2 E_{k_2} &= \frac{\rho_0}{\rho}(s_1 - s_2), \\ k_2 E_{k_2} - k_3 E_{k_3} &= \frac{\rho_0}{\rho}(s_2 - s_3), \\ E_S(k_1, k_2, k_3, S) &= T. \end{aligned} \tag{9.19}$$

Here, we should use results of measurements of stresses in a stationary medium with $||\varepsilon_{ij}|| = 0$ or distinguish those stresses from the stress tensor that are independent of the strain rate of the medium, i.e., eliminate σ_{ij}^{vis} from the stress tensor $||\sigma_{ij}||$.

To formulate a mathematical model describing relaxations of shear stresses in an isotropic viscoelastic Maxwell medium, we assume that no displacements of particles of a medium occur and the directions of the principal axes of the stress tensor $||\sigma_{ij}||$ remain unchanged. Respectively, the same is true for the metric effective elastic deformation tensor $||g_{ij}^{\text{eff}}||$. In other words, we postulate that, in the canonical representation,

$$\|\sigma_{ij}\| = \begin{bmatrix} \sigma_{11} & \sigma_{12} & \sigma_{13} \\ \sigma_{21} & \sigma_{22} & \sigma_{23} \\ \sigma_{31} & \sigma_{32} & \sigma_{33} \end{bmatrix} = U \begin{bmatrix} s_1 & 0 & 0 \\ 0 & s_2 & 0 \\ 0 & 0 & s_3 \end{bmatrix} U^*, \tag{9.20}$$

$$\|g_{ij}^{\text{eff}} = \begin{bmatrix} g_{11}^{\text{eff}} & g_{12}^{\text{eff}} & g_{13}^{\text{eff}} \\ g_{21}^{\text{eff}} & g_{22}^{\text{eff}} & g_{23}^{\text{eff}} \\ g_{31}^{\text{eff}} & g_{32}^{\text{eff}} & g_{33}^{\text{eff}} \end{bmatrix} = U \begin{bmatrix} [k_1^{\text{eff}}]^{-2} & 0 & 0 \\ 0 & [k_2^{\text{eff}}]^{-2} & 0 \\ 0 & 0 & [k_3^{\text{eff}}]^{-2} \end{bmatrix} U^* \tag{9.21}$$

the orthogonal matrix U is constant under relaxation. We assume that the principal stresses s_1, s_2, s_3 change in such a way that their differences decay. It is convenient to assume that k_1^{eff}, k_2^{eff}, k_3^{eff} change so that they come closer and closer together with time. Then the principal stresses

$$\begin{aligned} s_1 &= \frac{\rho_0}{k_2^{\text{eff}} k_3^{\text{eff}}} E_{k_1}(k_1^{\text{eff}}, k_2^{\text{eff}}, k_3^{\text{eff}}, S), \\ s_2 &= \frac{\rho_0}{k_3^{\text{eff}} k_1^{\text{eff}}} E_{k_2}(k_1^{\text{eff}}, k_2^{\text{eff}}, k_3^{\text{eff}}, S), \\ s_3 &= \frac{\rho_0}{k_1^{\text{eff}} k_2^{\text{eff}}} E_{k_3}(k_1^{\text{eff}}, k_2^{\text{eff}}, k_3^{\text{eff}}, S), \end{aligned} \tag{9.22}$$

will also approach each other.

To find k_i^{eff}, we use the equations

$$\begin{aligned} \frac{d \ln k_1}{dt} &= -\frac{1}{\tau} \ln \frac{k_1}{\sqrt[3]{k_1 k_2 k_3}}, \\ \frac{d \ln k_2}{dt} &= -\frac{1}{\tau} \ln \frac{k_2}{\sqrt[3]{k_1 k_2 k_3}}, \\ \frac{d \ln k_3}{dt} &= -\frac{1}{\tau} \ln \frac{k_3}{\sqrt[3]{k_1 k_2 k_3}}. \end{aligned} \tag{9.23}$$

Hereinafter, we omit the superscript "eff" for simplicity.

Throughout this section, we will consider only effective elastic extension coefficients k_i^{eff}.

It is obvious that the product $k_1 k_2 k_3$ remains unchanged because

$$\begin{aligned} \frac{d \ln(k_1 k_2 k_3)}{dt} &= \frac{d(\ln k_1 + \ln k_2 + \ln k_3)}{dt} \\ &= -\frac{1}{\tau}\left(\ln \frac{k_1}{\sqrt[3]{k_1 k_2 k_3}} + \ln \frac{k_2}{\sqrt[3]{k_1 k_2 k_3}} + \ln \frac{k_3}{\sqrt[3]{k_1 k_2 k_3}}\right) \\ &= -\frac{1}{\tau} \ln \frac{k_1 k_2 k_3}{(\sqrt[3]{k_1 k_2 k_3})^3} = -\frac{\ln 1}{\tau} = 0. \end{aligned}$$

- Within the framework of the suggested model of relaxation, the density $\rho = \rho_0/(k_1 k_2 k_3)$ remains constant.

Remark 9.2. The condition that the density is constant is necessary if there are no macroscopic displacements of material particles of the medium under consideration. We have shown that this condition is satisfied in our model.

However, if we try to apply our model to actual bodies, it turns out that the "relaxation time" τ is not necessarily constant. It depends on the stressed state and temperature of the medium. For example, for metals the relaxation time τ is several hours, months, or even years. If the stresses are larger than the so-called yield point, the relaxation time τ rapidly decreases. In processes with deformation of metals under explosive material, the relaxation time is of order 1–10 microseconds. For melts or fluids the relaxation time is of order 10^{-12} seconds. A fluid can be regarded as an elastic body with a very small relaxation time of shear stresses. This point of view was basic in the theory of viscosity coefficients for fluids developed by Ya. I. Frenkel.

9.5. The relaxation time τ is expressed in terms of parameters of the state of a medium as follows:

$$\tau = \tau_0 \exp\{-W(s_1, s_2, s_3, S)/T\}, \tag{9.24}$$

where S is the entropy, T is the temperature, s_1, s_2, s_3 are the principal stresses. In (9.24), W is the energy of activation of processes in the crystal lattice of grains of a polycrystal or in the polycrystal structure which lead to relaxation of shear stresses. Examples of such processes are motions of dislocation through the crystal, vacancy formation in crystals or displacements of some crystal grains relative to other grains in a polycrystalline medium. Naturally, the coefficient τ_0 may also depend on the stressed state and temperature.

In this book, we assume that τ is a known function of k_1, k_2, k_3, S defined in a phenomenological way, $\tau = \tau(k_1, k_2, k_3, S)$. In our relaxation model, the equation of state $E = E(k_1, k_2, k_3, S)$ remains unchanged under relaxations and for k_1, k_2, k_3 we take k_1^{eff}, k_2^{eff}, k_3^{eff} which can be determined in terms of the principal stresses and temperature by the Murnaghan formulas.

We will often omit the superscript "eff." But the relaxation process is not completely defined without the entropy condition. To formulate it, we will use the energy conservation law.

9.6. We discuss the so-called entropy condition. Consider the unit of mass of a medium with the heat influx q per unit time. Let the effective extension coefficients k_1, k_2, k_3 change in accordance with the above rule. The energy changes in accordance with the law

$$\frac{dE(k_1, k_2, k_3, S)}{dt} = E_{k_1}\frac{dk_1}{dt} + E_{k_2}\frac{dk_2}{dt} + E_{k_3}\frac{dk_3}{dt} + E_S\frac{dS}{dt} = q. \tag{9.25}$$

Since there are no actual displacements of material particles, the work of the external forces is not produced. Therefore, the work is equal to the change of the internal energy q. The change of k_1, k_2, k_3 is described by the relaxation equations

$$\frac{dk_i}{dt} = -k_i \frac{\ln k_i - \frac{1}{3}(\ln k_1 + \ln k_2 + \ln k_3)}{\tau}. \tag{9.26}$$

Using (9.25) and (9.26), we can write the equation describing the change of the entropy S:

$$\begin{aligned}
\frac{dS}{dt} &= \frac{1}{E_S}\left(q - E_{k_1}\frac{dk_1}{dt} - E_{k_2}\frac{dk_2}{dt} - E_{k_3}\frac{dk_3}{dt}\right) \\
&= \frac{1}{T}\left[q + \frac{1}{\tau}k_1 E_{k_1}\left(\ln k_1 - \frac{\ln k_1 + \ln k_2 + \ln k_3}{3}\right)\right. \\
&+ \frac{1}{\tau}k_2 E_{k_2}\left(\ln k_2 - \frac{\ln k_1 + \ln k_2 + \ln k_3}{3}\right) \\
&+ \left.\frac{1}{\tau}k_3 E_{k_3}\left(\ln k_3 - \frac{\ln k_1 + \ln k_2 + \ln k_3}{3}\right)\right] \\
&= \frac{q}{T} + \frac{1}{3\tau T}\big[(k_1 E_{k_1} - k_2 E_{k_2})(\ln k_1 - \ln k_2) \\
&+ (k_2 E_{k_2} - k_3 E_{k_3})(\ln k_2 - \ln k_3) \\
&+ (k_3 E_{k_3} - k_1 E_{k_1})(\ln k_3 - \ln k_1)\big].
\end{aligned} \tag{9.27}$$

It is natural to assume that the entropy increases under irreversible relaxations in an adiabatically isolated system ($q = 0$).

In order to obtain the entropy condition as a consequence of the above equations, it is necessary to require the following condition ($\tau > 0$ and $T = E_S > 0$):

$$\begin{aligned}
&(k_1 E_{k_1} - k_2 E_{k_2})(\ln k_1 - \ln k_2) + (k_2 E_{k_2} - k_3 E_{k_3})(\ln k_2 - \ln k_3) \\
&\quad + (k_3 E_{k_3} - k_1 E_{k_1})(\ln k_3 - \ln k_1) \geqslant 0.
\end{aligned} \tag{9.28}$$

Instead of (9.28), it is convenient to consider the following stronger restriction:

$$\begin{aligned}
\frac{k_1 E_{k_1} - k_2 E_{k_2}}{k_1^2 - k_2^2} &= \frac{1}{2}\frac{E_{k_1} - E_{k_2}}{k_1 - k_2} + \frac{1}{2}\frac{E_{k_1} + E_{k_2}}{k_1 + k_2} \geqslant 0, \\
\frac{k_2 E_{k_2} - k_3 E_{k_3}}{k_2^2 - k_3^2} &= \frac{1}{2}\frac{E_{k_2} - E_{k_3}}{k_2 - k_3} + \frac{1}{2}\frac{E_{k_2} + E_{k_3}}{k_2 + k_3} \geqslant 0, \\
\frac{k_3 E_{k_3} - k_1 E_{k_1}}{k_3^2 - k_1^2} &= \frac{1}{2}\frac{E_{k_3} - E_{k_1}}{k_3 - k_1} + \frac{1}{2}\frac{E_{k_3} + E_{k_1}}{k_3 + k_1} \geqslant 0,
\end{aligned} \tag{9.29}$$

which implies that

$$(k_i E_{k_i} - k_j E_{k_j})(\ln k_i - \ln k_j) = \frac{k_i E_{k_i} - k_j E_{k_j}}{k_i^2 - k_j^2}(k_i^2 - k_j^2)(\ln k_i - \ln k_j) \tag{9.30}$$

for each $i = 1, 2, 3$ because $(k_i^2 - k_j^2)(\ln k_i - \ln k_j) \geqslant 0$.

• *If the energy E satisfies* (9.29), *then the entropy condition holds.*

But what can we say about actual equations of state? These equations are based on results of experiments or are derived from the laws of statistical physics. Therefore, we cannot subject them to one or other restrictions convenient for our purposes.

We note that we introduced these restrictions in order to avoid contradictions between our mathematical model of relaxation of shear stresses and elementary facts of physics. We also note that the restrictions (9.29) can be justified by some arguments, although they are not directly connected with our relaxation model. Namely, as we will see, the inequalities (9.29) are necessary for the well-posedness of the Cauchy problem for the system of equations of the theory of elasticity (with or without relaxations of stresses). Naturally, there need be no doubt that the Cauchy problem for equations describing purely elastic processes in classical elastic media (for example, in metals) is well posed. This means that there can be no doubt that the entropy condition holds within the framework of our relaxation model.

9.7. We consider the case of small elastic deformations. Let the metric effective elastic deformation tensor $||g_{ij}^{\text{eff}}||$ be not necessarily diagonal. In our model, the relaxation of the metric effective elastic deformation tensor

$$G^{\text{eff}} = ||g_{ij}^{\text{eff}}|| = U \begin{bmatrix} 1/k_1^2 & 0 & 0 \\ 0 & 1/k_2^2 & 0 \\ 0 & 0 & 1/k_3^2 \end{bmatrix} U^* \tag{9.31}$$

is described by the equations

$$\frac{d}{dt}G^{\text{eff}} = U \begin{bmatrix} \frac{d}{dt}k_1^{-2} & 0 & 0 \\ 0 & \frac{d}{dt}k_2^{-2} & 0 \\ 0 & 0 & \frac{d}{dt}k_3^{-2} \end{bmatrix} U^* = -||\varphi_{ij}||$$

$$= U \begin{bmatrix} \frac{2}{\tau k_1^2}\ln\frac{k_1}{\sqrt[3]{k_1k_2k_3}} & 0 & 0 \\ 0 & \frac{2}{\tau k_2^2}\ln\frac{k_2}{\sqrt[3]{k_1k_2k_3}} & 0 \\ 0 & 0 & \frac{2}{\tau k_3^2}\ln\frac{k_3}{\sqrt[3]{k_1k_2k_3}} \end{bmatrix} U^*, \tag{9.32}$$

where the matrix $||\varphi_{ij}||$ describes the change rate of the tensor $||g_{ij}^{\text{eff}}||$ in the case where there are no displacements in the medium. It is easy to verify that, in the case of small deformations $g_{ik}^{\text{eff}} = \delta_{ik} - 2\varepsilon_{ik}^{\text{eff}}$ with small $\varepsilon_{ik}^{\text{eff}}$ and $k_i \approx 1 + \varepsilon_i^{\text{eff}}$, where $\varepsilon_i^{\text{eff}}$ are the eigenvalues of $||\varepsilon_{ik}^{\text{eff}}||$, we have the following approximate formulas:

$$
\begin{aligned}
\|\varphi_{ij}\| &\approx -\frac{2}{\tau} U \begin{bmatrix} \varepsilon_1^{\text{eff}} - \frac{\varepsilon_{kk}^{\text{eff}}}{3} & 0 & 0 \\ 0 & \varepsilon_2^{\text{eff}} - \frac{\varepsilon_{kk}^{\text{eff}}}{3} & 0 \\ 0 & 0 & \varepsilon_3^{\text{eff}} - \frac{\varepsilon_{kk}^{\text{eff}}}{3} \end{bmatrix} U^* \\
&= -\frac{2}{\tau} \left\| \varepsilon_{ij}^{\text{eff}} - \frac{1}{3}\delta_{ij}\varepsilon_{kk}^{\text{eff}} \right\| . \qquad (9.33)
\end{aligned}
$$

Therefore, in the case of small effective elastic deformations, the relaxation model is described by the linear system

$$
\frac{d\varepsilon_{ij}^{\text{eff}}}{dt} + \frac{1}{\tau}\left(\varepsilon_{ij}^{\text{eff}} - \frac{1}{3}\delta_{ij}\varepsilon_{kk}^{\text{eff}}\right) = 0. \qquad (9.34)
$$

As we already seen, the last system is equivalent to the Maxwell model (9.4). Thus, we have motivated the generalization of the Maxwell relaxation model to the case of finite (not necessarily small) deformations. Simultaneously, we have defined the metric effective elastic deformation tensor $\|g_{ij}^{\text{eff}}\|$.

§ 10. Equations for the Metric Effective Elastic Deformation Tensor

In Section 9, we derived equations describing relaxation of stresses in an isotropic motionless medium using the metric effective elastic deformation tensor. Our next goal is to describe the evolution of the metric effective elastic deformation tensor in a moving medium with relaxation.

10.1. Suppose that at some time t the state of a medium is characterized by the temperature $T(x_1, x_2, x_3)$ and the stresses $\sigma_{ij}(x_1, x_2, x_3)$ represented as the sum of elastic and viscous stresses: $\sigma_{ij} = \sigma_{ij}^{\text{el}} + \sigma_{ij}^{\text{vis}}$. Suppose that we know how to distinguish σ_{ij}^{el} from the stress tensor $\|\sigma_{ij}\|$ and consider only elastic stresses for σ_{ij}. We also assume that the medium is characterized by the equation of state $E = E(k_1, k_2, k_3, S)$, the density of the nonstressed state ρ_0 and the relaxation time $\tau = \tau(k_1, k_2, k_3, S)$.

Given σ_{ij} and T, we can find the metric effective elastic deformation tensor $g_{ij} = g_{ij}^{\text{eff}}$ and the entropy S from the equations (cf. (9.15))

$$\begin{aligned} &-2\rho_0\sqrt{\det\|g_{mn}\|}\, g_{ik}\frac{\partial E}{\partial g_{kj}} = \sigma_{ij}, \\ &\frac{\partial E}{\partial S} = T. \end{aligned} \tag{10.1}$$

We denote by $\overline{\sigma}_{ij}$, $\overline{T}$, $\overline{g}_{ij}$, $\overline{S}$ the values of σ_{ij}, T, g_{ij}, S at some point $(\overline{x}_1, \overline{x}_2, \overline{x}_3)$ respectively. We have

$$\|\overline{g}_{ij}\| = U \begin{bmatrix} \overline{k}_1^{-2} & 0 & 0 \\ 0 & \overline{k}_2^{-2} & 0 \\ 0 & 0 & \overline{k}_3^{-2} \end{bmatrix} U^* \qquad (UU^* = I, \quad \det U = +1). \tag{10.2}$$

We construct a matrix A_0 such that $A_0^* A_0 = \|\overline{g}_{ij}\|$. Such a matrix is not unique because for A_0 we can take any matrix of the form

$$A_0 = V^* \begin{bmatrix} 1/\overline{k}_1 & 0 & 0 \\ 0 & 1/\overline{k}_2 & 0 \\ 0 & 0 & 1/\overline{k}_3 \end{bmatrix} U^*, \tag{10.3}$$

where V is an arbitrary rotation matrix. We denote by C_0 the inverse matrix to A_0:

$$C_0 = A_0^{-1} = U \begin{bmatrix} \overline{k}_1 & 0 & 0 \\ 0 & \overline{k}_2 & 0 \\ 0 & 0 & \overline{k}_3 \end{bmatrix} V. \tag{10.4}$$

Under the transformation $A_0 = C_0^{-1}$ and any translation, points (x_1, x_2, x_3) go to points (x_{10}, x_{20}, x_{30}). In particular, $(\overline{x}_1, \overline{x}_2, \overline{x}_3)$ goes to $(\overline{x}_{10}, \overline{x}_{20}, \overline{x}_{30})$. Assuming that the corresponding deformation takes place in a short time period so that the relaxation processes cannot occur and that the entropy is preserved for each particle, we can conclude that the stresses at the point $(\overline{x}_{10}, \overline{x}_{20}, \overline{x}_{30})$ after this deformation are described by the standard (spherical) stress tensor $\sigma_{ij}(\overline{x}_{10}, \overline{x}_{20}, \overline{x}_{30}) = \delta_{ij} E_{k_1}(1, 1, 1, \overline{S})$. In a neighborhood of $(\overline{x}_{10}, \overline{x}_{20}, \overline{x}_{30})$, the medium is in the standard state (cf. Section 4). Indeed, the inverse transformation $C_0 = A_0^{-1}$ corresponds to the extension with coefficients k_1, k_2, k_3 along three orthogonal axes. The directed vectors of these transformed axes form columns of the orthogonal matrix U. The deformation is described by the metric deformation tensor (10.2). In the isentropic case, if the deformation occurs in a short time-interval (comparing with τ) during which the stresses cannot significantly change by relaxation processes, then the deformed medium (which was in the standard initial state before deformation) is characterized by the stress tensor

$$\overline{\sigma}_{ij}(\overline{x}_1, \overline{x}_2, \overline{x}_3)$$
$$= U \begin{bmatrix} \frac{\rho_0}{\overline{k}_2\overline{k}_3} E_{k_1}(\overline{k}_1, \overline{k}_2, \overline{k}_3, S) & 0 & 0 \\ 0 & \frac{\rho_0}{\overline{k}_3\overline{k}_1} E_{k_2}(\overline{k}_1, \overline{k}_2, \overline{k}_3, S) & 0 \\ 0 & 0 & \frac{\rho_0}{\overline{k}_1\overline{k}_2} E_{k_3}(\overline{k}_1, \overline{k}_2, \overline{k}_3, S) \end{bmatrix} U^*.$$

We see that the "reverse" elastic deformation described by the matrix A_0^{-1} transforms the standard state into the stressed state relative to $(\overline{x}_1, \overline{x}_2, \overline{x}_3)$. Therefore, A_0 can be regarded as a transformation "unloading" the medium in a neighborhood of the point $(\overline{x}_1, \overline{x}_2, \overline{x}_3)$, i.e., under this transformation, this neighborhood is transformed into the standard state.

Let

$$\begin{pmatrix} x_1 \\ x_2 \\ x_3 \end{pmatrix} = C_0 \begin{pmatrix} x_{10} \\ x_{20} \\ x_{30} \end{pmatrix} + \begin{pmatrix} l_1 \\ l_2 \\ l_3 \end{pmatrix}, \tag{10.5}$$

and let δx_1, δx_2, δx_3 be the displacement in time δt of a material particle with coordinates (x_1, x_2, x_3) at time t. Since the effective elastic deformation tensor $||\overline{g}_{ij}^{\text{eff}}||$ at the point $(\overline{x}_1, \overline{x}_2, \overline{x}_3)$ and time t coincides with the metric total deformation tensor described by the matrix $C_0 = A_0^{-1}$, i.e.,

$$||\overline{g}_{ij}^{\text{eff}}|| = ||\overline{g}_{ij}|| = A_0^* A_0,$$

in order to compute variations of $\overline{g}_{ij}^{\text{eff}}$, we can use the formulas for variations of the metric total deformation tensor $\overline{g}_{ij}$ obtained in Section 8:

$$\delta \overline{g}_{ij} = -\overline{g}_{ik} \frac{\partial(\delta x_k)}{\partial x_j} - \overline{g}_{kj} \frac{\partial(\delta x_k)}{\partial x_i}. \tag{10.6}$$

We emphasize that this is possible only if the relaxation process cannot occur in δt.

On the other hand, relaxation processes in a motionless medium are described by the equations $\delta \overline{g}_{ij} = -\varphi_{ij}\delta t$ (cf. Section 9). Therefore, it is natural to accept the following postulate.

• *Let a material particle have the coordinates $\overline{x}_1$, $\overline{x}_2$, $\overline{x}_3$ at time t. Then the variation $\delta \overline{g}_{ij}$ caused by elastic deformation and relaxation in time δt is expressed as the sum*

$$\delta \overline{g}_{ij} = [\delta \overline{g}_{ij}]_{\text{el}} + (\delta \overline{g}_{ij})_{\text{rel}} = \left[-\overline{g}_{ik} \frac{\partial(\delta x_k)}{\partial x_j} - \overline{g}_{kj} \frac{\partial(\delta x_k)}{\partial x_i}\right] + (-\varphi_{ij}\delta t). \tag{10.7}$$

10.2. The equalities (10.7) should be valid for $\overline{g}_{ik}$ at any point. Therefore,

$$\delta g_{ij}^{\text{eff}}(x_1, x_2, x_3, t) = \frac{\partial g_{ij}^{\text{eff}}}{\partial t}\delta t + \frac{\partial g_{ij}^{\text{eff}}}{\partial x_r}\delta x_r = -g_{ik}^{\text{eff}}\frac{\partial(\delta x_k)}{\partial x_j} - g_{kj}^{\text{eff}}\frac{\partial(\delta x_k)}{\partial x_i} - \varphi_{ij}\delta t. \tag{10.8}$$

Let the displacements δx_k be determined by the velocity $u_k = u_k(x_1, x_2, x_3, t)$, i.e., $\delta x_k = u_k \delta t$. Then from the equalities (10.8) for the variations of $g_{ij}^{\text{eff}} = \overline{g}_{ij}$ we obtain the equations

$$\frac{\partial g_{ij}^{\text{eff}}}{\partial t} + u_r\frac{\partial g_{ij}^{\text{eff}}}{\partial x_r} + g_{ik}^{\text{eff}}\frac{\partial u_k}{\partial x_j} + g_{kj}^{\text{eff}}\frac{\partial u_k}{\partial x_i} = -\varphi_{ij} \tag{10.9}$$

which differ from the equations for the metric total deformation tensor (cf. Section 8) only by the right-hand sides $-\varphi_{ij}$. To compute $-\varphi_{ij}$, we use the rules in Section 9. For $\|g_{ij}^{\text{eff}}\|$ of the form (9.31) we find φ_{ij} by formula (9.32). We also recall (cf. (9.15)) that $\rho = \rho_0\sqrt{\det\|g_{ij}^{\text{eff}}\|}$, where the density of the nonstressed state ρ_0 is constant in our case.

The equations of linear elasticity used in the case of small deformations and displacements (cf. Section 19) have the form

$$\begin{aligned}
&\frac{\partial \varepsilon_{ij}}{\partial t} - \frac{1}{2}\left(\frac{\partial u_i}{\partial x_j} + \frac{\partial u_j}{\partial x_i}\right) = 0,\\
&\rho_0\frac{\partial u_i}{\partial t} - \frac{\partial \sigma_{ij}}{\partial x_j} = 0,\\
&\sigma_{ij} = \lambda\delta_{ij}(\varepsilon_{11} + \varepsilon_{22} + \varepsilon_{33}) + 2\mu\varepsilon_{ij}.
\end{aligned}$$

To describe relaxation, we consider the first equation with the nonzero right-hand side

$$\widetilde{\varphi}_{ij} = -\frac{1}{2\mu\tau}\left(\sigma_{ij} - \frac{\delta_{ij}}{3}\sigma_{kk}\right) = -\frac{1}{\tau}\left(\varepsilon_{ij} - \frac{\delta_{ij}}{3}e_{kk}\right).$$

As a result, we obtain the Maxwell viscoelastic equations. We generalize this model to the case of large deformations.

In what follows we need the equality

$$\frac{\partial \rho}{\partial g_{ij}^{\text{eff}}} g_{jk}^{\text{eff}} = \frac{1}{2}\rho\delta_{ik} \tag{10.10}$$

or, in matrix notation,

$$\left\|\frac{\partial \rho}{\partial g_{ij}} g_{jk}\right\| = \frac{\rho}{2}I \tag{10.11}$$

obtained in Section 8. Multiplying equations (10.9) by $\dfrac{\partial\rho}{\partial g^{\text{eff}}_{kj}}$ and taking the sum over k, j, we find

$$\frac{\partial\rho}{\partial g^{\text{eff}}_{kj}}\left(\frac{\partial g^{\text{eff}}_{kj}}{\partial t}+u_r\frac{\partial g^{\text{eff}}_{kj}}{\partial x_r}\right)+\frac{\partial\rho}{\partial g^{\text{eff}}_{kj}}g^{\text{eff}}_{kl}\frac{\partial u_l}{\partial x_j}+\frac{\partial\rho}{\partial g^{\text{eff}}_{kj}}g^{\text{eff}}_{lj}\frac{\partial u_l}{\partial x_k}=-\varphi_{kj}\frac{\partial\rho}{\partial g^{\text{eff}}_{kj}},$$

where the left-hand side can be transformed as follows:

$$\frac{\partial\rho}{\partial t}+u_r\frac{\partial\rho}{\partial x_r}+\frac{\rho}{2}\delta_{lj}\frac{\partial u_l}{\partial x_j}+\frac{\rho}{2}\delta_{kl}\frac{\partial u_k}{\partial x_l}\equiv\frac{\partial\rho}{\partial t}+u_r\frac{\partial\rho}{\partial x_r}+\rho\frac{\partial u_l}{\partial x_l}\equiv\frac{\partial\rho}{\partial t}+\frac{\partial(\rho u_i)}{\partial x_i}.$$

Thus, we obtain the equation for the density

$$\frac{\partial\rho}{\partial t}+\frac{\partial(\rho u_i)}{\partial x_i}=-\varphi_{kj}\frac{\partial\rho}{\partial g^{\text{eff}}_{kj}}. \tag{10.12}$$

The term $\varphi_{kj}\dfrac{\partial\rho}{\partial g^{\text{eff}}_{kj}}$ on the right-hand side of (10.12) is the trace of the matrix product of $||\varphi_{kj}||$ and $\left\|\dfrac{\partial\rho}{\partial g^{\text{eff}}_{kj}}\right\|=\dfrac{\rho}{2}[G^{\text{eff}}]^{-1}$. Using (9.31) and (9.32), we find

$$\frac{2}{\tau}\frac{\rho}{2}U\begin{bmatrix}\dfrac{1}{k_1^2}\ln\dfrac{k_1}{\sqrt[3]{k_1k_2k_3}} & 0 & 0\\ 0 & \dfrac{1}{k_2^2}\ln\dfrac{k_2}{\sqrt[3]{k_1k_2k_3}} & 0\\ 0 & 0 & \dfrac{1}{k_3^2}\ln\dfrac{k_3}{\sqrt[3]{k_1k_2k_3}}\end{bmatrix}U^*U\begin{bmatrix}k_1^2 & 0 & 0\\ 0 & k_2^2 & 0\\ 0 & 0 & k_3^2\end{bmatrix}U^*$$

$$=[U^*]^{-1}\begin{bmatrix}\dfrac{\rho}{\tau}\ln\dfrac{k_1}{\sqrt[3]{k_1k_2k_3}} & 0 & 0\\ 0 & \dfrac{\rho}{\tau}\ln\dfrac{k_2}{\sqrt[3]{k_1k_2k_3}} & 0\\ 0 & 0 & \dfrac{\rho}{\tau}\ln\dfrac{k_3}{\sqrt[3]{k_1k_2k_3}}\end{bmatrix}U^*.$$

Since the traces of similar matrices are equal, we have

$$\begin{aligned}\operatorname{tr}\left\|\varphi_{kj}\frac{\partial\rho}{\partial g^{\text{eff}}_{jl}}\right\|&=\frac{\rho}{\tau}\ln\frac{k_1}{\sqrt[3]{k_1k_2k_3}}+\frac{\rho}{\tau}\ln\frac{k_2}{\sqrt[3]{k_1k_2k_3}}+\frac{\rho}{\tau}\ln\frac{k_3}{\sqrt[3]{k_1k_2k_3}}\\&=\frac{\rho}{\tau}\ln\frac{k_1k_2k_3}{(\sqrt[3]{k_1k_2k_3})^3}=\frac{\rho}{\tau}\ln 1=0.\end{aligned} \tag{10.13}$$

Thus, we have proved that

$$\varphi_{kj}\frac{\partial\rho}{\partial g_{kj}^{\text{eff}}}=0,\qquad \frac{\partial\rho}{\partial t}+\frac{\partial(\rho u_i)}{\partial x_i}=0. \tag{10.14}$$

10.3. The equation for density (the so-called equation of continuity) has the same form as in the case of a medium without relaxation. This can be caused by the requirement of our relaxation model (cf. Section 9) that the density remains constant under change of the metric effective elastic deformation tensor g_{ij}^{eff} in the case of pure relaxation (a motionless medium).

Remark 10.1. The equation for density is a consequence of the conservation of mass (cf. Section 14). We derived this equation on the basis of the relaxation arguments in order to confirm the reasonableness of our model.

10.4. In Section 15, we will show that within the framework of our model, the entropy increases if we deal with relaxation processes. In the case of pure relaxation, this fact about the entropy was established in Section 9. To treat the general case, we need to use the energy conservation law. The following identity is also useful for this purpose:

$$-E_{g_{ij}}\varphi_{ij}=-\frac{1}{3\tau}\Big[(k_1E_{k_1}-k_2E_{k_2})\ln\frac{k_1}{k_2}+(k_2E_{k_2}-k_3E_{k_3})\ln\frac{k_2}{k_3}+(k_3E_{k_3}-k_1E_{k_1})\ln\frac{k_3}{k_1}\Big].$$

It follows from the following chain of equalities:

$$-E_{g_{ij}}\varphi_{ij}=-\operatorname{tr}\big(\|E_{g_{ji}}\|\,\|\varphi_{ij}\|\big)=-\operatorname{tr}U\begin{bmatrix}\frac{k_1^3}{2}E_{k_1} & 0 & 0\\ 0 & \frac{k_2^3}{2}E_{k_2} & 0\\ 0 & 0 & \frac{k_3^3}{2}E_{k_3}\end{bmatrix}U^*$$

$$\times U\begin{bmatrix}\frac{2}{\tau k_1^2}\ln\frac{k_1}{\sqrt[3]{k_1k_2k_3}} & 0 & 0\\ 0 & \frac{2}{\tau k_2^2}\ln\frac{k_2}{\sqrt[3]{k_1k_2k_3}} & 0\\ 0 & 0 & \frac{2}{\tau k_3^2}\ln\frac{k_3}{\sqrt[3]{k_1k_2k_3}}\end{bmatrix}U^*$$

$$= -\operatorname{tr}\begin{bmatrix} \frac{k_1}{\tau}E_{k_1}\ln\frac{k_1}{\sqrt[3]{k_1k_2k_3}} & 0 & 0 \\ 0 & \frac{k_2}{\tau}E_{k_2}\ln\frac{k_2}{\sqrt[3]{k_1k_2k_3}} & 0 \\ 0 & 0 & \frac{k_3}{\tau}E_{k_3}\ln\frac{k_3}{\sqrt[3]{k_1k_2k_3}} \end{bmatrix}$$

$$= -\frac{1}{\tau}\left(k_1E_{k_1}\ln\frac{k_1}{\sqrt[3]{k_1k_2k_3}} + k_2E_{k_2}\ln\frac{k_2}{\sqrt[3]{k_1k_2k_3}} + k_3E_{k_3}\ln\frac{k_3}{\sqrt[3]{k_1k_2k_3}}\right)$$

$$= -\frac{1}{3\tau}\left[(k_1E_{k_1} - k_2E_{k_2})\ln\frac{k_1}{k_2} + (k_2E_{k_2} - k_3E_{k_3})\ln\frac{k_2}{k_3} + (k_3E_{k_3} - k_1E_{k_1})\ln\frac{k_3}{k_1}\right],$$

Remark 10.2. In our consideration, it was not required that the medium is transformed to the stressed state simultaneously by the same transformation at all points of the medium. In fact, *we construct a canonical deformation of a neighborhood of each point of the medium.*

Up to this moment, we assumed that material particles of an elastic medium in the nonstressed state are described by the Cartesian coordinates. In Section 11, we will establish a criterion which allows us to recognize if it is possible for known $g_{ij}^{\text{eff}}(x_1, x_2, x_3)$, to find a mapping of the Euclidean space with the Cartesian coordinates $x_{10} = x_{10}(x_1, x_2, x_3)$ $(x_1 = x_1(x_{10}, x_{20}, x_{30}))$, $x_{20} = x_{20}(x_1, x_2, x_3)$ $(x_2 = x_2(x_{10}, x_{20}, x_{30}))$, $x_{30} = x_{30}(x_1, x_2, x_3)$ $(x_3 = x_3(x_{10}, x_{20}, x_{30}))$ such that the tensors $g_{ij} = g_{ij}(x_1, x_2, x_3) = \dfrac{\partial x_{k0}}{\partial x_i}\dfrac{\partial x_{k0}}{\partial x_j}$ and g_{ij}^{eff} coincide. This criterion yields the solvability conditions for the equations for the initial coordinates $x_{k0} = x_{k0}(x_1, x_2, x_3)$ with given $g_{ij}(x_1, x_2, x_3)$. It turns out that for the solvability of these equations it is necessary that the entries g_{ij} satisfy some system of second-order differential equations. Naturally, it is not necessary that the tensor g_{ij}^{eff} satisfies such conditions.

§ 11. Compatibility Conditions

In this section, we derive conditions on $g_{ij}(x_1, x_2, x_3)$ under which there exists a solution $x_{i0} = x_{i0}(x_1, x_2, x_3)$ to the system

$$g_{ij}(x_1, x_2, x_3) = \frac{\partial x_{k0}}{\partial x_i}\frac{\partial x_{k0}}{\partial x_j} \equiv \frac{\partial x_{10}}{\partial x_i}\frac{\partial x_{10}}{\partial x_j} + \frac{\partial x_{20}}{\partial x_i}\frac{\partial x_{20}}{\partial x_j} + \frac{\partial x_{30}}{\partial x_i}\frac{\partial x_{30}}{\partial x_j}, \tag{11.1}$$

called the *compatibility conditions*. For the sake of simplicity, we replace x_{10}, x_{20}, x_{30} with ξ_1, ξ_2, ξ_3 and consider the system

$$g_{ij}(x_1, x_2, x_3) = \frac{\partial \xi_k}{\partial x_i}\frac{\partial \xi_k}{\partial x_j}, \tag{11.2}$$

where the matrix $G = \|g_{ij}\|$ is assumed to be symmetric and positive definite and the entries of G are sufficiently smooth functions. The proof is rather cumbersome and can be omitted at first reading.

In this section, we give different formulation of the compatibility conditions. In the case of small deformations (linear elasticity), the compatibility conditions have the simple form and are called the *Saint-Venant conditions*.

11.1. We start by considering the following generalization of (11.2):

$$\frac{\partial \xi_k}{\partial x_i} h_{kl} \frac{\partial \xi_l}{\partial x_j} = g_{ij}(x_1, x_2, x_3), \tag{11.3}$$

or, in the matrix notation,

$$G = \begin{bmatrix} \frac{\partial \xi_1}{\partial x_1} & \frac{\partial \xi_2}{\partial x_1} & \frac{\partial \xi_3}{\partial x_1} \\ \frac{\partial \xi_1}{\partial x_2} & \frac{\partial \xi_2}{\partial x_2} & \frac{\partial \xi_3}{\partial x_2} \\ \frac{\partial \xi_1}{\partial x_3} & \frac{\partial \xi_2}{\partial x_3} & \frac{\partial \xi_3}{\partial x_3} \end{bmatrix} H \begin{bmatrix} \frac{\partial \xi_1}{\partial x_1} & \frac{\partial \xi_1}{\partial x_2} & \frac{\partial \xi_1}{\partial x_3} \\ \frac{\partial \xi_2}{\partial x_1} & \frac{\partial \xi_2}{\partial x_2} & \frac{\partial \xi_2}{\partial x_3} \\ \frac{\partial \xi_3}{\partial x_1} & \frac{\partial \xi_3}{\partial x_2} & \frac{\partial \xi_3}{\partial x_3} \end{bmatrix},$$

where $H = \|h_{kl}\|$ is a constant symmetric positive definite matrix. Since H is constant, we have

$$\frac{\partial h_{kl}}{\partial x_1} = \frac{\partial h_{kl}}{\partial x_2} = \frac{\partial h_{kl}}{\partial x_3} = 0. \tag{11.4}$$

Later we will show that the solvability of (11.3) for at least one constant matrix H implies that (11.3) is solvable for $H = I$ ($h_{kl} = \delta_{kl}$), i.e., the original system (11.2) is solvable.

Remark 11.1. For given $g_{ij}(x_1, x_2, x_3)$ and $\xi_k = \xi_k(x_1, x_2, x_3)$ ($k = 1, 2, 3$) with $\det \left\| \frac{\partial \xi_k}{\partial x_i} \right\| \neq 0$ we can construct a unique symmetric matrix H satisfying (11.3) by the formula

$$H = \begin{bmatrix} \frac{\partial \xi_1}{\partial x_1} & \frac{\partial \xi_2}{\partial x_1} & \frac{\partial \xi_3}{\partial x_1} \\ \frac{\partial \xi_1}{\partial x_2} & \frac{\partial \xi_2}{\partial x_2} & \frac{\partial \xi_3}{\partial x_2} \\ \frac{\partial \xi_1}{\partial x_3} & \frac{\partial \xi_2}{\partial x_3} & \frac{\partial \xi_3}{\partial x_3} \end{bmatrix}^{-1} G \begin{bmatrix} \frac{\partial \xi_1}{\partial x_1} & \frac{\partial \xi_1}{\partial x_2} & \frac{\partial \xi_1}{\partial x_3} \\ \frac{\partial \xi_2}{\partial x_1} & \frac{\partial \xi_2}{\partial x_2} & \frac{\partial \xi_2}{\partial x_3} \\ \frac{\partial \xi_3}{\partial x_1} & \frac{\partial \xi_3}{\partial x_2} & \frac{\partial \xi_3}{\partial x_3} \end{bmatrix}^{-1}. \tag{11.5}$$

From the representation (11.5) and the condition $\det \left\| \dfrac{\partial \xi_k}{\partial x_i} \right\| \neq 0$ it follows that the matrix H is symmetric and positive definite.

Differentiating equations (11.3), we obtain the system

$$\begin{aligned}
\frac{\partial g_{ik}}{\partial x_m} &= h_{jl}\left(\frac{\partial^2 \xi_j}{\partial x_i \partial x_m}\frac{\partial \xi_l}{\partial x_k} + \frac{\partial^2 \xi_l}{\partial x_k \partial x_m}\frac{\partial \xi_j}{\partial x_i}\right) + \frac{\partial h_{jl}}{\partial x_m}\frac{\partial \xi_j}{\partial x_i}\frac{\partial \xi_l}{\partial x_k}, \\
\frac{\partial g_{im}}{\partial x_k} &= h_{jl}\left(\frac{\partial^2 \xi_j}{\partial x_i \partial x_k}\frac{\partial \xi_l}{\partial x_m} + \frac{\partial^2 \xi_l}{\partial x_k \partial x_m}\frac{\partial \xi_j}{\partial x_i}\right) + \frac{\partial h_{jl}}{\partial x_k}\frac{\partial \xi_j}{\partial x_i}\frac{\partial \xi_l}{\partial x_m}, \\
\frac{\partial g_{km}}{\partial x_i} &= h_{jl}\left(\frac{\partial^2 \xi_j}{\partial x_k \partial x_i}\frac{\partial \xi_l}{\partial x_m} + \frac{\partial^2 \xi_l}{\partial x_m \partial x_i}\frac{\partial \xi_j}{\partial x_k}\right) + \frac{\partial h_{jl}}{\partial x_i}\frac{\partial \xi_j}{\partial x_k}\frac{\partial \xi_l}{\partial x_m}.
\end{aligned} \tag{11.6}$$

Then

$$\begin{aligned}
\frac{\partial g_{ik}}{\partial x_m} + \frac{\partial g_{im}}{\partial x_k} - \frac{\partial g_{km}}{\partial x_i} &= h_{jl}\left(2\frac{\partial^2 \xi_l}{\partial x_k \partial x_m}\frac{\partial \xi_j}{\partial x_i} + \underline{\frac{\partial^2 \xi_j}{\partial x_i \partial x_m}\frac{\partial \xi_l}{\partial x_k} - \frac{\partial^2 \xi_l}{\partial x_m \partial x_i}\frac{\partial \xi_j}{\partial x_k}}\right) \\
&+ \left(\frac{\partial h_{jl}}{\partial x_m}\frac{\partial \xi_j}{\partial x_i}\frac{\partial \xi_l}{\partial x_k} + \frac{\partial h_{jl}}{\partial x_k}\frac{\partial \xi_j}{\partial x_i}\frac{\partial \xi_l}{\partial x_m} - \frac{\partial h_{jl}}{\partial x_i}\frac{\partial \xi_j}{\partial x_k}\frac{\partial \xi_l}{\partial x_m}\right).
\end{aligned}$$

The underlined terms on the right-hand side of this formula can be eliminated since, in view of the symmetry property $h_{jl} = h_{lj}$, we have

$$\begin{aligned}
h_{jl}\left(\frac{\partial^2 \xi_j}{\partial x_i \partial x_m}\frac{\partial \xi_l}{\partial x_k} - \frac{\partial^2 \xi_l}{\partial x_m \partial x_i}\frac{\partial \xi_j}{\partial x_k}\right) &= h_{lj}\frac{\partial^2 \xi_j}{\partial x_i \partial x_m}\frac{\partial \xi_l}{\partial x_k} - h_{jl}\frac{\partial^2 \xi_l}{\partial x_m \partial x_i}\frac{\partial \xi_j}{\partial x_k} \\
&\equiv h_{jl}\frac{\partial^2 \xi_l}{\partial x_i \partial x_m}\frac{\partial \xi_j}{\partial x_k} - h_{jl}\frac{\partial^2 \xi_l}{\partial x_i \partial x_m}\frac{\partial \xi_j}{\partial x_k} = 0.
\end{aligned}$$

Then

$$\begin{aligned}
&\frac{1}{2}\left(\frac{\partial g_{ik}}{\partial x_m} + \frac{\partial g_{im}}{\partial x_k} - \frac{\partial g_{km}}{\partial x_i}\right) - h_{jl}\frac{\partial \xi_j}{\partial x_i}\frac{\partial^2 \xi_l}{\partial x_k \partial x_m} \\
&= \frac{1}{2}\left(\frac{\partial h_{jl}}{\partial x_m}\frac{\partial \xi_j}{\partial x_i}\frac{\partial \xi_l}{\partial x_k} + \frac{\partial h_{jl}}{\partial x_k}\frac{\partial \xi_j}{\partial x_i}\frac{\partial \xi_l}{\partial x_m} - \frac{\partial h_{jl}}{\partial x_i}\frac{\partial \xi_j}{\partial x_k}\frac{\partial \xi_l}{\partial x_m}\right).
\end{aligned} \tag{11.7}$$

The last identity holds for any $\xi_k = \xi_k(x_1, x_2, x_3)$ and the corresponding h_{jl}.

Since $h_{jl}(x_1, x_2, x_3)$ are constant, from (11.7) we obtain the equality

$$\frac{1}{2}\left(\frac{\partial g_{ik}}{\partial x_m} + \frac{\partial g_{im}}{\partial x_k} - \frac{\partial g_{km}}{\partial x_i}\right) - h_{jl}\frac{\partial \xi_j}{\partial x_i}\frac{\partial^2 \xi_l}{\partial x_k \partial x_m} = 0. \tag{11.8}$$

We note that the converse assertion is also true: if (11.8) holds, then h_{ij} are constants. Indeed, from (11.8) we find

$$\frac{\partial h_{jl}}{\partial x_m}\frac{\partial \xi_j}{\partial x_i}\frac{\partial \xi_l}{\partial x_k} + \frac{\partial h_{jl}}{\partial x_k}\frac{\partial \xi_j}{\partial x_i}\frac{\partial \xi_l}{\partial x_m} - \frac{\partial h_{jl}}{\partial x_i}\frac{\partial \xi_j}{\partial x_k}\frac{\partial \xi_l}{\partial x_m} = 0.$$

Adding these equalities with the equalities

$$\frac{\partial h_{jl}}{\partial x_i}\frac{\partial \xi_j}{\partial x_m}\frac{\partial \xi_l}{\partial x_k}+\frac{\partial h_{jl}}{\partial x_k}\frac{\partial \xi_j}{\partial x_m}\frac{\partial \xi_l}{\partial x_i}-\frac{\partial h_{jl}}{\partial x_m}\frac{\partial \xi_j}{\partial x_k}\frac{\partial \xi_l}{\partial x_i}=0$$

obtained by permutation of i and m, we find

$$\frac{\partial h_{jl}}{\partial x_m}\left(\frac{\partial \xi_j}{\partial x_i}\frac{\partial \xi_l}{\partial x_k}-\frac{\partial \xi_j}{\partial x_k}\frac{\partial \xi_l}{\partial x_i}\right)+\frac{\partial h_{jl}}{\partial x_k}\left(\frac{\partial \xi_j}{\partial x_i}\frac{\partial \xi_l}{\partial x_m}+\frac{\partial \xi_j}{\partial x_m}\frac{\partial \xi_l}{\partial x_i}\right)$$
$$+\frac{\partial h_{jl}}{\partial x_i}\left(-\frac{\partial \xi_j}{\partial x_k}\frac{\partial \xi_l}{\partial x_m}+\frac{\partial \xi_j}{\partial x_m}\frac{\partial \xi_l}{\partial x_k}\right)=0. \tag{11.9}$$

Since $h_{jl}=h_{lj}$, the first and third terms on the left-hand side of (11.9) vanish. After simple transformations of the second term, we find

$$2\frac{\partial h_{jl}}{\partial x_k}\frac{\partial \xi_j}{\partial x_i}\frac{\partial \xi_l}{\partial x_m}=0, \tag{11.10}$$

or, in the matrix notation,

$$2\begin{bmatrix}\frac{\partial \xi_1}{\partial x_1} & \frac{\partial \xi_2}{\partial x_1} & \frac{\partial \xi_3}{\partial x_1}\\ \frac{\partial \xi_1}{\partial x_2} & \frac{\partial \xi_2}{\partial x_2} & \frac{\partial \xi_3}{\partial x_2}\\ \frac{\partial \xi_1}{\partial x_3} & \frac{\partial \xi_2}{\partial x_3} & \frac{\partial \xi_3}{\partial x_3}\end{bmatrix}\begin{bmatrix}\frac{\partial h_{11}}{\partial x_k} & \frac{\partial h_{12}}{\partial x_k} & \frac{\partial h_{13}}{\partial x_k}\\ \frac{\partial h_{21}}{\partial x_k} & \frac{\partial h_{22}}{\partial x_k} & \frac{\partial h_{23}}{\partial x_k}\\ \frac{\partial h_{31}}{\partial x_k} & \frac{\partial h_{32}}{\partial x_k} & \frac{\partial h_{33}}{\partial x_k}\end{bmatrix}\begin{bmatrix}\frac{\partial \xi_1}{\partial x_1} & \frac{\partial \xi_1}{\partial x_2} & \frac{\partial \xi_1}{\partial x_3}\\ \frac{\partial \xi_2}{\partial x_1} & \frac{\partial \xi_2}{\partial x_2} & \frac{\partial \xi_2}{\partial x_3}\\ \frac{\partial \xi_3}{\partial x_1} & \frac{\partial \xi_3}{\partial x_2} & \frac{\partial \xi_3}{\partial x_3}\end{bmatrix}=0.$$

Since the matrix $\left\|\frac{\partial \xi_i}{\partial x_j}\right\|$ is nonsingular, we conclude that $\frac{\partial h_{jl}}{\partial x_k}=0$, i.e., h_{jl} are constant.

Thus, we have proved the following assertion.

• *Let* det $\left\|\frac{\partial \xi_i}{\partial x_j}\right\| \neq 0$, *and let H be a matrix in the system* (11.3). *Then the matrix H is constant if and only if the condition* (11.8) *holds.*

11.2. Now, excluding h_{jl} from (11.3), we obtain linear equations for ξ_1, ξ_2, ξ_3 with coefficients depending on g_{ik} and their derivatives. We denote by $B=\|b_{rt}\|$ and $F=\|f_{si}\|$ the inverse matrices to $\left\|\frac{\partial \xi_l}{\partial x_r}\right\|$ and G respectively, i.e., $\frac{\partial \xi_l}{\partial x_r}b_{rt}=\delta_{lt}$ and $f_{si}g_{ir}=\delta_{sr}$. Using the relations

$$h_{jl}\frac{\partial \xi_j}{\partial x_i}=h_{jr}\frac{\partial \xi_j}{\partial x_i}\delta_{rl}=h_{jr}\frac{\partial \xi_j}{\partial x_i}\frac{\partial \xi_r}{\partial x_p}b_{pl}=g_{ip}b_{pl},$$

we can write the condition (11.8) in the form

$$g_{ip}b_{pl}\frac{\partial^2 \xi_l}{\partial x_k \partial x_m}=\frac{1}{2}\left(\frac{\partial g_{ik}}{\partial x_m}+\frac{\partial g_{im}}{\partial x_k}-\frac{\partial g_{km}}{\partial x_i}\right), \tag{11.11}$$

where $i = 1,2,3$. For fixed k and m we consider these equations as a single vector equation. Applying the matrix $B^{-1}G^{-1} = B^{-1}F$ with entries

$$\frac{\partial \xi_p}{\partial x_s} f_{si} \equiv \frac{\partial \xi_p}{\partial x_1} f_{1i} + \frac{\partial \xi_p}{\partial x_2} f_{2i} + \frac{\partial \xi_p}{\partial x_3} f_{3i}$$

to both sides of this vector equality, we obtain the required linear system

$$\frac{\partial^2 \xi_p}{\partial x_k \partial x_m} = \frac{1}{2} f_{si} \left(\frac{\partial g_{ik}}{\partial x_m} + \frac{\partial g_{im}}{\partial x_k} - \frac{\partial g_{km}}{\partial x_i} \right) \frac{\partial \xi_p}{\partial x_s} \tag{11.12}$$

for ξ_1, ξ_2, ξ_3.

Thus, equation (11.12) has a solution ξ_1, ξ_2, ξ_3 if the following conditions hold:

(a) $h_{jl} \dfrac{\partial \xi_j}{\partial x_i} \dfrac{\partial^2 \xi_l}{\partial x_k \partial x_m} = \dfrac{1}{2} \left(\dfrac{\partial g_{ik}}{\partial x_m} + \dfrac{\partial g_{im}}{\partial x_k} - \dfrac{\partial g_{km}}{\partial x_i} \right),$ [cf. (11.8)]

(b) $g_{ij} = h_{kl} \dfrac{\partial \xi_k}{\partial x_i} \dfrac{\partial \xi_l}{\partial x_j},$ [cf. (11.3)]

(c) $\det \left\| \dfrac{\partial \xi_k}{\partial x_i} \right\| \neq 0.$

Conversely, if there exists a solution to the linear system (11.12) satisfying conditions (b) and (c), then the condition (a) is also satisfied. This can be easily verified by making the transformations in the converse order.

We introduce the notation for the coefficients of the linear system (11.12):

$$\Gamma^s_{km} = \frac{1}{2} f_{si} \left(\frac{\partial g_{ik}}{\partial x_m} + \frac{\partial g_{im}}{\partial x_k} - \frac{\partial g_{km}}{\partial x_i} \right), \quad \|f_{si}\| = F = G^{-1}. \tag{11.13}$$

It is obvious that $\Gamma^s_{km} = \Gamma^s_{mk}$.

Thus, we have proved the following assertion.

• *The matrix H from the system $g_{ij} = h_{kl} \dfrac{\partial \xi_k}{\partial x_i} \dfrac{\partial \xi_l}{\partial x_j}$ for ξ_1, ξ_2, ξ_3 such that* $\det \left\| \dfrac{\partial \xi_k}{\partial x_j} \right\| \neq 0$ *is a constant matrix if and only if $\xi_p(x_1, x_2, x_3)$ satisfy the equations*

$$\frac{\partial^2 \xi_p}{\partial x_k \partial x_m} = \Gamma^s_{km} \frac{\partial \xi_p}{\partial x_s} \equiv \Gamma^1_{km} \frac{\partial \xi_p}{\partial x_1} + \Gamma^2_{km} \frac{\partial \xi_p}{\partial x_2} + \Gamma^3_{km} \frac{\partial \xi_p}{\partial x_3}.$$

Further, we will omit the subscript p because we have the same equation for ξ_1, ξ_2, ξ_3. Therefore, we consider the single equation

$$\frac{\partial^2 \xi}{\partial x_k \partial x_m} = \Gamma^s_{km} \frac{\partial \xi}{\partial x_s}. \tag{11.14}$$

11.3. We reduce the study of the second-order equations (11.14) to the case of the first-order equation. To do this, we assume that ξ_1, ξ_2, ξ_3 is a solution to (11.4) and denote by η_1, η_2, η_3 the partial derivatives $\frac{\partial \xi}{\partial x_1}, \frac{\partial \xi}{\partial x_2}, \frac{\partial \xi}{\partial x_3}$. Then η_1, η_2, η_3 satisfy the first-order equations

$$\frac{\partial \eta_k}{\partial x_m} = \Gamma^s_{km} \eta_s. \tag{11.15}$$

Conversely, if η_1, η_2, η_3 form a sufficiently smooth solution to equations (11.15), then, by the symmetry property $\Gamma^s_{km} = \Gamma^s_{mk}$, we have

$$\frac{\partial \eta_k}{\partial x_m} - \frac{\partial \eta_m}{\partial x_k} = \Gamma^s_{km} \eta_s - \Gamma^s_{mk} \eta_s = 0.$$

Integrating the exact differential

$$d\xi = \eta_1 dx_1 + \eta_2 dx_2 + \eta_3 dx_3$$

along an arbitrary path, we see that $\xi(x_1, x_2, x_3)$ satisfies the conditions

$$\frac{\partial \xi}{\partial x_i} = \eta_i, \qquad \frac{\partial^2 \xi}{\partial x_k \partial x_m} = \Gamma^s_{km} \frac{\partial \xi}{\partial x_s}.$$

Further, we consider a solution η_1, η_2, η_3 to (11.15) taking some prescribed values at a fixed point $(\overline{x}_1, \overline{x}_2, \overline{x}_3)$:

$$\overline{\eta}_i = \eta_i(\overline{x}_1, \overline{x}_2, \overline{x}_3). \tag{11.16}$$

Then there exists a solution ξ_1, ξ_2, ξ_3 to (11.14)

$$\frac{\partial^2 \xi_p}{\partial x_k \partial x_m} = \Gamma^s_{km} \frac{\partial \xi_p}{\partial x_s}$$

satisfying the conditions

$$\xi_p(\overline{x}_1, \overline{x}_2, \overline{x}_3) = \overline{\xi}_p, \qquad \left[\frac{\partial \xi_p}{\partial x_k}\right]\Bigg|_{\substack{x_1=\overline{x}_1\\ x_2=\overline{x}_2\\ x_3=\overline{x}_3}} = \overline{\eta}_{pk}. \tag{11.17}$$

The converse assertion is also true: the existence of a solution to (11.14), (11.17) implies the solvability of the first-order system (11.15) with the initial condition (11.16). Indeed, for arbitrary $\overline{\eta}_1, \overline{\eta}_2, \overline{\eta}_3$ we choose $\overline{\eta}_{pk}$ such that $\overline{\eta}_{11} = \overline{\eta}_1$, $\overline{\eta}_{21} = \overline{\eta}_2$, $\overline{\eta}_{31} = \overline{\eta}_3$ ($\overline{\eta}_{p2}$ and $\overline{\eta}_{p3}$ are arbitrary) and construct $\xi_p(x_1, x_2, x_3)$ satisfying (11.14), (11.17) as above. Then the required η_k are determined by the formula $\eta_k = \frac{\partial \xi_1}{\partial x_k}$.

If there exists at least one triple $\xi_1 = \xi_1^*(x_1, x_2, x_3)$, $\xi_2 = \xi_2^*(x_1, x_2, x_3)$, $\xi_3 = \xi_3^*(x_1, x_2, x_3)$ with nonzero Jacobian in some domain that satisfies the second-order differential equations (11.14), then the first-order system (11.15) is solvable in the same domain. Indeed, to construct a solution ξ_1, ξ_2, ξ_3 to (11.14) satisfying the condition

$$\left[\frac{\partial \xi_p}{\partial x_k}\right]\Bigg|_{\substack{x_1=\overline{x}_1\\x_2=\overline{x}_2\\x_3=\overline{x}_3}} = \overline{\eta}_{pk},$$

we can put $\xi_i(x_1,x_2,x_3) = \alpha_{ij}\xi_j^*(x_1,x_2,x_3)$. Then

$$\left[\frac{\partial \xi_i}{\partial x_k}\right]\Bigg|_{\substack{x_1=\overline{x}_1\\x_2=\overline{x}_2\\x_3=\overline{x}_3}} = \alpha_{ij}\left[\frac{\partial \xi_j^*}{\partial x_k}\right]\Bigg|_{\substack{x_1=\overline{x}_1\\x_2=\overline{x}_2\\x_3=\overline{x}_3}} = \overline{\eta}_{ik}.$$

It is possible to choose such α_{ij} because the matrix $\left\|\frac{\partial \xi_j^*}{\partial x_k}\right\|$ is nonsingular at any point $(\overline{x}_1,\overline{x}_2,\overline{x}_3)$. Since, adding any constant to any solution $\xi_p(x_1,x_2,x_3)$, we obtain the same equations, it is easy to show that $\xi_p(\overline{x}_1,\overline{x}_2,\overline{x}_3)=\overline{\xi}_p$.

Thus, we have justified the following assertion.

• *The verification of the solvability conditions for the second-order system* (11.14) *is reduced to the equivalent but much simpler question on the solvability conditions for the linear first-order system of nine equations* ($k=1,2,3$; $m=1,2,3$) (*cf.* (11.15))

$$\frac{\partial \eta_k}{\partial x_m} = \Gamma^s_{km}\eta_s$$

or, in the matrix notation,

$$\begin{aligned}
\frac{\partial \eta_k}{\partial x_1} &= \Gamma^1_{k1}(x_1,x_2,x_3)\eta_1 + \Gamma^2_{k1}(x_1,x_2,x_3)\eta_2 + \Gamma^3_{k1}(x_1,x_2,x_3)\eta_3,\\
\frac{\partial \eta_k}{\partial x_2} &= \Gamma^1_{k2}(x_1,x_2,x_3)\eta_1 + \Gamma^2_{k2}(x_1,x_2,x_3)\eta_2 + \Gamma^3_{k2}(x_1,x_2,x_3)\eta_3,\\
\frac{\partial \eta_k}{\partial x_3} &= \Gamma^1_{k3}(x_1,x_2,x_3)\eta_1 + \Gamma^2_{k3}(x_1,x_2,x_3)\eta_2 + \Gamma^3_{k3}(x_1,x_2,x_3)\eta_3.
\end{aligned} \tag{11.18}$$

11.4. Now, we derive the solvability conditions for the linear system (11.15). Assuming that the solution η_1,η_2,η_3 is sufficiently smooth, we compute the second-order derivatives

$$\frac{\partial^2 \eta_k}{\partial x_i \partial x_j} = \Gamma^s_{ki}\frac{\partial \eta_s}{\partial x_j} + \frac{\partial \Gamma^s_{ki}}{\partial x_j}\eta_s = \Gamma^s_{ki}\Gamma^r_{sj}\eta_r + \frac{\partial \Gamma^r_{ki}}{\partial x_j}\eta_r = \left(\Gamma^s_{ki}\Gamma^r_{sj} + \frac{\partial \Gamma^r_{ki}}{\partial x_j}\right)\eta_r.$$

Then

$$0 = \frac{\partial^2 \eta_k}{\partial x_i \partial x_j} - \frac{\partial^2 \eta_k}{\partial x_j \partial x_i} = \left(\Gamma^s_{ki}\Gamma^r_{sj} - \Gamma^s_{kj}\Gamma^r_{si} + \frac{\partial \Gamma^r_{ki}}{\partial x_j} - \frac{\partial \Gamma^r_{kj}}{\partial x_i}\right)\eta_r.$$

This equality holds, in particular, at the point $(\overline{x}_1, \overline{x}_2, \overline{x}_3)$ where the solution takes the prescribed values: $\eta_1 = \overline{\eta}_1$, $\eta_2 = \overline{\eta}_2$, $\eta_3 = \overline{\eta}_3$. The solvability means that an initial point $(\overline{x}_1, \overline{x}_2, \overline{x}_3)$ and the values $\overline{\eta}_1$, $\overline{\eta}_2$, $\overline{\eta}_3$ are arbitrary . Consequently, the following conditions are necessary:

$$\Gamma^s_{ki}\Gamma^r_{sj} - \Gamma^s_{kj}\Gamma^r_{si} + \frac{\partial \Gamma^r_{ki}}{\partial x_j} - \frac{\partial \Gamma^r_{kj}}{\partial x_i} = 0. \tag{11.19}$$

We show that *the necessary conditions* (11.19) *are also sufficient.*

For given $(\overline{x}_1, \overline{x}_2, \overline{x}_3)$ and $\eta_i(\overline{x}_1, \overline{x}_2, \overline{x}_3) = \overline{\eta}_i$ we find a solution $\eta_i(x_1, \overline{x}_2, \overline{x}_3)$ to the ordinary differential (with respect to x_1) equations

$$\frac{\partial \eta_i}{\partial x_1} = \Gamma^s_{i1}(x_1, \overline{x}_2, \overline{x}_3)\eta_s.$$

Then η_1, η_2, η_3 are defined on the line $x_2 = \overline{x}_2$, $x_3 = \overline{x}_3$. Similarly, from the ordinary differential equations

$$\frac{\partial \eta_i}{\partial x_2} = \Gamma^s_{i2}(\overline{x}_1, x_2, \overline{x}_3)\eta_s$$

we find η_1, η_2, η_3 on the line $x_1 = \overline{x}_1$, $x_3 = \overline{x}_3$ passing through the initial point. Then we try to construct η_1, η_2, η_3 on the plane $x_3 = \overline{x}_3$ (more exactly, in a domain of the plane $x_3 = \overline{x}_3$ containing the initial point). For this purpose, we solve for each x_2 the initial-value problem

$$\begin{aligned} &\frac{\partial \eta'_i}{\partial x_1} = \Gamma^s_{i1}(x_1, x_2, \overline{x}_3)\eta'_s, \\ &\eta'_i(\overline{x}_1, x_2, \overline{x}_3) = \eta_i(\overline{x}_1, x_2, \overline{x}_3) \end{aligned} \tag{11.20}$$

or for any fixed x_1 the initial-value problem

$$\begin{aligned} &\frac{\partial \eta''_i}{\partial x_2} = \Gamma^s_{i2}(x_1, x_2, \overline{x}_3)\eta''_s, \\ &\eta''_i(x_1, \overline{x}_2, \overline{x}_3) = \eta_i(x_1, \overline{x}_2, \overline{x}_3). \end{aligned} \tag{11.21}$$

If $g_{ik}(x_1, x_2, x_3)$ and, consequently, the coefficients $\Gamma^s_{ij}(x_1, x_2, x_3)$ are sufficiently smooth, the solutions η'_i and η''_i are smooth functions of x_1, x_2, x_3.

We show that $\eta'_i = \eta''_i$. For this purpose, we differentiate the equation in (11.20) with respect to x_2. Using the equality

$$\frac{\partial \Gamma^s_{i1}}{\partial x_2} = \frac{\partial \Gamma^s_{i2}}{\partial x_1} + \Gamma^r_{i2}\Gamma^s_{r1} - \Gamma^r_{i1}\Gamma^s_{r2}$$

from the solvability condition (11.19), we find

$$\frac{\partial^2 \eta'_i}{\partial x_1 \partial x_2} = \frac{\partial \Gamma^s_{i1}}{\partial x_2}\eta'_s + \Gamma^s_{i1}\frac{\partial \eta'_s}{\partial x_2} = \left(\frac{\partial \Gamma^s_{i2}}{\partial x_1} + \Gamma^r_{i2}\Gamma^s_{r1} - \Gamma^r_{i1}\Gamma^s_{r2}\right)\eta'_s + \Gamma^r_{i1}\frac{\partial \eta'_r}{\partial x_2}. \tag{11.22}$$

Using the equation in (11.20), we transform (11.22) as follows:

$$\begin{aligned}\frac{\partial^2 \eta_i'}{\partial x_1 \partial x_2} &= \frac{\partial \Gamma_{i2}^s}{\partial x_1}\eta_s' + \Gamma_{i2}^r \Gamma_{r1}^s \eta_s' - \Gamma_{i1}^r \Gamma_{r2}^s \eta_s' + \Gamma_{i1}^r \frac{\partial \eta_r'}{\partial x_2} \\ &= \frac{\partial \Gamma_{i2}^s}{\partial x_1}\eta_s' + \Gamma_{i2}^r \frac{\partial \eta_r'}{\partial x_1} + \Gamma_{i1}^r \Gamma_{r2}^s \eta_s' + \Gamma_{i1}^r \frac{\partial \eta_r'}{\partial x_2}.\end{aligned}$$

Rearranging terms, we obtain the system of linear homogeneous ordinary differential (with respect to x_1) equations

$$\frac{\partial}{\partial x_1}\left(\frac{\partial \eta_i'}{\partial x_2} - \Gamma_{i2}^s \eta_s'\right) + \Gamma_{i1}^r \left(\frac{\partial \eta_i'}{\partial x_2} - \Gamma_{i2}^s \eta_s'\right) = 0 \tag{11.23}$$

for $\dfrac{\partial \eta_i'}{\partial x_2} - \Gamma_{i2}^s \eta_s'$. We defined η_i' at $x_1 = \overline{x}_1$ such that the following equality holds: $\dfrac{\partial \eta_i'}{\partial x_2} - \Gamma_{i2}^s \eta_s' = 0$ ($x_1 = \overline{x}_1$, $x_3 = \overline{x}_3$). Therefore, from (11.23) it follows that this equality holds everywhere in the considered domain of the plane $x_3 = \overline{x}_3$ (the uniqueness theorem!). Hence the equality $\eta_i' = \eta_i''$ is proved.

11.5. Introduce the notation $\eta_i(x_1, x_2, \overline{x}_3) \equiv \eta_i' = \eta_i''$. Then η_i satisfies the equations

$$\frac{\partial \eta_i}{\partial x_1} = \Gamma_{i1}^s \eta_s, \quad \frac{\partial \eta_i}{\partial x_2} = \Gamma_{i2}^s \eta_s, \quad x_3 = \overline{x}_3. \tag{11.24}$$

We extend η_i to the whole space (x_1, x_2, x_3) (more exactly, to a neighborhood of the initial point $(\overline{x}_1, \overline{x}_2, \overline{x}_3)$). To do this, we define η_i as a solution to the equations

$$\frac{\partial \eta_i}{\partial x_3} = \Gamma_{i3}^s \eta_s \tag{11.25}$$

with prescribed values at $x_3 = \overline{x}_3$. It remains to show that the equations in (11.24)

$$\frac{\partial \eta_i}{\partial x_1} = \Gamma_{i1}^s \eta_s, \quad \frac{\partial \eta_i}{\partial x_2} = \Gamma_{i2}^s \eta_s,$$

hold everywhere, not only on the plane $x_3 = \overline{x}_3$. We consider only the first equation in (11.24) (the second equation can be treated similarly). Differentiating (11.25) with respect to x_1, we obtain the equality

$$\frac{\partial^2 \eta_i}{\partial x_3 \partial x_1} - \frac{\partial \Gamma_{i3}^s}{\partial x_1}\eta_s - \Gamma_{i3}^s \frac{\partial \eta_s}{\partial x_1} = 0 \tag{11.26}$$

and transform it with the help of the following equation from the solvability condition (11.19):

$$\frac{\partial \Gamma_{i3}^s}{\partial x_1} = \frac{\partial \Gamma_{i1}^s}{\partial x_3} + \Gamma_{i1}^r \Gamma_{r3}^s - \Gamma_{i3}^r \Gamma_{r1}^s.$$

We find

$$\begin{aligned}
0 &= \frac{\partial^2 \eta_i}{\partial x_3 \partial x_1} - \frac{\partial \Gamma_{i3}^s}{\partial x_1}\eta_s - \Gamma_{i3}^s \frac{\partial \eta_s}{\partial x_1} \\
&= \frac{\partial^2 \eta_i}{\partial x_3 \partial x_1} - \left(\frac{\partial \Gamma_{i1}^s}{\partial x_3} + \Gamma_{i1}^r \Gamma_{r3}^s - \Gamma_{i3}^r \Gamma_{r1}^s\right)\eta_s - \Gamma_{i3}^s \frac{\partial \eta_s}{\partial x_1} \\
&= \left(\frac{\partial^2 \eta_i}{\partial x_3 \partial x_1} - \frac{\partial \Gamma_{i1}^s}{\partial x_3}\eta_s - \Gamma_{i1}^s \frac{\partial \eta_s}{\partial x_3}\right) + \left(\Gamma_{i1}^s \frac{\partial \eta_s}{\partial x_3} - \Gamma_{i1}^r \Gamma_{r3}^s \eta_s\right) \\
&\quad + \left(\Gamma_{i3}^r \Gamma_{r1}^s \eta_s - \Gamma_{i3}^s \frac{\partial \eta_s}{\partial x_1}\right) \qquad (11.27) \\
&= \frac{\partial}{\partial x_3}\left(\frac{\partial \eta_i}{\partial x_1} - \Gamma_{i1}^s \eta_s\right) + \Gamma_{i1}^s \left(\frac{\partial \eta_s}{\partial x_3} - \Gamma_{i3}^k \eta_k\right) - \Gamma_{i3}^r \left(\frac{\partial \eta_r}{\partial x_1} - \Gamma_{r1}^s \eta_s\right).
\end{aligned}$$

Since $\dfrac{\partial \eta_s}{\partial x_3} - \Gamma_{s3}^k \eta_k = 0$, we arrive at the following system of linear ordinary differential (with respect to x_3) equations:

$$\frac{\partial}{\partial x_3}\left(\frac{\partial \eta_i}{\partial x_1} - \Gamma_{i1}^s \eta_s\right) - \Gamma_{i3}^r \left(\frac{\partial \eta_r}{\partial x_1} - \Gamma_{r1}^s \eta_s\right) = 0 \qquad (11.28)$$

for $\dfrac{\partial \eta_i}{\partial x_1} - \Gamma_{i1}^s \eta_s$. By the uniqueness theorem, $\dfrac{\partial \eta_i}{\partial x_1} - \Gamma_{i1}^s \eta_s = 0$ for all x_3 provided that these equalities hold on the initial plane $x_3 = \overline{x}_3$. But the initial condition is satisfied by construction.

• *Thus, we have proved that the condition* (11.19) *is sufficient for the solvability of the linear system* (11.14) *or* (11.15).

11.6. Thus, at was shown, if the system (11.2) with respect to ξ_1, ξ_2, ξ_3 is solvable, then the solution ξ_1, ξ_2, ξ_3 satisfies the linear system (11.12) (or (11.14)); moreover, $\det \left\| \dfrac{\partial \xi_k}{\partial x_i} \right\| \neq 0$. The last relation holds since $\left\| \dfrac{\partial \xi_k}{\partial x_i} \right\|$ is nonsingular because of the equality $\left(\det \left\| \dfrac{\partial \xi_k}{\partial x_i} \right\|\right)^2 = \det G > 0$ which follows from (11.2). Thus, the solvability of (11.2) implies the solvability of the linear system (11.12) (or (11.14)).

Let us show that the solvability of the linear system (11.14) implies the solvability of the system (11.2). As was already shown, if the system (11.14) is solvable, then there exists a solution ξ_1, ξ_2, ξ_3 satisfying the initial condition (11.17) at the point $(\overline{x}_1, \overline{x}_2, \overline{x}_3)$.

In particular, we can choose $\overline{\eta}_{pk}$ such that

$$g_{ij}(\overline{x}_1, \overline{x}_2, \overline{x}_3) = \overline{\eta}_{pi}\overline{\eta}_{pj} = \delta_{pk}\overline{\eta}_{pi}\overline{\eta}_{kj} = \delta_{pk} \left.\frac{\partial \xi_p}{\partial x_i}\frac{\partial \xi_p}{\partial x_j}\right|_{\overline{x}_1, \overline{x}_2, \overline{x}_3}.$$

Indeed, it suffices to write $G = \|g_{ij}\|$ in the canonical form

$$G = U \begin{bmatrix} 1/k_1^2 & 0 & 0 \\ 0 & 1/k_2^2 & 0 \\ 0 & 0 & 1/k_3^2 \end{bmatrix} U^* \qquad (UU^* = I,\ \det U = +1)$$

and set

$$\|\overline{\eta}_{pi}\| = V^* \begin{bmatrix} 1/k_1 & 0 & 0 \\ 0 & 1/k_2 & 0 \\ 0 & 0 & 1/k_3 \end{bmatrix} U^*, \tag{11.29}$$

where V is an arbitrary orthogonal matrix.

As was mentioned, for a solution $\xi_1(x_1, x_2, x_3)$, $\xi_2(x_1, x_2, x_3)$, $\xi_3(x_1, x_2, x_3)$ satisfying the initial data we can construct a constant symmetric matrix H such that $g_{ij} = h_{kl}\dfrac{\partial \xi_k}{\partial x_i}\dfrac{\partial \xi_k}{\partial x_j}$. By construction, at the point $(\overline{x}_1, \overline{x}_2, \overline{x}_3)$, we have

$$g_{ij} = \delta_{kl}\frac{\partial \xi_k}{\partial x_i}\frac{\partial \xi_l}{\partial x_j} \qquad (h_{kl} = \delta_{kl}).$$

Consequently, the equalities

$$h_{kl} = \delta_{kl}, \quad g_{ij} = \delta_{kl}\frac{\partial \xi_k}{\partial x_i}\frac{\partial \xi_l}{\partial x_j} = \frac{\partial \xi_k}{\partial x_i}\frac{\partial \xi_k}{\partial x_j}$$

hold everywhere. As was mentioned above, the last equality with a positive definite matrix G leads to the condition $\det \left\| \dfrac{\partial \xi_k}{\partial x_j} \right\| \neq 0$.

• *The solvability of the equations (cf (11.2))* $g_{ij} = \dfrac{\partial \xi_k}{\partial x_i}\dfrac{\partial \xi_k}{\partial x_j}$ *is equivalent to the solvability of the linear system (cf.* (11.14))

$$\frac{\partial^2 \xi_p}{\partial x_k \partial x_m} = \Gamma^s_{km}\frac{\partial \xi_p}{\partial x_s}$$

which, in turn, is solvable if and only if (cf. (11.19))

$$R^r_{ki,j} \equiv \Gamma^s_{ki}\Gamma^r_{sj} - \Gamma^s_{kj}\Gamma^r_{si} + \frac{\partial \Gamma^r_{ki}}{\partial x_j} - \frac{\partial \Gamma^r_{kj}}{\partial x_i} = 0, \tag{11.30}$$

where (cf. (11.13))

$$\Gamma^s_{km} = \frac{1}{2} f_{si}\left(\frac{\partial g_{ik}}{\partial x_m} + \frac{\partial g_{im}}{\partial x_k} - \frac{\partial g_{km}}{\partial x_i}\right), \quad \|f_{si}\| = F = G^{-1}.$$

• The equalities (11.30) are called the *compatibility conditions.*

The validity of the compatibility conditions (11.30) is sufficient and necessary for the representation of the metric deformation tensor G in the form $G = A^*A$, where A is the distortion matrix,

$$A = \begin{bmatrix} a_{11} & a_{12} & a_{13} \\ a_{21} & a_{22} & a_{23} \\ a_{31} & a_{32} & a_{33} \end{bmatrix} = \begin{bmatrix} \frac{\partial x_{10}}{\partial x_1} & \frac{\partial x_{10}}{\partial x_2} & \frac{\partial x_{10}}{\partial x_3} \\ \frac{\partial x_{20}}{\partial x_1} & \frac{\partial x_{20}}{\partial x_2} & \frac{\partial x_{20}}{\partial x_3} \\ \frac{\partial x_{30}}{\partial x_1} & \frac{\partial x_{30}}{\partial x_2} & \frac{\partial x_{30}}{\partial x_3} \end{bmatrix},$$

which is the Jacobi matrix of some nondegenerate mapping

$$x_{10} = x_{10}(x_1, x_2, x_3), \quad x_{20} = x_{20}(x_1, x_2, x_3), \quad x_{30} = x_{30}(x_1, x_2, x_3).$$

Without loss of generality, we can assume that $\det A > 0$. Otherwise, we change the sign of x_{10}, x_{20}, x_{30}, which leads to the change of the sign for all a_{ij} and, consequently, for $\det A$.

11.7. It is convenient to introduce the notation

$$R_{ki,jl} = g_{lr} R^r_{ki,j}, \tag{11.31}$$

where

$$R^r_{ki,j} = f_{rl} R_{ki,jl},$$

and write the compatibility condition (11.30) in terms of $R_{ki,jl}$ instead of $R^r_{ki,j}$.

A simple but cumbersome calculation shows that

$$R_{ij,kl} = \frac{1}{2}\left(\frac{\partial^2 g_{li}}{\partial x_j \partial x_k} - \frac{\partial^2 g_{ik}}{\partial x_j \partial x_l} - \frac{\partial^2 g_{lj}}{\partial x_i \partial x_k} + \frac{\partial^2 g_{jk}}{\partial x_i \partial x_l}\right)$$
$$+ g_{\alpha\beta}\left(\Gamma^\beta_{il}\Gamma^\alpha_{jk} - \Gamma^\beta_{jl}\Gamma^\alpha_{ik}\right) \tag{11.32}$$

which implies $R_{ij,kl} = R_{kl,ij}$ and $R_{ij,kl} = -R_{ij,lk} = -R_{ji,kl}$. In particular, $R_{ii,kl} = R_{ij,kk} = 0$. We note that only following components $R_{ij,kl}$ may be different from zero:

$$\begin{aligned} &R_{12,12} = -R_{21,12} = R_{21,21}, \qquad &&R_{12,13} = -R_{21,13} = R_{21,31}, \\ &R_{23,23} = -R_{32,23} = R_{32,32}, &&R_{23,21} = -R_{32,21} = R_{32,12}, \\ &R_{31,31} = -R_{13,31} = R_{13,13}, &&R_{31,32} = -R_{13,32} = R_{13,23}. \end{aligned}$$

Thus, the compatibility condition is reduced to the following six equalities:

$$\begin{aligned} &R_{12,12} = 0, \quad R_{12,13} = 0, \quad R_{23,23} = 0, \\ &R_{23,21} = 0, \quad R_{31,31} = 0, \quad R_{31,32} = 0. \end{aligned} \tag{11.33}$$

11.8. Here, we consider the compatibility conditions in the case of small deformations (linear elasticity). This means that, in the representation $g_{ik} = \delta_{ik} - 2\varepsilon_{ik}$, the components ε_{ik} of the Almansi strain tensor and their derivatives with respect to x_1, x_2, x_3 are assumed to be small. Then the compatibility conditions (11.30) take the form

$$0 = -R_{ij,kl} = \frac{\partial^2 \varepsilon_{li}}{\partial x_j \partial x_k} - \frac{\partial^2 \varepsilon_{ik}}{\partial x_j \partial x_l} - \frac{\partial^2 \varepsilon_{lj}}{\partial x_i \partial x_k} + \frac{\partial^2 \varepsilon_{jk}}{\partial x_i \partial x_l} + \dots, \tag{11.34}$$

where the dots denote the second-order terms. Since we consider only small deformations, we should assume that the displacements of ζ_i (together with its derivatives) are also small: $x_i = x_{i0} + \zeta_i(x_1, x_2, x_3)$. Therefore,

$$\begin{aligned}\delta_{ij} - 2\varepsilon_{ij} = g_{ij} &= \frac{\partial x_{k0}}{\partial x_i}\frac{\partial x_{k0}}{\partial x_j} = \frac{\partial(x_k - \zeta_k)}{\partial x_i}\frac{\partial(x_k - \zeta_k)}{\partial x_j} \\ &= \delta_{ij} - \frac{\partial \zeta_i}{\partial x_j} - \frac{\partial \zeta_j}{\partial x_i} + \frac{\partial \zeta_k}{\partial x_i}\frac{\partial \zeta_k}{\partial x_j} = \delta_{ij} - \frac{\partial \zeta_i}{\partial x_j} - \frac{\partial \zeta_j}{\partial x_i} + \dots\end{aligned}$$

and, consequently,

$$\varepsilon_{ij} = \frac{1}{2}\left(\frac{\partial \zeta_i}{\partial x_j} + \frac{\partial \zeta_j}{\partial x_i}\right) + \dots \tag{11.35}$$

In the case of small deformations (linear elasticity), ε_{ij} is expressed in terms of displacements of ζ_i by the formula

$$\varepsilon_{ij} = \frac{1}{2}\left(\frac{\partial \zeta_i}{\partial x_j} + \frac{\partial \zeta_j}{\partial x_i}\right). \tag{11.36}$$

It is easy to verify that ε_{ij} satisfy the equations

$$\frac{\partial^2 \varepsilon_{li}}{\partial x_j \partial x_k} - \frac{\partial^2 \varepsilon_{ik}}{\partial x_j \partial x_l} - \frac{\partial^2 \varepsilon_{lj}}{\partial x_i \partial x_k} + \frac{\partial^2 \varepsilon_{jk}}{\partial x_i \partial x_l} = 0 \tag{11.37}$$

where the left-hand side coincides with the leading terms in the compatibility condition (11.34).

• The linear equations (11.37) are called the *Saint-Venant compatibility conditions for small deformations.*

Let us show the necessity of the Saint-Venant compatibility conditions (11.37). Applying the operator curl to the equation

$$\begin{pmatrix}\varepsilon_{11}\\ \varepsilon_{12}\\ \varepsilon_{13}\end{pmatrix} = \frac{1}{2}\begin{pmatrix}\dfrac{\partial \zeta_1}{\partial x_1}\\ \dfrac{\partial \zeta_1}{\partial x_2}\\ \dfrac{\partial \zeta_1}{\partial x_3}\end{pmatrix} + \frac{1}{2}\frac{\partial}{\partial x_1}\begin{pmatrix}\zeta_1\\ \zeta_2\\ \zeta_3\end{pmatrix},$$

we find

$$\operatorname{curl}\begin{pmatrix}\varepsilon_{11}\\ \varepsilon_{12}\\ \varepsilon_{13}\end{pmatrix} = \begin{pmatrix}\dfrac{\partial\varepsilon_{12}}{\partial x_3}-\dfrac{\partial\varepsilon_{13}}{\partial x_2}\\ \dfrac{\partial\varepsilon_{13}}{\partial x_1}-\dfrac{\partial\varepsilon_{11}}{\partial x_3}\\ \dfrac{\partial\varepsilon_{11}}{\partial x_2}-\dfrac{\partial\varepsilon_{12}}{\partial x_1}\end{pmatrix} = \frac{1}{2}\frac{\partial}{\partial x_1}\begin{pmatrix}\dfrac{\partial\zeta_2}{\partial x_3}-\dfrac{\partial\zeta_3}{\partial x_2}\\ \dfrac{\partial\zeta_3}{\partial x_1}-\dfrac{\partial\zeta_1}{\partial x_3}\\ \dfrac{\partial\zeta_1}{\partial x_2}-\dfrac{\partial\zeta_2}{\partial x_1}\end{pmatrix} = \frac{1}{2}\frac{\partial}{\partial x_1}\operatorname{curl}\begin{pmatrix}\zeta_1\\ \zeta_2\\ \zeta_3\end{pmatrix}$$

or, shortly,

$$\operatorname{curl}_j \varepsilon_{1j} = \frac{1}{2}\frac{\partial}{\partial x_1}\operatorname{curl}_j \zeta_j. \tag{11.38}$$

The notation curl_j means that the operator curl is applied to the vector with components marked by the subscript j. Similarly,

$$\operatorname{curl}_j \varepsilon_{2j} = \frac{1}{2}\frac{\partial}{\partial x_2}\operatorname{curl}_j \zeta_j, \quad \operatorname{curl}_j \varepsilon_{3j} = \frac{1}{2}\frac{\partial}{\partial x_3}\operatorname{curl}_j \zeta_j. \tag{11.39}$$

Applying curl_i to $\operatorname{curl}_j \varepsilon_{ij} = \dfrac{1}{2}\dfrac{\partial}{\partial x_i}(\operatorname{curl}_j \zeta_j)$, we find

$$\operatorname{curl}_i \operatorname{curl}_j \varepsilon_{ij} = \frac{1}{2}\operatorname{curl}_i \frac{\partial}{\partial x_i}(\operatorname{curl}_j \zeta_j) = 0 \tag{11.40}$$

since curl of gradient is zero. Writing (11.40) componentwise, we obtain six conditions two of which are as follows:

$$\begin{aligned}&\frac{\partial^2\varepsilon_{11}}{\partial x_2^2} - 2\frac{\partial^2\varepsilon_{12}}{\partial x_1\partial x_2} + \frac{\partial^2\varepsilon_{22}}{\partial x_1^2} = 0,\\ &\frac{\partial^2\varepsilon_{12}}{\partial x_3^2} - \frac{\partial^2\varepsilon_{13}}{\partial x_3\partial x_2} - \frac{\partial^2\varepsilon_{32}}{\partial x_1\partial x_3} + \frac{\partial^2\varepsilon_{33}}{\partial x_1\partial x_2} = 0,\end{aligned} \tag{11.41}$$

whereas the remaining ones are obtained from (11.41) by cyclic permutation of subscripts: $(1,2,3) \to (2,3,1) \to (3,1,2)$. These six equalities coincide with the Saint-Venant compatibility conditions (11.37). The converse assertion is also true.

• *If* $\varepsilon_{ij}(x_1, x_2, x_3)$ *satisfy the compatibility condition* (11.40), *then there exist* $\zeta_1(x_1, x_2, x_3)$, $\zeta_2(x_1, x_2, x_3)$, $\zeta_3(x_1, x_2, x_3)$ *such that* (*cf.* (11.36))

$$\varepsilon_{ij} = \frac{1}{2}\left(\frac{\partial\zeta_i}{\partial x_j} + \frac{\partial\zeta_j}{\partial x_i}\right).$$

The required $\zeta_i(x_1, x_2, x_3)$ can be found from the auxiliary equations

$$\frac{\partial\zeta_i}{\partial x_j} = \varepsilon_{ij}(x_1, x_2, x_3) + \Phi_{ij}(x_1, x_2, x_3),$$

where $\Phi_{ij} = -\Phi_{ji}$ are a solution of the equations

$$\frac{\partial \Phi_{ik}}{\partial x_l} = \frac{\partial \varepsilon_{il}}{\partial x_k} - \frac{\partial \varepsilon_{kl}}{\partial x_i}.$$

Let us check the integrability condition for these equations:

$$\frac{\partial \Phi_{ik}}{\partial x_1} = \frac{\partial \varepsilon_{i1}}{\partial x_k} - \frac{\partial \varepsilon_{k1}}{\partial x_i},$$
$$\frac{\partial \Phi_{ik}}{\partial x_2} = \frac{\partial \varepsilon_{i2}}{\partial x_k} - \frac{\partial \varepsilon_{k2}}{\partial x_i},$$
$$\frac{\partial \Phi_{ik}}{\partial x_3} = \frac{\partial \varepsilon_{i3}}{\partial x_k} - \frac{\partial \varepsilon_{k3}}{\partial x_i}.$$

It is easy to see that the integrability of these equations is provided by the conditions

$$\frac{\partial}{\partial x_m}\left(\frac{\partial \varepsilon_{il}}{\partial x_k} - \frac{\partial \varepsilon_{kl}}{\partial x_i}\right) - \frac{\partial}{\partial x_l}\left(\frac{\partial \varepsilon_{im}}{\partial x_k} - \frac{\partial \varepsilon_{km}}{\partial x_i}\right) = 0$$

which coincide with the Saint-Venant compatibility conditions (11.37). The system

$$\frac{\partial \zeta_i}{\partial x_1} = \varepsilon_{i1} + \Phi_{i1},$$
$$\frac{\partial \zeta_i}{\partial x_2} = \varepsilon_{i2} + \Phi_{i2},$$
$$\frac{\partial \zeta_i}{\partial x_3} = \varepsilon_{i3} + \Phi_{i3}$$

is integrable because

$$\frac{\partial(\varepsilon_{i2} + \Phi_{i2})}{\partial x_3} - \frac{\partial(\varepsilon_{i3} + \Phi_{i3})}{\partial x_2} = 0,$$
$$\frac{\partial(\varepsilon_{i3} + \Phi_{i3})}{\partial x_1} - \frac{\partial(\varepsilon_{i1} + \Phi_{i1})}{\partial x_3} = 0,$$
$$\frac{\partial(\varepsilon_{i1} + \Phi_{i1})}{\partial x_2} - \frac{\partial(\varepsilon_{i2} + \Phi_{i2})}{\partial x_1} = 0.$$

We prove this only for the first equality (the remaining equalities are verified in a similar way):

$$\frac{\partial(\varepsilon_{i2} + \Phi_{i2})}{\partial x_3} - \frac{\partial(\varepsilon_{i3} + \Phi_{i3})}{\partial x_2} = \left(\frac{\partial \varepsilon_{i2}}{\partial x_3} - \frac{\partial \varepsilon_{i3}}{\partial x_2}\right) + \frac{\partial \Phi_{i2}}{\partial x_3} - \frac{\partial \Phi_{i3}}{\partial x_2}$$
$$= \left(\frac{\partial \varepsilon_{i2}}{\partial x_3} - \frac{\partial \varepsilon_{i3}}{\partial x_2}\right) + \left(\frac{\partial \varepsilon_{i3}}{\partial x_2} - \frac{\partial \varepsilon_{23}}{\partial x_i}\right) - \left(\frac{\partial \varepsilon_{i2}}{\partial x_3} - \frac{\partial \varepsilon_{32}}{\partial x_i}\right) = 0.$$

• *Thus, the question if the Saint-Venant compatibility conditions* (11.37) *are necessary and sufficient for the solvability of the linear system* (11.36)

$$\frac{1}{2}\left(\frac{\partial \zeta_i}{\partial x_j} + \frac{\partial \zeta_j}{\partial x_i}\right) = \varepsilon_{ij},$$

is much simpler than the same question for the metric deformation tensor in the case of finite deformations.

§ 12. Description of Relaxation of Shear Stresses by Equations for Effective Distortion

12.1. In Section 9, we defined the metric effective elastic deformation tensor $||g_{ij}^{\text{eff}}||$. In an isotropic medium with relaxation of shear stresses, the tensor $||g_{ij}^{\text{eff}}||$ is determined from the stress tensor $||\sigma_{ij}||$ on the basis of the laws of elasticity. It is assumed that stresses act during a time-interval less than the relaxation time τ. In Section 10, we explained how to compute relaxation of stresses using equations for the metric effective elastic deformation tensor $||g_{ij}^{\text{eff}}||$. Inserting additional special terms into these equations, we can describe relaxation of the deviator of the metric deformation tensor and, consequently, relaxation of the deviator of the stress tensor. We prefer to use the equations for $||g_{ij}^{\text{eff}}||$ but not the equations for stresses because our choice provides us with more simple and clear computations.

To describe an actual deformation, we can use the metric total deformation tensor $G = ||g_{ij}||$, as well as the distortion tensor $A = ||a_{ij}||$ connected with G by the equality $G = A^*A$. As was mentioned in Section 6, for a given A we can find G in a unique way, whereas the reconstruction of A from G is not necessarily unique. Nevertheless, it is often more convenient to start with the metric deformation tensor G and determine the distortion tensor A from G. For example, such a situation can occur if, instead of $G = ||g_{ik}||$, we deal with the metric effective elastic deformation tensor $G^{\text{eff}} = ||g_{ik}^{\text{eff}}||$ which connects with the stressed state but not with deformations. If the metric effective elastic deformation tensor G^{eff} admits the canonical representation

$$G^{\text{eff}} = U\begin{bmatrix} 1/k_1^2 & 0 & 0 \\ 0 & 1/k_2^2 & 0 \\ 0 & 0 & 1/k_3^2 \end{bmatrix} U^* \qquad (UU^* = I,\ \det U = +1)$$

at some point, then for the effective distortion tensor A^{eff} we can take any matrix of the form

$$A^{\text{eff}} = V^*\begin{bmatrix} 1/k_1 & 0 & 0 \\ 0 & 1/k_2 & 0 \\ 0 & 0 & 1/k_3 \end{bmatrix} U^*,$$

where V is an arbitrary orthogonal matrix which is not necessarily the same at distinct points (x_1, x_2, x_3). If $G^{\text{eff}} = G$ corresponds to an actual deformation described by a mapping

$$x_1 = x_1(x_{10}, x_{20}, x_{30}), \quad x_2 = x_2(x_{10}, x_{20}, x_{30}), \quad x_3 = x_3(x_{10}, x_{20}, x_{30})$$

and no relaxation of stresses occurs under this deformation, we can restore the initial coordinates (x_{10}, x_{20}, x_{30}) by solving the partial differential equations

$$\frac{\partial x_{k0}}{\partial x_i}\frac{\partial x_{k0}}{\partial x_j} = g_{ij}.$$

As can be seen from the proof of the solvability of this system, the arbitrariness degree is not so large and means the arbitrary choice of the rotation matrix V at only one point. It is easy to recognize if given g_{ik} form the metric deformation tensor for some actual deformation. It suffices to verify the compatibility conditions. As a rule, g_{ik}^{eff} do not satisfy the compatibility conditions. Therefore, the representation of g_{ik}^{eff} in terms of the effective distortion a_{ik}^{eff} is rather arbitrary (it is impossible to connect the values of V at distinct points).

In the study of nonstationary process, we deal with the values of the tensors $||\sigma_{ik}||$ and $||g_{ik}^{\text{eff}}||$ at distinct time moments. Similarly, the effective distortion tensor $||a_{ik}^{\text{eff}}||$ should depend on time. As is known (cf. Section 8), if the motion of a medium is described by the velocity field $u_i = u_i(x_1, x_2, x_3, t)$, then the evolution of the actual distortion $A = ||a_{ik}|| = \left\|\frac{\partial x_{i0}}{\partial x_k}\right\|$ is described by the differential equations

$$\frac{\partial a_{ij}}{\partial t} + u_k \frac{\partial a_{ij}}{\partial x_k} + a_{ik}\frac{\partial u_k}{\partial x_j} = 0 \tag{12.1}$$

or, in the matrix notation,

$$\frac{\partial}{\partial t}A + u_k \frac{\partial}{\partial x_k}A + AW = 0, \qquad W = ||w_{ik}||, \ w_{ik} = \frac{\partial u_i}{\partial x_k}. \tag{12.2}$$

From (12.2) we can obtain equations for G by combining the equations for the matrix A and its transposed matrix A^*. Namely,

$$\begin{aligned}
\frac{dG}{dt} + W^*G + GW &= \frac{d}{dt}(A^*A) + W^*(A^*A) + (A^*A)W \\
&= A^*\left(\frac{dA}{dt} + AW\right) + \left(\frac{dA^*}{dt} + W^*A\right)A \\
&= A^*\left(\frac{\partial A}{\partial t} + u_k\frac{\partial A}{\partial x_k} + AW\right) + \left(\frac{\partial A}{\partial t} + u_k\frac{\partial A}{\partial x_k} + AW\right)^* A \\
&= A^* \cdot 0 + 0 \cdot A = 0.
\end{aligned} \tag{12.3}$$

To describe relaxation processes, we used, instead of G, the metric effective elastic deformation tensor G^{efff} whose evolution is described (cf. Section 10) by the equation

$$\frac{d}{dt}G^{\text{eff}} + G^{\text{eff}}W + W^*G^{\text{eff}} = -\Phi = -\|\varphi_{ij}\|, \tag{12.4}$$

where Φ describes the relaxation rate.

12.2. It is natural to try to derive from (12.2) an equation for the effective distortion A^{eff}. For this purpose, we consider equation (12.1) but with special right-hand side Ψ:

$$\frac{d}{dt}A^{\text{eff}} + A^{\text{eff}}W = -\Psi. \tag{12.5}$$

Assuming the representation $G^{\text{eff}} = (A^{\text{eff}})^* A^{\text{eff}}$, we obtain the equations

$$\begin{aligned}
&\frac{d}{dt}G^{\text{eff}} + G^{\text{eff}}W + W^*G^{\text{eff}} \\
&\quad = (A^{\text{eff}})^* \left(\frac{d}{dt}A^{\text{eff}} + W^* A^{\text{eff}}\right) + \left(\frac{d}{dt}A^{\text{eff}} + W^* A^{\text{eff}}\right)^* A^{\text{eff}} \\
&\quad = -(A^{\text{eff}})^*\Psi - \Psi^* A^{\text{eff}}
\end{aligned} \tag{12.6}$$

which coincide with equations (12.4) if the right-hand side Ψ is such that

$$(A^{\text{eff}})^*\Psi + \Psi^* A^{\text{eff}} = \Phi. \tag{12.7}$$

We note that if

$$(A^{\text{eff}})^* A^{\text{eff}} = G^{\text{eff}}, \tag{12.8}$$

at some time moment and the evolution of A^{eff} is described by equation (12.5), then the equality (12.8) holds at all later time moments.

Developing our relaxation model, we suggested in Section 10 to define Φ and Ψ by the following rule. If

$$G^{\text{eff}} = U \begin{bmatrix} 1/k_1^2 & 0 & 0 \\ 0 & 1/k_2^2 & 0 \\ 0 & 0 & 1/k_3^2 \end{bmatrix} U^*,$$

where $UU^* = I$, $\det U = +1$, then we set

$$-\Phi = \frac{2}{\tau} U \begin{bmatrix} \frac{1}{k_1^2}\ln\frac{k_1}{\sqrt[3]{k_1k_2k_3}} & 0 & 0 \\ 0 & \frac{1}{k_2^2}\ln\frac{k_2}{\sqrt[3]{k_1k_2k_3}} & 0 \\ 0 & 0 & \frac{1}{k_3^2}\ln\frac{k_3}{\sqrt[3]{k_1k_2k_3}} \end{bmatrix} U^*. \tag{12.9}$$

If

$$A^{\text{eff}} = V^* \begin{bmatrix} 1/k_1 & 0 & 0 \\ 0 & 1/k_2 & 0 \\ 0 & 0 & 1/k_3 \end{bmatrix} U^*,$$

where $UU^* = VV^* = I$, $\det U = \det V = +1$, then we set

$$-\Psi = \frac{1}{\tau} V^* \begin{bmatrix} \frac{1}{k_1} \ln \frac{k_1}{\sqrt[3]{k_1 k_2 k_3}} & 0 & 0 \\ 0 & \frac{1}{k_2} \ln \frac{k_2}{\sqrt[3]{k_1 k_2 k_3}} & 0 \\ 0 & 0 & \frac{1}{k_3} \ln \frac{k_3}{\sqrt[3]{k_1 k_2 k_3}} \end{bmatrix} U^*. \quad (12.10)$$

It is easy to see that for Φ and Ψ defined by formulas (12.9) and (12.10) respectively the condition (12.7) is satisfied.

• *It is important to emphasize that our method for describing relaxation of shear stresses based on the use of additional special terms in the equations requires no new physical ideas. In fact, our arguments are the same as before when we introduced the notion of the metric effective elastic deformation tensor* $\|g_{ij}^{\text{eff}}\|$ *and derived equations describing the behavior of* $\|g_{ij}^{\text{eff}}\|$. *We only suggest to introduce, with a certain degree of arbitrariness, new quantities that used in order to restore the same tensor* G^{eff} *as in the model described in Section* 10.

12.3. Now, we formulate the **compatibility conditions** in terms of the distortion A. By A we can mean the "actual" distortion A determined as the Jacobi matrix

$$A = \begin{bmatrix} a_{11} & a_{12} & a_{13} \\ a_{21} & a_{22} & a_{23} \\ a_{31} & a_{32} & a_{33} \end{bmatrix} = \begin{bmatrix} \frac{\partial x_{10}}{\partial x_1} & \frac{\partial x_{10}}{\partial x_2} & \frac{\partial x_{10}}{\partial x_3} \\ \frac{\partial x_{20}}{\partial x_2} & \frac{\partial x_{20}}{\partial x_2} & \frac{\partial x_{20}}{\partial x_2} \\ \frac{\partial x_{30}}{\partial x_1} & \frac{\partial x_{30}}{\partial x_2} & \frac{\partial x_{30}}{\partial x_3} \end{bmatrix} \qquad \left(a_{ij} = \frac{\partial x_{i0}}{\partial x_j} \right)$$

of a mapping

$$\begin{aligned} x_1 &= x_1(x_{10}, x_{20}, x_{30}) \qquad & (x_{10} = x_{10}(x_1, x_2, x_3)), \\ x_2 &= x_2(x_{10}, x_{20}, x_{30}) \qquad & (x_{20} = x_{20}(x_1, x_2, x_3)), \\ x_3 &= x_3(x_{10}, x_{20}, x_{30}) \qquad & (x_{30} = x_{30}(x_1, x_2, x_3)) \end{aligned}$$

as the Jacobi matrix or the "effective" distortion A^{eff} found with a certain degree of arbitrariness from the metric effective elastic deformation tensor G^{eff}.

• We say that the *compatibility conditions* hold if for given $a_{ij}(x_1, x_2, x_3)$ there exists a solution $x_{i0} = x_{i0}(x_1, x_2, x_3)$ to the system

$$a_{ij}(x_1, x_2, x_3) = \frac{\partial x_{i0}}{\partial x_j}. \qquad (12.11)$$

It turns out that the question about the compatibility conditions expressed in terms of the distortion is much simpler than that in the case where we use the metric deformation tensor. Indeed, the system (12.11) is solvable if and only if the following nine equalities hold:

$$\frac{\partial a_{j2}}{\partial x_3} - \frac{\partial a_{j3}}{\partial x_2} = 0, \qquad \frac{\partial a_{j3}}{\partial x_1} - \frac{\partial a_{j1}}{\partial x_3} = 0, \qquad \frac{\partial a_{j1}}{\partial x_2} - \frac{\partial a_{j2}}{\partial x_1} = 0,$$

where $j = 1, 2, 3$.

12.4. Based on the notion of distortion, we define the Burgers tensor.

• The tensor

$$B = \|B_{ij}\| = \begin{bmatrix} B_{11} & B_{12} & B_{13} \\ B_{21} & B_{22} & B_{23} \\ B_{31} & B_{32} & B_{33} \end{bmatrix} \tag{12.12}$$

with elements

$$\begin{aligned} -B_{i1} &= \frac{\partial a_{i3}}{\partial x_2} - \frac{\partial a_{i2}}{\partial x_3}, \\ -B_{i2} &= \frac{\partial a_{i1}}{\partial x_3} - \frac{\partial a_{i3}}{\partial x_1}, \\ -B_{i3} &= \frac{\partial a_{i2}}{\partial x_1} - \frac{\partial a_{i1}}{\partial x_2} \end{aligned} \tag{12.13}$$

is called the *Burgers tensor* constructed from the distortion $A = \|a_{ij}\|$.

Using the notion of the Burgers tensor, we can express the compatibility condition as follows.

• We say that the *compatibility condition* holds if the Burgers tensor B constructed from the distortion A vanishes.

Formulas (12.13) for the components of the Burgers tensor can be written as follows:

$$B_{ij} = -e_{j\alpha\beta}\frac{\partial a_{i\beta}}{\partial x_\alpha}, \tag{12.14}$$

where $e_{j\alpha\beta}$ is equal to 1 if $(j\alpha\beta)$ is an even permutation and to 0 otherwise.

12.5. It is of interest to reformulate the compatibility conditions with the tensor G (cf. Section 11) in terms of the Burgers tensor.

• The tensor G satisfies *the compatibility conditions* $R_{ij,kl} = 0$ if and only if it admits the representation $G = A^*A$, where A is some distortion with the zero Burgers tensor $B = 0$.

We can interpret the Burgers tensor as follows. We first assume that $A = \|a_{ij}(x_1, x_2, x_3)\|$ describes the "actual" distortion connected with the "initial coordinates" $x_{i0} = x_{i0}(x_1, x_2, x_3)$ by the equalities $\dfrac{\partial x_{i0}}{\partial x_j} = a_{ij}$. Then we can restore the "initial coordinates" by integrating the exact differential $dx_{i0} = a_{i1}dx_1 + a_{i2}dx_2 + a_{i3}dx_3$ along an arbitrary path. Moreover, x_{i0} is uniquely determined in any connected domain if $\overline{x}_{10}$, $\overline{x}_{20}$, $\overline{x}_{30}$ are known at some fixed point $(\overline{x}_1, \overline{x}_2, \overline{x}_3)$. The uniqueness follows from the fact that the integral of the exact differential along any closed path $\mathcal{L}$ vanishes:

$$\oint_{\mathcal{L}} a_{i1}dx_1 + a_{i2}dx_2 + a_{i3}dx_3 = 0. \tag{12.15}$$

By the Stokes theorem, for any surface (film) Π given by the parametric equations $x_1 = x_1(\alpha, \beta)$, $x_2 = x_2(\alpha, \beta)$, $x_3 = x_3(\alpha, \beta)$ and bounded by a contour $\mathcal{L}$ (i.e., the film is spanned by $\mathcal{L}$), the integral along $\mathcal{L}$ can be expressed in terms of the integral over Π:

$$\begin{aligned}
&\oint_{\mathcal{L}} a_{i1}dx_1 + a_{i2}dx_2 + a_{i3}dx_3 \\
&\quad = \iint_{\Pi} \left[\left(\frac{\partial a_{i3}}{\partial x_2} - \frac{\partial a_{i2}}{\partial x_3} \right) \frac{\partial(x_2, x_3)}{\partial(\alpha, \beta)} + \left(\frac{\partial a_{i1}}{\partial x_3} - \frac{\partial a_{i3}}{\partial x_1} \right) \frac{\partial(x_3, x_1)}{\partial(\alpha, \beta)} \right. \\
&\qquad + \left. \left(\frac{\partial a_{i2}}{\partial x_1} - \frac{\partial a_{i1}}{\partial x_2} \right) \frac{\partial(x_1, x_2)}{\partial(\alpha, \beta)} \right] d\alpha d\beta \\
&\quad = \iint_{\Pi} -B_{i1}\, dx_2dx_3 - B_{i2}\, dx_3dx_1 - B_{i3}\, dx_1dx_2 = 0.
\end{aligned}$$

The independence of the "initial coordinate" x_{i0} of the integration path follows from the fact that the Burgers tensor (more exactly, its ith row) vanishes.

In the case $B \neq 0$, the attempt to restore the "initial coordinates" is doomed to failure since for any contour $\mathcal{L}$ we obtain some vector (b_1, b_2, b_3) with components

$$\begin{aligned}
-b_i &= \oint_{\mathcal{L}} a_{i1}dx_1 + a_{i2}dx_2 + a_{i3}dx_3 \\
&= -\iint_{\Pi} B_{i1}\, dx_2dx_3 + B_{i2}\, dx_3dx_1 + B_{i3}\, dx_1dx_2
\end{aligned} \tag{12.16}$$

which is, in general, differs from zero.

• The vector (b_1, b_2, b_3) defined by formulas (12.16) is called the *Burgers vector* corresponding to the contour $\mathcal{L}$.

The components of the Burgers tensor are connected with components of the Burgers vector. If we consider contours bounding infinitesimal small area elements on the planes parallel to the coordinate planes, then $B_{i1}dx_2dx_3$ coincide with the components of the Burgers vector corresponding to the rectangle on the plane x_1 = const with vertices (x_2, x_3), $(x_2 + dx_2, x_3)$, $(x_2 + dx_2, x_3 + dx_3)$, $(x_2, x_3 + dx_3)$. Thus, the Burgers tensor characterizes the distribution of "nonuniqueness" preventing the reconstruction of the "initial coordinates" x_{j0}.

Remark 12.2. The notions of the distortion, Burgers tensor, and Burgers vector are fundamental object of the theory of dislocations which deals with defects of crystal lattice. They will be discussed from the point of view of this theory in Section 13. Within the framework of our approach, these notions appear as constructions connected with the stress filed in an isotropic viscoelastic medium with relaxation in order to describe the phenomenological model of relaxation of shear stresses (a nonlinear version of the Maxwell model).

12.6. We note that formulas (12.13) imply the following identities between the components of the Burgers tensor:

$$\frac{\partial B_{i1}}{\partial x_1} + \frac{\partial B_{i2}}{\partial x_2} + \frac{\partial B_{i3}}{\partial x_3} = 0. \tag{12.17}$$

This implies the equality

$$\iint\limits_{\tilde{\Pi}} B_{i1}\, dx_2 dx_3 + B_{i2}\, dx_3 dx_1 + B_{i3}\, dx_1 dx_2 = 0$$

on any closed surface $\tilde{\Pi}$ bounding some three-dimensional domain and the independence of the integral

$$\iint\limits_{\Pi} B_{i1}\, dx_2 dx_3 + B_{i2}\, dx_3 dx_1 + B_{i3}\, dx_1 dx_2$$

of the choice of a film Π spanned by $\mathcal{L}$.

Let us derive equations describing the evolution of the Burgers tensor with time. For this purpose, we differentiate with respect to x_n the equation (one of equations in (12.5))

$$\frac{\partial a_{im}}{\partial t} + u_k \frac{\partial a_{im}}{\partial x_k} + a_{ik} \frac{\partial u_k}{\partial x_m} = -\psi_{im}$$

and subtract the equation

$$\frac{\partial a_{in}}{\partial t} + u_k \frac{\partial a_{in}}{\partial x_k} + a_{ik} \frac{\partial u_k}{\partial x_n} = -\psi_{in}$$

differentiated with respect to x_m. We obtain the equation

$$\frac{\partial}{\partial t}\left(\frac{\partial a_{im}}{\partial x_n}-\frac{\partial a_{in}}{\partial x_m}\right)+u_r\frac{\partial}{\partial x_r}\left(\frac{\partial a_{im}}{\partial x_n}-\frac{\partial a_{in}}{\partial x_m}\right)$$
$$+\left[\frac{\partial a_{im}}{\partial x_k}\frac{\partial u_k}{\partial x_n}+\frac{\partial a_{ik}}{\partial x_n}\frac{\partial u_k}{\partial x_m}-\frac{\partial a_{in}}{\partial x_k}\frac{\partial u_k}{\partial x_m}-\frac{\partial a_{ik}}{\partial x_m}\frac{\partial u_k}{\partial x_n}\right]$$
$$=-\left(\frac{\partial \psi_{mi}}{\partial x_n}-\frac{\partial \psi_{ni}}{\partial x_m}\right) \tag{12.18}$$

which can be written in an elegant form if all the derivatives of the components of distortion are arranged in such a way that the resulting equation contains only the components of the Burgers tensor. Indeed, it suffices to explain how to handle with the terms in the square brackets for some m, n. We consider only the case $m=3$, $n=2$ and write explicitly the summation with respect to k:

$$\left(\frac{\partial a_{im}}{\partial x_k}\frac{\partial u_k}{\partial x_n}+\frac{\partial a_{ik}}{\partial x_n}\frac{\partial u_k}{\partial x_m}-\frac{\partial a_{in}}{\partial x_k}\frac{\partial u_k}{\partial x_m}-\frac{\partial a_{ik}}{\partial x_m}\frac{\partial u_k}{\partial x_n}\right)$$
$$=\frac{\partial a_{i3}}{\partial x_1}\frac{\partial u_1}{\partial x_2}+\frac{\partial a_{i3}}{\partial x_2}\frac{\partial u_2}{\partial x_2}+\frac{\partial a_{i3}}{\partial x_3}\frac{\partial u_3}{\partial x_2}+\frac{\partial a_{i1}}{\partial x_2}\frac{\partial u_1}{\partial x_3}+\frac{\partial a_{i2}}{\partial x_2}\frac{\partial u_2}{\partial x_3}+\frac{\partial a_{i3}}{\partial x_2}\frac{\partial u_3}{\partial x_3}$$
$$-\frac{\partial a_{i2}}{\partial x_1}\frac{\partial u_1}{\partial x_3}-\frac{\partial a_{i2}}{\partial x_2}\frac{\partial u_2}{\partial x_3}-\frac{\partial a_{i2}}{\partial x_3}\frac{\partial u_3}{\partial x_3}-\frac{\partial a_{i1}}{\partial x_3}\frac{\partial u_1}{\partial x_2}-\frac{\partial a_{i2}}{\partial x_3}\frac{\partial u_2}{\partial x_2}-\frac{\partial a_{i3}}{\partial x_3}\frac{\partial u_3}{\partial x_2}$$
$$=-\left(\frac{\partial a_{i1}}{\partial x_3}-\frac{\partial a_{i3}}{\partial x_1}\right)\frac{\partial u_1}{\partial x_2}-\left(\frac{\partial a_{i2}}{\partial x_1}-\frac{\partial a_{i1}}{\partial x_3}\right)\frac{\partial u_1}{\partial x_3}$$
$$+\left(\frac{\partial a_{i3}}{\partial x_2}-\frac{\partial a_{i2}}{\partial x_3}\right)\left(\frac{\partial u_2}{\partial x_2}+\frac{\partial u_3}{\partial x_3}\right)$$
$$=-\left(\frac{\partial u_2}{\partial x_2}+\frac{\partial u_3}{\partial x_3}\right)B_{i1}+B_{i2}\frac{\partial u_1}{\partial x_2}+B_{i3}\frac{\partial u_1}{\partial x_3}$$
$$=-\left(\frac{\partial u_1}{\partial x_1}+\frac{\partial u_2}{\partial x_2}+\frac{\partial u_3}{\partial x_3}\right)B_{i1}+B_{i1}\frac{\partial u_1}{\partial x_1}+B_{i2}\frac{\partial u_1}{\partial x_2}+B_{i3}\frac{\partial u_1}{\partial x_3}$$
$$=-\frac{\partial u_k}{\partial x_k}B_{i1}+B_{ik}\frac{\partial u_1}{\partial x_k}.$$

Finally,

$$\frac{\partial B_{i1}}{\partial t}+u_k\frac{\partial B_{i1}}{\partial x_k}-\frac{\partial u_k}{\partial x_k}B_{i1}+B_{ik}\frac{\partial u_1}{\partial x_k}=\chi_{i1},$$

where

$$\chi_{i1}=-\left(\frac{\partial \psi_{3i}}{\partial x_2}-\frac{\partial \psi_{2i}}{\partial x_3}\right).$$

Similarly, we have

$$\frac{\partial B_{ij}}{\partial t}+u_k\frac{\partial B_{ij}}{\partial x_k}-\frac{\partial u_k}{\partial x_k}B_{ij}+B_{ik}\frac{\partial u_j}{\partial x_k}=\chi_{ij}.$$

These equations can be written in the following elegant matrix form:

$$\frac{dB}{dt} - \left(\frac{\partial u_1}{\partial x_1} + \frac{\partial u_2}{\partial x_2} + \frac{\partial u_3}{\partial x_3}\right) B + BW^* = \chi. \tag{12.19}$$

We recall that

$$W = \begin{bmatrix} \frac{\partial u_1}{\partial x_1} & \frac{\partial u_1}{\partial x_2} & \frac{\partial u_1}{\partial x_3} \\ \frac{\partial u_2}{\partial x_1} & \frac{\partial u_2}{\partial x_2} & \frac{\partial u_2}{\partial x_3} \\ \frac{\partial u_3}{\partial x_1} & \frac{\partial u_3}{\partial x_2} & \frac{\partial u_3}{\partial x_3} \end{bmatrix}, \qquad W^* = \begin{bmatrix} \frac{\partial u_1}{\partial x_1} & \frac{\partial u_2}{\partial x_1} & \frac{\partial u_3}{\partial x_1} \\ \frac{\partial u_1}{\partial x_2} & \frac{\partial u_2}{\partial x_2} & \frac{\partial u_3}{\partial x_2} \\ \frac{\partial u_1}{\partial x_3} & \frac{\partial u_2}{\partial x_3} & \frac{\partial u_3}{\partial x_3} \end{bmatrix}.$$

Equations (12.18) form a system of linear ordinary differential equations along the trajectories of particles $\frac{dx_k}{dt} = u_k$. The right-hand sides χ_{ij} of (12.18) form the matrix χ whose ith column is the vector obtained by applying the operator "curl" to the ith column of the relaxation matrix Ψ and taking the opposite sign.

By the properties of solutions to linear ordinary equations (the uniqueness theorem), if $B_{ij} = 0$ at the initial time moment and $\Psi = \|\psi_{ij}\| = 0$ (hence $\chi_{ij} = 0$), then $B_{ij} = 0$ at all later time moments. In other words, if the compatibility conditions hold at the initial time moment, then they hold at all later times if the deformation is elastic (the rate of inelastic relaxation processes ψ_{ij} is zero).

Problem 12.1. Using the equations (cf. Section 8)

$$\begin{aligned} &\frac{\partial \rho c_{ij}}{\partial t} + \frac{\partial (u_k \rho c_{ij})}{\partial x_k} - \frac{\partial u_i}{\partial x_k} \rho c_{kj} = 0, \\ &\frac{\partial \rho}{\partial t} + \frac{\partial (u_k \rho)}{\partial x_k} = 0, \quad \|c_{ij}\| = C = A^{-1} \end{aligned} \tag{12.20}$$

with the initial data at $t = t_0$

$$c_{ij} = \delta_{ij}, \quad \rho = \rho_0 = \text{const}$$

prove the following relation for $t \geqslant t_0$:

$$\frac{\partial (\rho c_{ij})}{\partial x_i} = 0, \quad \rho_0 \rho^2 = \det(\rho C). \tag{12.21}$$

It is easy to see that $c_{ij}(x_1, x_2, x_3) = \frac{\partial x_i}{\partial x_{j0}}$ and, consequently, $B_{ij}(x_1, x_2, x_3, t) = 0$. One can show that for any matrix $C = A^{-1}$ the condition (12.21) with some $\rho = \rho_0$ is equivalent to the condition $B = 0$, i.e., (12.21) is another form of the compatibility conditions.

§ 13. Defects of Crystal Lattice. Review of Notions (Dislocations)

As a rule, an isotropic continuum medium is a polycrystalline structure of pressed crystal grains. Sometime, it is useful to consider ideal (or almost ideal, i.e., with small defects) crystalline bodies. In this case, the notions of dislocation, Burgers tensor, and distortion naturally appear. This motivates the choice of these terms in our consideration, although we introduced them in order to develop the theory of nonlinear viscoelastic Maxwell media from the purely phenomenological point of view.

An ideal crystalline medium can be thought of as a parallelepipedic lattice with atoms, groups of atoms or ions at knots. The lattice knots can be represented as points with integer coordinates z_1, z_2, z_3 relative to some frame of vectors

$$\begin{pmatrix} r_{11} \\ r_{21} \\ r_{31} \end{pmatrix}, \quad \begin{pmatrix} r_{12} \\ r_{22} \\ r_{32} \end{pmatrix}, \quad \begin{pmatrix} r_{13} \\ r_{23} \\ r_{33} \end{pmatrix}.$$

The Cartesian coordinates of these points are computed by the formulas

$$\begin{pmatrix} x_{10} \\ x_{20} \\ x_{30} \end{pmatrix} = z_1 \begin{pmatrix} r_{11} \\ r_{21} \\ r_{31} \end{pmatrix} + z_2 \begin{pmatrix} r_{12} \\ r_{22} \\ r_{32} \end{pmatrix} + z_3 \begin{pmatrix} r_{13} \\ r_{23} \\ r_{33} \end{pmatrix} = \begin{bmatrix} r_{11} & r_{12} & r_{13} \\ r_{21} & r_{22} & r_{23} \\ r_{31} & r_{32} & r_{33} \end{bmatrix} \begin{pmatrix} z_1 \\ z_2 \\ z_3 \end{pmatrix}.$$

The subscript 0 in the notation x_{i0} means that to emphasie that the lattice has the above structure in the standard stressed state (cf. Section 4). In practice, the vector with components r_{ij} is very small in actual crystals.

The same lattice can be obtained from different frames. Indeed, it is determined by any frame with the matrix

$$\begin{bmatrix} r'_{11} & r'_{12} & r'_{13} \\ r'_{21} & r'_{22} & r'_{23} \\ r'_{31} & r'_{32} & r'_{33} \end{bmatrix} = \begin{bmatrix} r_{11} & r_{12} & r_{13} \\ r_{21} & r_{22} & r_{23} \\ r_{31} & r_{32} & r_{33} \end{bmatrix} \begin{bmatrix} y_{11} & y_{12} & y_{13} \\ y_{21} & y_{22} & y_{23} \\ y_{31} & y_{32} & y_{33} \end{bmatrix}$$

obtained from the matrix determining the initial frame by multiplying from the right by an arbitrary (unimodular) matrix $\|y_{ij}\|$ with integer entries such that $\det \|y_{ij}\| = 1$. Indeed,

$$\begin{bmatrix} r_{11} & r_{12} & r_{13} \\ r_{21} & r_{22} & r_{23} \\ r_{31} & r_{32} & r_{33} \end{bmatrix} \begin{pmatrix} z_1 \\ z_2 \\ z_3 \end{pmatrix} = \begin{bmatrix} r'_{11} & r'_{12} & r'_{13} \\ r'_{21} & r'_{22} & r'_{23} \\ r'_{31} & r'_{32} & r'_{33} \end{bmatrix} \begin{pmatrix} z'_1 \\ z'_2 \\ z'_3 \end{pmatrix},$$

where the coordinates z_1', z_2', z_3' are expressed in terms of z_1, z_2, z_3 by the formula

$$\begin{pmatrix} z_1' \\ z_2' \\ z_3' \end{pmatrix} = \begin{bmatrix} y_{11} & y_{12} & y_{13} \\ y_{21} & y_{22} & y_{23} \\ y_{31} & y_{32} & y_{33} \end{bmatrix}^{-1} \begin{pmatrix} z_1 \\ z_2 \\ z_3 \end{pmatrix}.$$

They are integers because z_1, z_2, z_3 and the entries of the inverse matrix are integers (the inverse matrix is also unimodular and its entries are integers). It is obvious that any point with integer coordinates (z_1', z_2', z_3') relative to the "new" frame has the following integer coordinates relative to the "old" frame:

$$\begin{pmatrix} z_1 \\ z_2 \\ z_3 \end{pmatrix} = \begin{bmatrix} y_{11} & y_{12} & y_{13} \\ y_{21} & y_{22} & y_{23} \\ y_{31} & y_{32} & y_{33} \end{bmatrix} \begin{pmatrix} z_1' \\ z_2' \\ z_3' \end{pmatrix}.$$

Thus, the lattices constructed from different frames coincide.

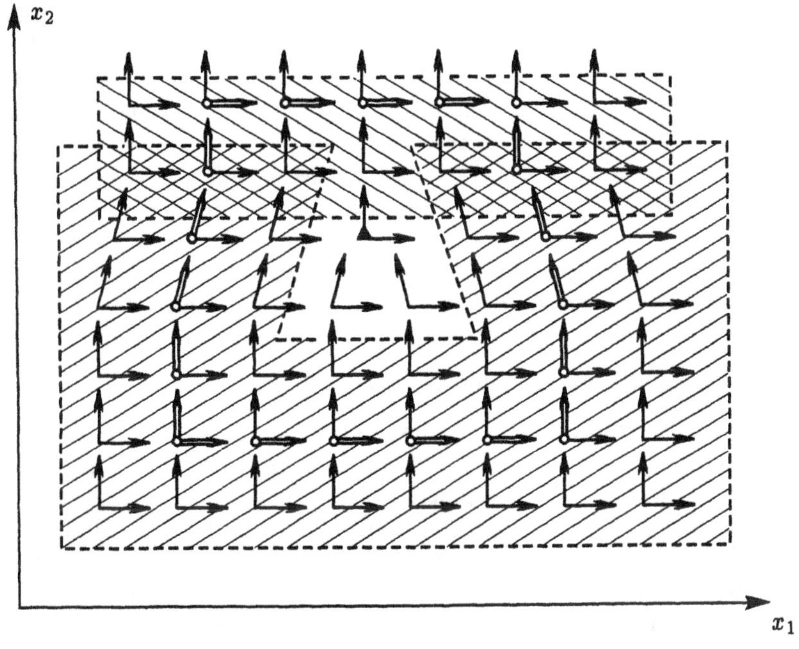

Fig. 3

After the deformation defined by the mapping $x_1 = x_1(x_{10}, x_{20}, x_{30})$, $x_2 = x_2(x_{10}, x_{20}, x_{30})$, $x_3 = x_3(x_{10}, x_{20}, x_{30})$, the image of knots of the crystal lattice form a discrete system that is not, in general, a parallelepipedic lattice. However, if the mapping is smooth, then in sufficiently small domains (so that the derivatives

$\frac{\partial x_i}{\partial x_{j0}}$ can be assumed to be constant in these domains) but containing a large number of points of the deformed lattice we can assume with large degree of accuracy that the obtained discrete set is parallelepipedic. It is obtained from the initial lattice by a translation and the linear transformation $\|c_{ij}\| = \left\|\frac{\partial x_i}{\partial x_{j0}}\right\|$. It seems that if the positions of points of the lattice before and after deformation are known, we can find the corresponding mapping and, consequently, its Jacobian. But this is not true because we cannot definitely say what frame in the initial lattice is the preimage of our fixed frame in the deformed lattice. Jacobian can be found up to an integer unimodular (right) factor.

However, if the deformation $x_i = x_i(x_{10}, x_{20}, x_{30}, t)$, $t_0 \leqslant t \leqslant t_1$, $x_{i0} = x_i(x_{10}, x_{20}, x_{30}, t_0)$ is continuous in a time-interval $t_0 \leqslant t \leqslant t_1$, some arguments based on energy cost can be suggested.

Suppose that, under a deformation, one admissible frame goes to another admissible frame of the same lattice. Then the total work is zero since the initial and final states are characterized by the internal energy and, consequently, are the same from this point of view (we consider only "elastic energy" and ignore heat influx corresponding to the energy of vibration of particles in a neighborhood of their equilibrium state). In spite of this fact, some intermediate states of the lattice can be significantly distorted, which requires huge amount of energy. Such situations are characterized by the so-called energy barrier. As a rule, under actual deformations of a crystalline medium, energy barriers do not occur.

Therefore, for a fixed frame in the lattice distorted by the mapping $x_i = x_i(x_{10}, x_{20}, x_{30})$, for its preimage in the initial lattice it is reasonable to take the frame with the lowest energy barrier. Thereby we can compute the Jacobian $C = \|c_{ij}\| = \left\|\frac{\partial x_i}{\partial x_{j0}}\right\|$ and the distortion $A = C^{-1}$.

In an actual crystalline medium, the position of atoms and ions generating a lattice can be found by the X-ray structural analysis. Hence the distortion is experimentally definable. But, in practice, the structure of an actual crystal is not ideal. Respectively, we discuss some possible defects in crystal lattices.

Consider a two-dimensional set of points of the plane (x_1, x_2) as in Fig. 3. It can be thought of as the section by the plane $x_3 =$ const of the three-dimensional block of identical two-dimensional layers dislocated by the same vector parallel to the x_3-axis. In the picture, we shade two intersecting regions obtained by deformation from regions in the initial lattice (cf. Figs. 4 and 5). With each point in Fig. 3 a frame of two vectors is associated, and their preimages are distinguished in Figs. 4 and 5.

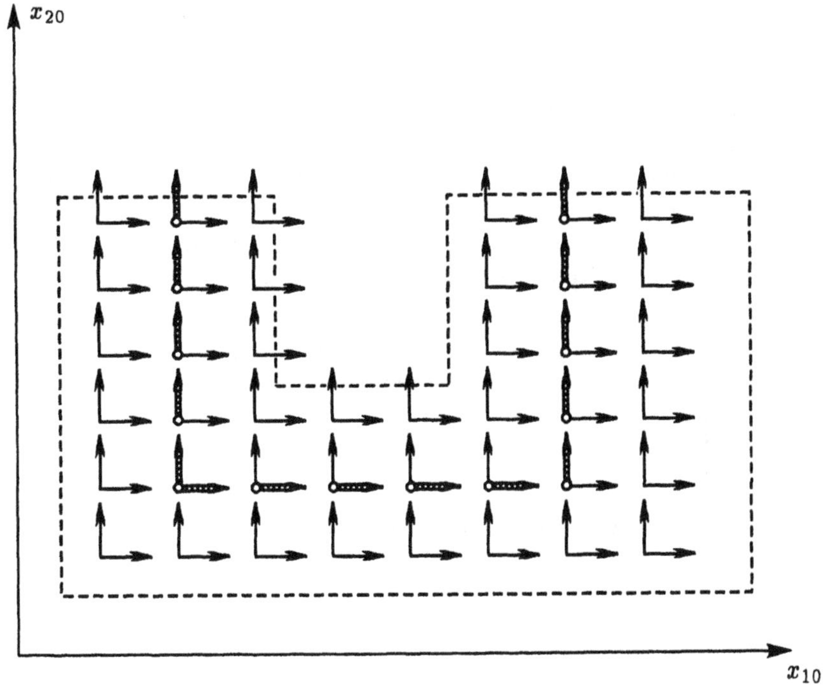

Fig. 4

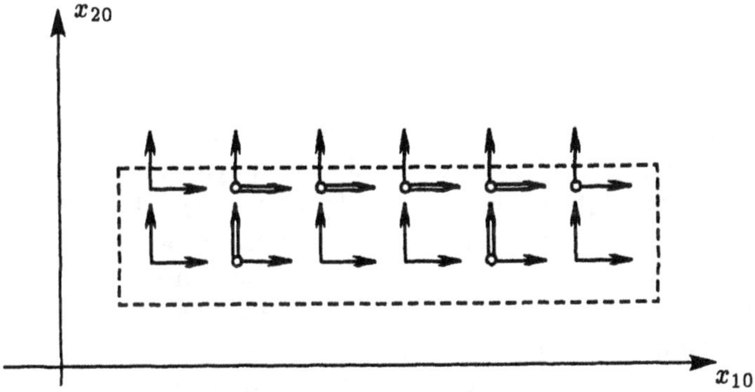

Fig. 5

The distinguished vectors form a closed path in Fig. 3. The preimages of the corresponding parts of this path are presented in Figs. 4 and 5.

In spite of the fact that points in Fig. 3 do not form a lattice, the distortion can be determined almost everywhere except for a neighborhood of the defect distinguished in Fig. 3.

In any case, the distortion is defined everywhere in a neighborhood of the distinguished closed path. If for an arbitrary point (x_1, x_2) we try to restore the coordinates of its preimage (x_{10}, x_{20}), we meet certain difficulties because there is no uniform preimage in the form of a parallelepipedic lattice. Moving along the contour in the counterclockwise direction, we move from point to point in the plane (x_{10}, x_{20}) alternatively in the horizontal and in the vertical direction by one edge (vector of frame) of the ideal lattice in the standard stressed state.

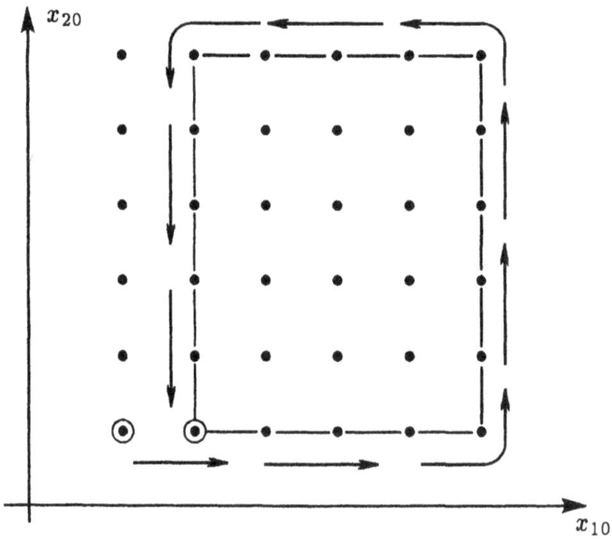

Fig. 6

It is clear that moving along the contour in the plane (x_1, x_2), we pass along the path on the plane (x_{10}, x_{20}) which is not already closed since its upper and lower segments contain different number of edges (cf. Figs. 4 and 5). This nonclosed contour is presented in Fig. 6. To make it closed, we should apply a vector, called a *Burgers vector*, to the terminal point and direct it toward the initial point. In our case, the *Burgers vector* coincides with the horizontal edge of the standard frame but is directed in the opposite direction. It is clear that the components b_i of the Burgers vector are found from the equalities

$$\sum(a_{i1}\Delta x_1 + a_{i2}\Delta x_2 + a_{i3}\Delta x_3) + b_i = 0,$$

where the sum is taken over all the edges of the contour and a_{ij} are components of the distortion computed as the partial derivatives $\dfrac{\partial x_{i0}}{\partial x_j} = a_{ij}$. However, it should be kept in mind that in different parts of the contour we use different copies of this nondeformed space and different preimages presented in Figs. 4 and 5.

Breakdown of the structure of a lattice, as in the above example, is called *dislocation*. In our example, we deal with the so-called *edge dislocation*. Let us explain this term.

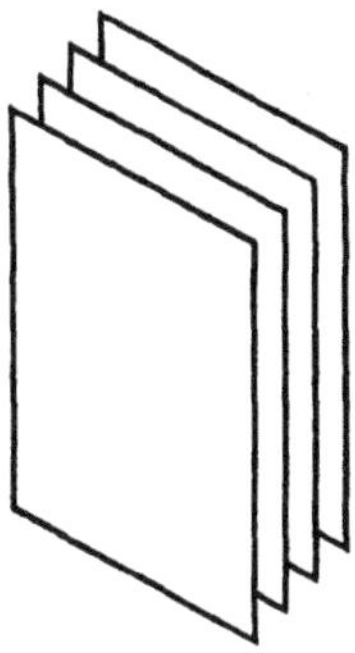

Fig. 7

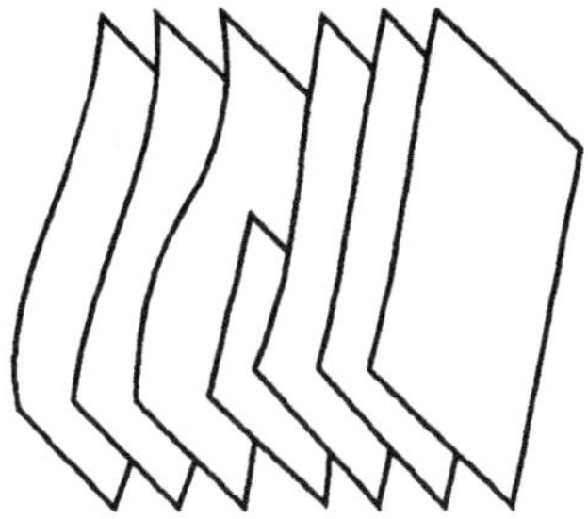

Fig. 8

A crystal without dislocations can be presented as a block of sheets — parallel plane layers of points of the lattice (Fig. 7). A crystal with edge dislocation is a block with one incomplete sheet (Fig. 8). The boundary of this sheet is a defect of the lattice, called the edge dislocation line.

Another simple type of dislocation is a screw dislocation. In this case, sheets of the block are inseparable and, like sheets of the Riemann surface of the function $\mathrm{Ln}\, z$, form a screw surface presented in Fig. 9. Going around the point O, common for all sheets of the block, along some contour consisting of edges of the lattice, we pass to the second sheet. During the next round, we get to the second sheet, then we pass to the third sheet and so on. The dislocation concentrated at O is called a *screw dislocation*. The Burgers vector is perpendicular to the round plane. (In the case of an edge dislocation, the Burgers vector lies in the round plane.)

In the general case, a dislocation is a sufficiently arbitrary space curve along which the Burgers vector relative to the round of this curve along some closed contour remains constant or is equal to some vector of translation of the lattice. (The contour should go around the dislocation only one time.)

The condition that the Burgers vector is preserved along a dislocation means that there are no endpoints of the dislocation line inside the crystal. The dislocation line either is closed or goes to the boundary of crystal or bifurcates in other dislocation.

Usually, in a crystal there are a number of dislocations whose lines are close in the sense of macroscopic size, but are at a distant in comparison with distances between neighboring points of the crystal lattice. In this situation, an average approach to the study of defects is more reasonable. In this case, we assume that a contour consists of small intervals that join neighboring points of the crystal lattice. Then the integral

$$\oint a_{i1}dx_1 + a_{i2}dx_2 + a_{i3}dx_3$$

along this contour defined by the distortion a_{ij} can be considered instead of the sum $\sum(a_{i1}\Delta x_1 + a_{i2}\Delta x_2 + a_{i3}\Delta x_3)$. In our opinion, this remark completely explains the definition of the Burgers vector (b_1, b_2, b_3) relative to the contour $\mathcal{L}$ corresponding to the distortion $A = ||a_{ij}||$:

$$-b_i = \oint_{\mathcal{L}} a_{ij}dx_j.$$

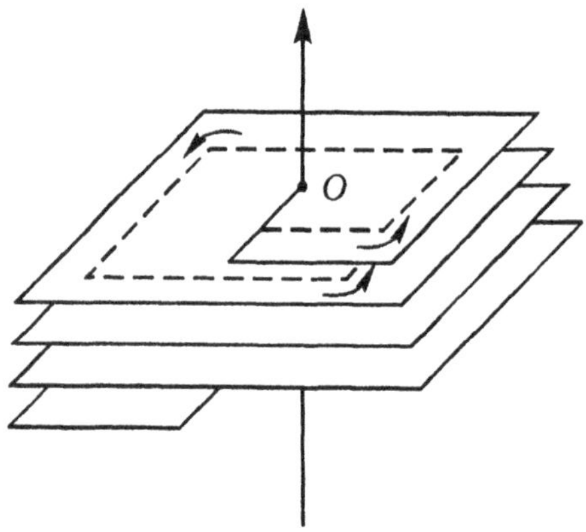

Fig. 9

This formula was taken as the definition in Section 12, where the distortion was formally (and nonuniquely) introduced from the stressed state and temperature of the medium. For a crystalline medium with dislocations, the notions of the distortion and Burgers vector are very descriptive. In such a medium, the Burgers tensor with components

$$B_{i1} = \frac{\partial a_{i2}}{\partial x_3} - \frac{\partial a_{i3}}{\partial x_2}, \quad B_{i2} = \frac{\partial a_{i3}}{\partial x_1} - \frac{\partial a_{i1}}{\partial x_3}, \quad B_{i3} = \frac{\partial a_{i1}}{\partial x_2} - \frac{\partial a_{i2}}{\partial x_1},$$

is usually referred to as the *dislocation density tensor*. In the continual theory of dislocations, equations for distortion are written in the form

$$\frac{\partial a_{ij}}{\partial t} + u_k \frac{\partial a_{ij}}{\partial x_k} + a_{ik} \frac{\partial u_k}{\partial x_j} = -\psi_{ij},$$

which was used in Section 12 to describe the evolution of the effective distortion. The right-hand sides ψ_{ij} in the continual theory of dislocations are defined on the basis of some arguments reasonable from the point of view of microscopic processes in the crystal lattice. Within the framework of our approach to the study of Maxwell media with relaxation, we constructed interpolation formulas for ψ_{ij} based on purely phenomenological arguments.

The above facts of the theory of dislocation motivate the terminology in Section 12.

Chapter III

Differential Equations of Dynamical Processes

§ 14. Conservation Laws

We already used the momentum, moment of momentum, mass, and energy conservation laws while treating the structure of a stressed state (Section 2), relationships between stresses and deformations (Sections 4 and 8), and a relaxation model (cf. Sections 9 and 10). Now, we write these laws (except for the moment of momentum one) as integral identities and deduce differential equations which, together with equations for the deformation (distortion) tensor, form a complete system of equations describing processes in an elastic or Maxwell relaxation medium.

Recall that from the equations for the metric (total or effective) deformation tensor (cf. Sections 8 and 10) we deduced the equation describing the evolution of the density

$$\frac{\partial \rho}{\partial t} + \frac{\partial(\rho u_i)}{\partial x_i} = 0. \tag{14.1}$$

As was mentioned, we need this to control equations describing the behavior of g_{ij} (g_{ij}^{eff}) because the equation for the density expresses the mass conservation law which is usually formulated as the integral identity

$$\iiint\limits_{\partial\Omega} \rho\, dx_1\, dx_2\, dx_3 + \rho u_1\, dx_2\, dx_3\, dt + \rho u_2\, dx_3\, dx_1\, dt + \rho u_3\, dx_1\, dx_2\, dt = 0$$

over a closed three-dimensional hypersurface $\partial\Omega$, the boundary of an arbitrary four-dimensional domain Ω.

This identity is usually established in a cylinder Ω [1] with three-dimensional bases $\partial_0\Omega$ ($t = t_0$) and $\partial_1\Omega$ ($t = t_1$) and the generators parallel to the t-axis. The lateral "surface" of such a cylinder is denoted by $\partial'\Omega$. Then the mass conservation law takes the form

$$\iiint\limits_{\partial_1\Omega} \rho\, dx_1\, dx_2\, dx_3 - \iiint\limits_{-\partial_0\Omega} \rho\, dx_1\, dx_2\, dx_3$$

$$= \iiint\limits_{-\partial'\Omega} \rho u_1\, dx_2\, dx_3\, dt + \rho u_2\, dx_1\, dx_3\, dt + \rho u_3\, dx_1\, dx_2\, dt.$$

We put the sign "$-$" at the pieces $\partial_0\Omega$ and $\partial'\Omega$ of $\partial\Omega$ in order to emphasize that the orientation is changed, which leads to the change of signs of the corresponding integrals. The integrals

$$\iiint\limits_{\partial_1\Omega} \rho\, dx_1\, dx_2\, dx_3, \qquad \iiint\limits_{-\partial_0\Omega} \rho\, dx_1\, dx_2\, dx_3$$

of the density ρ of the medium represent the mass of the medium concentrated in $\partial_1\Omega$ and $\partial_0\Omega$ respectively. We assume that the difference between these integrals is equal to the mass flow through the lateral surface of Ω. The expression

$$u_1\, dx_2\, dx_3 + u_2\, dx_3\, dx_1 + u_3\, dx_1\, dx_2$$

is the increment in volume per unit time, and the expression

$$(\rho u_1\, dx_2\, dx_3 + \rho u_2\, dx_3\, dx_1 + \rho u_3\, dx_1\, dx_2)\, dt$$

is the increment in mass during the increment in time dt. The integral identity

$$\iiint\limits_{\partial\Omega} \rho\, dx_1\, dx_2\, dx_3 + \rho u_1\, dx_2\, dx_3\, dt + \rho u_2\, dx_3\, dx_1\, dt + \rho u_3\, dx_1\, dx_2\, dt = 0$$

is valid for any (smooth or discontinuous) density ρ and components u_1, u_2, u_3 of the velocity. Therefore, this identity can be used in obtaining relations on discontinuities, i.e., on shock waves and contact surfaces. However, we are interested in the case of smooth distributions described by sufficiently good differentiable functions $\rho = \rho(x_1, x_2, x_3, t)$, $u_i = u_i(x_1, x_2, x_3, t)$. By the Gauss

[1] In fact, any domain can be approximated by a step function such that each step is a cylinder. Therefore, the conservation law in an arbitrary domain follows from that in cylinders. If the domain and integrands are not too pathological, the proof of this fact is a standard exercise from the calculus and is left to the reader.

divergence theorem, from the integral identity over any hypersurface $\partial\Omega$ for such functions we obtain the differential equality (14.1), called the *differential form of the continuity equation.* Note that the integral identity, called the *integral form of the continuity equation,* has a wider area of applications.

As was mentioned, the continuity equation was established while obtaining the differential equations for the deformation tensor and the distortion tensor (cf. Section 10). Now, we consider the momentum and energy conservation laws. They lead to new differential equations.

The momentum of the medium concentrated in the bases $\partial_1\Omega$ $(t = t_1)$ and $\partial_0\Omega$ $(t = t_0)$ of Ω is described by the vectors with components

$$\iiint\limits_{\partial_1\Omega} \rho u_i \, dx_1 \, dx_2 \, dx_3, \quad i = 1,2,3 \qquad (t = t_1),$$

$$\iiint\limits_{-\partial_0\Omega} \rho u_i \, dx_1 \, dx_2 \, dx_3, \quad i = 1,2,3 \qquad (t = t_0).$$

The momentum conservation law can be written in the form

$$\iiint\limits_{\partial_1\Omega} \rho u_i \, dx_1 \, dx_2 \, dx_3 - \iiint\limits_{-\partial_0\Omega} \rho u_i \, dx_1 \, dx_2 \, dx_3$$

$$= \iiint\limits_{-\partial'\Omega} \rho u_i (u_1 \, dx_2 \, dx_3 + u_2 \, dx_3 \, dx_1 + u_3 \, dx_1 \, dx_2) dt$$

$$+ \iiint\limits_{-\partial'\Omega} (-\sigma_{i1} \, dx_2 \, dx_3 - \sigma_{i2} \, dx_3 \, dx_1 - \sigma_{i3} \, dx_1 \, dx_2) \, dt$$

$$+ \iiiint\limits_{\Omega} \rho F_i \, dx_1 \, dx_2 \, dx_3 \, dt, \quad i = 1,2,3.$$

The last expression means that the increment in momentum is equal to the sum of the momentum of the medium flowing through the lateral boundary of Ω (the first term on the right-hand side), the momentum of the surface stresses on the lateral boundary (the second term), and the momentum of the mass forces applied to the interior of Ω (the third term). The components of the mass force density (per unit mass) are denoted by F_i.

In an arbitrary four-dimensional domain Ω, the *momentum conservation law* takes the form

$$\iiint\limits_{\partial\Omega} \rho u_i \, dx_1 \, dx_2 \, dx_3 + (\rho u_i u_1 - \sigma_{i1}) \, dx_2 \, dx_3 \, dt + (\rho u_i u_2 - \sigma_{i2}) \, dx_3 \, dx_1 \, dt$$

$$+ (\rho u_i u_3 - \sigma_{i3}) \, dx_1 \, dx_2 \, dt = \iiiint\limits_{\Omega} \rho F_i \, dx_1 \, dx_2 \, dx_3 \, dt, \quad i = 1,2,3.$$

These equalities can be used in the study of smooth or discontinuous functions describing the flow of a medium. By the Gauss divergence theorem, for smooth functions from the integral equalities we obtain the differential equations

$$\frac{\partial(\rho u_i)}{\partial t} + \frac{\partial(\rho u_i u_j - \sigma_{ij})}{\partial x_j} = \rho F_i, \quad i = 1, 2, 3, \tag{14.2}$$

called the *differential form of the momentum conservation law.*

The total energy concentrated in the bases $\partial_0\Omega$ $(t = t_0)$ and $\partial_1\Omega$ $(t = t_1)$ of Ω is represented as follows:

$$\iiint\limits_{-\partial_0\Omega} \rho\Big(E + \frac{u_i u_i}{2}\Big)\, dx_1\, dx_2\, dx_3 \quad (t = t_0),$$

$$\iiint\limits_{\partial_1\Omega} \rho\Big(E + \frac{u_i u_i}{2}\Big)\, dx_1\, dx_2\, dx_3 \quad (t = t_1).$$

We recall that E is the internal energy per unit mass, ρE is the internal energy per unit volume, and $\rho\dfrac{u_i u_i}{2} = \rho\dfrac{u_1^2 + u_2^2 + u_3^2}{2}$ is the kinetic energy per unit volume. In the case of a cylinder, the conservation law takes the form

$$\begin{aligned}
&\iiint\limits_{\partial_1\Omega} \rho\Big(E + \frac{u_i u_i}{2}\Big)\, dx_1\, dx_2\, dx_3 - \iiint\limits_{-\partial_0\Omega} \rho\Big(E + \frac{u_i u_i}{2}\Big)\, dx_1\, dx_2\, dx_3 \\
&\quad = \iiint\limits_{-\partial'\Omega} \rho\Big(E + \frac{u_i u_i}{2}\Big)(u_1\, dx_2\, dx_3 + u_2\, dx_3\, dx_1 + u_3\, dx_1\, dx_2)\, dt \\
&\quad + \iiint\limits_{-\partial'\Omega} u_i(-\sigma_{i1}\, dx_2\, dx_3 - \sigma_{i2}\, dx_3\, dx_1 - \sigma_{i3}\, dx_1\, dx_2)\, dt \\
&\quad + \iiiint\limits_{\Omega} \rho(u_1 F_1 + u_2 F_2 + u_3 F_3)\, dx_1\, dx_2\, dx_3\, dt \\
&\quad + \iiiint\limits_{\Omega} Q\, dx_1\, dx_2\, dx_3\, dt \\
&\quad + \iiint\limits_{-\partial'\Omega} (q_1\, dx_2\, dx_3 + q_2\, dx_3\, dx_1 + q_3\, dx_1\, dx_2)\, dt.
\end{aligned}$$

The left-hand side of this equality is the increment in energy during the time interval $[t_0, t_1]$. On the right-hand side, the first term describes the energy of the medium flowing through the lateral boundary. The second term represents the work done by the surface stresses in producing displacements $u_i\, dt$ of particles passing through the lateral boundary. The third term is the work done by the mass forces with density ρF_i in producing displacements of interior points. The last two terms describe the heat influx. Here, Q is the volume density of heat sources and q_i, $i = 1,2,3$, are the components of the heat flux vector corresponding to the influx through the lateral boundary.

In an arbitrary domain Ω, the *energy conservation law* takes the form [2)]

$$\iiint_{\partial\Omega} \rho\left(E + \frac{u_i u_i}{2}\right) dx_1\, dx_2\, dx_3$$
$$+ \left[\rho u_1 \left(E + \frac{u_i u_i}{2}\right) - u_k \sigma_{k1} + q_1\right] dx_2\, dx_3\, dt$$
$$+ \left[\rho u_2 \left(E - \frac{u_i u_i}{2}\right) - u_k \sigma_{k2} + q_2\right] dx_3\, dx_1\, dt$$
$$+ \left[\rho u_3 \left(E + \frac{u_i u_i}{2}\right) - u_k \sigma_{k3} + q_3\right] dx_1\, dx_2\, dt$$
$$= \iiiint_{\Omega} [\rho u_k F_k + Q]\, dx_1\, dx_2\, dx_3\, dt.$$

If the functions occurring in this integral identity are smooth, then from the Gauss divergence theorem we obtain the differential equation

$$\frac{\partial[\rho(E + u_i u_i/2)]}{\partial t} + \frac{\partial[\rho u_j(E + u_i u_i/2) - u_k \sigma_{kj} + q_j]}{\partial x_j} = \rho u_k F_k + Q, \quad (14.3)$$

called the *differential form of the energy conservation law.*

The heat flux vector q_j, $i = 1,2,3$, arises because of the heat transfer. By the Fourier law, this vector is proportional to the temperature gradient ($q_j = -\varkappa \dfrac{\partial T}{\partial x_j}$) since the flow is from regions of higher to lower temperature ($\varkappa \geqslant 0$).

We have formulated the heat conduction law only for an isotropic body. In general, this law has the form

$$q_j = -\varkappa_{ji} \frac{\partial T}{\partial x_i},$$

where $\varkappa_{ij}$ is the heat conductivity coefficient tensor. In what follows, we consider heat conduction processes only in the case of an isotropic body and do not take into account any external heat sources ($Q = 0$).

Sometimes it is convenient to introduce the density of the total heat influx

$$\widetilde{Q} = Q + \frac{\partial}{\partial x_j}\left(\varkappa_{ji} \frac{\partial T}{\partial x_i}\right)$$

which is the sum of the power of external sources Q and the power of the influx caused by the heat conduction. Using $\widetilde{Q}$, we write the differential equation for

[2)] As in the case of the mass (or momentum) conservation law, this formula can be deduced from the energy conservation law in cylinders.

the energy conservation law in the form

$$\frac{\partial[\rho(E+u_iu_i/2)]}{\partial t}+\frac{\partial[\rho u_j(E+u_iu_i/2)-u_k\sigma_{ki}]}{\partial x_j}=\rho u_kF_k+\widetilde{Q}.$$

§ 15. Consequences of Conservation Laws

In Section 14, we demonstrated how conservation laws can be used to derive differential equations that describe the motion of media under deformations. In this section, we derive differential equations that are consequences of conservation laws.

We start with the continuity equation

$$\frac{\partial\rho}{\partial t}+\frac{\partial(\rho u_j)}{\partial x_j}=0$$

and write it in the following form:

$$\frac{\partial\rho}{\partial t}+u_j\frac{\partial\rho}{\partial x_j}+\rho\frac{\partial u_j}{\partial x_j}=\frac{d\rho}{dt}+\rho\frac{\partial u_j}{\partial x_j}=0,$$

$$\frac{d\ln\rho}{dt}=-\frac{\partial u_j}{\partial x_j}\equiv-\left(\frac{\partial u_1}{\partial x_1}+\frac{\partial u_2}{\partial x_2}+\frac{\partial u_3}{\partial x_3}\right).$$

The momentum conservation law

$$\frac{\partial(\rho u_i)}{\partial t}+\frac{\partial(\rho u_iu_j-\sigma_{ij})}{\partial x_j}=\rho F_i$$

can be written in the form

$$u_i\left[\frac{\partial\rho}{\partial t}+\frac{\partial(\rho u_j)}{\partial x_j}\right]+\rho\left(\frac{\partial u_i}{\partial t}+u_j\frac{\partial u_i}{\partial x_j}\right)-\frac{\partial\sigma_{ij}}{\partial x_j}=\rho F_i,$$

and, with the help of the continuity equation, in the form

$$\frac{\partial u_i}{\partial t}+u_j\frac{\partial u_i}{\partial x_j}-\frac{1}{\rho}\frac{\partial\sigma_{ij}}{\partial x_j}\equiv\frac{du_i}{dt}-\frac{1}{\rho}\frac{\partial\sigma_{ij}}{\partial x_j}=F_i.$$

In the case of an isotropic stressed state which occurs in a fluid ($\sigma_{ij}=-p\delta_{ij}$), the equation

$$\frac{du_i}{dt}+\frac{1}{\rho}\frac{\partial p}{\partial x_j}=F_i$$

is called the *Euler equation*.

The expression on the left-hand side of the energy conservation law

$$\frac{\partial[\rho(E+u_k u_k/2)]}{\partial t}+\frac{\partial[\rho u_j(E+u_k u_k/2)-u_k\sigma_{kj}]}{\partial x_j}$$
$$=\frac{\partial}{\partial x_j}\left(\varkappa\frac{\partial T}{\partial x_j}\right)+Q+\rho u_k F_k,$$

can be written as the sum of 4 terms:

$$\frac{\partial[\rho(E+u_k u_k/2)]}{\partial t}+\frac{\partial[\rho u_j(E+u_k u_k/2)-u_k\sigma_{kj}]}{\partial x_j}=$$
$$\left\{u_k\left[\frac{\partial(\rho u_k)}{\partial t}+\frac{\partial(\rho u_j u_k-\sigma_{kj})}{\partial x_j}\right]\right\}+\left\{\left(E-\frac{u_k u_k}{2}\right)\cdot\left[\frac{\partial\rho}{\partial t}+\frac{\partial(\rho u_j)}{\partial x_j}\right]\right\}$$
$$+\left\{\rho\left(\frac{\partial E}{\partial t}+u_j\frac{\partial E}{\partial x_j}\right)\right\}-\left\{\sigma_{kj}\frac{\partial u_k}{\partial x_j}\right\}.$$

The first two distinguished terms can be simplified with the help of the momentum conservation law and the continuity equation as follows:

$$u_k\left[\frac{\partial(\rho u_k)}{\partial t}+\frac{\partial(\rho u_j u_k-\sigma_{kj})}{\partial x_j}\right]=u_k\rho F_k,$$
$$\left(E-\frac{1}{2}u_k u_k\right)\left[\frac{\partial\rho}{\partial t}+\frac{\partial(\rho u_j)}{\partial x_j}\right]=0,$$

which allows us to write the equation for internal energy:

$$\rho\left(\frac{\partial E}{\partial t}+u_j\frac{\partial E}{\partial x_j}\right)\equiv\rho\frac{dE}{dt}=\left\{\frac{\partial}{\partial x_j}\left(\varkappa\frac{\partial T}{\partial x_j}\right)+Q\right\}+\sigma_{kj}\frac{\partial u_k}{\partial x_j}=\tilde{Q}+\sigma_{kj}\frac{\partial u_k}{\partial x_j}.$$

The left-hand side $\rho\dfrac{dE}{dt}$ of this equation expresses the rate of the increment in the internal energy (per unit volume) of a fixed element of a medium. The first term $\tilde{Q}$ on the right-hand side is the heat for the increment in internal energy, and the second term is called the *work of the internal surface forces*. We already considered this notion in Section 5 and used it in Section 8 in the study of the stressed state of an elastic medium.

The equality

$$\rho\frac{dE}{dt}=\tilde{Q}+\sigma_{km}\frac{\partial u_k}{\partial x_m}$$

is sufficient in order to justify the possibility of identification of the sum $\sigma_{km}\dfrac{\partial u_k}{\partial x_m}$ and the density of power of the internal surface stresses. We use the decomposition of σ_{km} into "elastic" and "viscous" parts: $\sigma_{km}=\sigma^{\mathrm{el}}_{km}+\sigma^{\mathrm{vis}}_{km}$, as was postulated in Section 5, and the Murnaghan formulas (cf. Sections 7 and 8)

$$\sigma^{\mathrm{el}}_{km} = -2\rho g^{\mathrm{eff}}_{kj}\frac{\partial E}{\partial g^{\mathrm{eff}}_{jm}} = -2\rho g_{kj}\frac{\partial E}{\partial g_{jm}},$$

which express the elastic stresses in terms of metric effective elastic deformation tensor

$$\rho\frac{dE}{dt} = \widetilde{Q} + \sigma^{\mathrm{vis}}_{km}\frac{\partial u_k}{\partial x_m} + \sigma^{\mathrm{el}}_{km}\frac{\partial u_k}{\partial x_m} = \widetilde{Q} + \sigma^{\mathrm{vis}}_{km}\frac{\partial u_k}{\partial x_m} - 2\rho g_{kj}\frac{\partial E}{\partial g_{jm}}\frac{\partial u_k}{\partial x_m}$$

$$= \widetilde{Q} + \sigma^{\mathrm{vis}}_{km}\frac{\partial u_k}{\partial x_m} - \rho\frac{\partial E}{\partial g_{jm}}\frac{\partial u_k}{\partial x_m}g_{kj} - \rho\frac{\partial E}{\partial g_{mj}}g_{km}\frac{\partial u_k}{\partial x_j}$$

$$= \widetilde{Q} + \sigma^{\mathrm{vis}}_{km}\frac{\partial u_k}{\partial x_m} - \rho\frac{\partial E}{\partial g_{mj}}\left(g_{mk}\frac{\partial u_k}{\partial x_j} + \frac{\partial u_k}{\partial x_m}g_{kj}\right).$$

Making these transformations, we take into account the symmetry property $\left(g_{km} = g_{mk},\ \dfrac{\partial E}{\partial g_{mj}} = \dfrac{\partial E}{\partial g_{jm}}\right)$ and omit the superscript "eff" in the notation g^{eff}_{ik}. Recall that the behavior of the metric effective elastic deformation tensor is described by the equations (cf. Section 10)

$$\frac{dg_{mj}}{dt} + g_{mk}\frac{\partial u_k}{\partial x_j} + \frac{\partial u_k}{\partial x_m}g_{kj} = -\varphi_{mj}.$$

Using these equations, we can reduce the equation for energy to the form

$$\rho\frac{dE}{dt} = \rho\left(E_S\frac{dS}{dt} + E_{g_{mj}}\frac{dg_{mj}}{dt}\right) = \widetilde{Q} + \sigma^{\mathrm{vis}}_{km}\frac{\partial u_k}{\partial x_m} + \rho\frac{\partial E}{\partial g_{mj}}\left(\frac{dg_{mj}}{dt} + \varphi_{mj}\right),$$

which implies the entropy equation ($E_S = T$)

$$\frac{dS}{dt} \equiv \frac{\partial S}{\partial t} + u_i\frac{\partial S}{\partial x_i} = \frac{\widetilde{Q}}{\rho T} + \frac{1}{\rho T}\sigma^{\mathrm{vis}}_{km}\frac{\partial u_k}{\partial x_m} + \frac{1}{T}E_{g_{mj}}\varphi_{mj}.$$

On the right-hand side, the first term represents the rate of entropy changed in view of the heat influx, the second one in view of the viscous friction, and the third represents the rate of the change of entropy caused by relaxation processes. We note that relaxation and friction are irreversible dissipative processes and each of them leads to the increase of entropy. The mathematical statement of this postulate is expressed by the inequalities

$$E_{g_{mj}}\varphi_{mj} > 0, \quad \sigma^{\mathrm{vis}}_{km}\frac{\partial u_k}{\partial x_m} > 0,$$

which were already used in Sections 5 and 10 where we constructed equations governing irreversible processes.

Multiplying the equation for entropy

$$\frac{\partial S}{\partial t} + u_j \frac{\partial S}{\partial x_j} = \frac{1}{\rho T}\left[Q + \frac{\partial}{\partial x_j}\left(\varkappa \frac{\partial T}{\partial x_j}\right)\right] + \frac{1}{T} E_{g_{km}} \varphi_{km} + \frac{1}{\rho T} \sigma_{km}^{\text{vis}} \frac{\partial u_k}{\partial x_m}$$

by the density ρ and adding to the left-hand side the expression (which is equal to zero by the continuity equation)

$$S\left[\frac{\partial \rho}{\partial t} + \frac{\partial(\rho u_j)}{\partial x_k}\right],$$

we obtain the entropy equation in the form

$$\frac{\partial(\rho S)}{\partial t} + \frac{\partial(\rho S u_j)}{\partial x_j} = \frac{Q}{T} + \frac{1}{T}\frac{\partial}{\partial x_j}\left(\varkappa \frac{\partial T}{\partial x_j}\right) + \frac{\rho}{T} E_{g_{km}} \varphi_{km} + \frac{1}{T} \sigma_{km}^{\text{vis}} \frac{\partial u_k}{\partial x_m}$$
$$= \frac{Q}{T} + \frac{\partial}{\partial x_j}\left(\frac{\varkappa}{T}\frac{\partial T}{\partial x_j}\right) + \frac{\varkappa}{T^2}\frac{\partial T}{\partial x_j}\frac{\partial T}{\partial x_j} + \frac{\rho}{T} E_{g_{km}} \varphi_{km} + \frac{1}{T} \sigma_{km}^{\text{vis}} \frac{\partial u_k}{\partial x_m},$$

which is equivalent to the following integral identity similar to a conservation law:

$$\iint_{\partial\Omega} \rho S\, dx_1\, dx_2\, dx_3 + \left(\rho S u_1 - \frac{\varkappa}{T}\frac{\partial T}{\partial x_1}\right) dx_2\, dx_3\, dt$$
$$+ \left(\rho S u_2 - \frac{\varkappa}{T}\frac{\partial T}{\partial x_2}\right) dx_3\, dx_1\, dt + \left(\rho S u_3 - \frac{\varkappa}{T}\frac{\partial T}{\partial x_3}\right) dx_1\, dx_2\, dt$$
$$= \iiint_{\Omega} \left(\frac{Q}{T} + \frac{\varkappa}{T^2}\frac{\partial T}{\partial x_j}\frac{\partial T}{\partial x_j} + \frac{\rho}{T} E_{g_{km}} \varphi_{km} + \frac{1}{T} \sigma_{km}^{\text{vis}} \frac{\partial u_k}{\partial x_m}\right) dx_1\, dx_2\, dx_3\, dt.$$

The vector with components $\rho S u_i - \dfrac{\varkappa}{T}\dfrac{\partial T}{\partial x_j}$ describes the entropy flux caused by the mass and heat (the term $-\dfrac{\varkappa}{T}\dfrac{\partial T}{\partial x_j}$) transfer, and the terms on the right-hand side are distributed "entropy sources" arising from irreversible processes.

Thus, the term

$$\frac{\varkappa}{T^2}\frac{\partial T}{\partial x_j}\frac{\partial T}{\partial x_j} \equiv \frac{\varkappa}{T^2}\left[\left(\frac{\partial T}{\partial x_1}\right)^2 + \left(\frac{\partial T}{\partial x_2}\right)^2 + \left(\frac{\partial T}{\partial x_3}\right)^2\right]$$

describes the entropy production caused by the temperature equalization in view of the heat transfer. The positivity of this term is justified by the fact that the heat conductivity coefficient $\varkappa$ is positive. The entropy balance law (more exactly, "entropy growth law") is a consequence of the equations considered earlier only for smooth functions describing the flow of the medium. This fact significantly differs from the conservation of mass, momentum, and energy which are applicable to a wider field.

It is possible to generalize this law to flows with shock waves where distributions of physical quantities are discontinuous. However, such generalizations should be formulated in the form of inequalities. In Section 21 below we will discuss this question (in the case where a gas is not heat conductive).

§ 16. Deformation of a Layer and a Bar

Starting with this section, we consider the behavior of media with properties described by the above mathematical models. The simplest problem is to clarify the dependence between deformation of an extending (or compressing) layer of a medium and the forces necessary for this deformation to be realized. A more complicated version is the deformation of a bar whose lateral bounds are free from stresses.

Let a deformation of the layer $0 \leqslant x_1 \leqslant h$ be caused by the forced motion of the boundary plane $h = h(t)$ $(h(0) = h_0)$. The other boundary plane $x_1 = 0$ is fixed.

If the layer is made of some homogeneous isotropic material, it is natural to assume that the stress tensor is the same at all points of the layer and remains unchanged under rotations of the layer around the x_1-axis at any angle. Hence the components σ_{21} and σ_{31} of the stress vector on the area element with normal directed along the x_1-axis vanish. The stresses at the area elements that are perpendicular to the x_2-axis and x_3-axis should also be perpendicular to these area elements because they should be transformed into themselves under any reflections with respect to the coordinate axes (the stressed state should remain unchanged in the case under consideration). Therefore, $\sigma_{12} = \sigma_{32} = 0$ and $\sigma_{13} = \sigma_{23} = 0$. Since the axes x_2 and x_3 play the same role, we obtain the equality $\sigma_{22} = \sigma_{33} = s_2$. Thus, the stress tensor has the form

$$[\sigma_{ij}] = \begin{bmatrix} s_1 & 0 & 0 \\ 0 & s_2 & 0 \\ 0 & 0 & s_3 \end{bmatrix}, \qquad s_2 = s_3.$$

Assume that the layer is made from a material, the properties of which are described by the Maxwell model of a viscoelastic medium with relaxation of shear stresses. We also assume that stresses of internal friction can be neglected. The metric effective elastic deformation tensor G^{eff} computed by the stress tensor and the temperature of the medium should possess the same symmetry property as the stress tensor. Thus,

$$G^{\text{eff}} = \begin{bmatrix} 1/k_1^2 & 0 & 0 \\ 0 & 1/k_2^2 & 0 \\ 0 & 0 & 1/k_3^2 \end{bmatrix}, \qquad k_2 = k_3.$$

The effective extension coefficients k_1 and k_2 and the entropy S are determined from the equations

$$\begin{aligned}
&\frac{\rho_0}{k_2^2}\, E_{k_1}(k_1, k_2, k_2, S) = s_1, \\
&E_S(k_1, k_2, k_2, S) = T, \\
&\frac{\rho_0}{k_1 k_2}\, E_{k_2}(k_1, k_2, k_3, S)\big|_{k_3=k_2} = s_2.
\end{aligned}$$

The effective distortion $A^{\text{eff}} = A$ was introduced (cf. Section 12) as a solution to the matrix equation $G^{\text{eff}} = A^* A$. In the case under consideration, this solution takes the form

$$A = V^* \begin{bmatrix} 1/k_1 & 0 & 0 \\ 0 & 1/k_2 & 0 \\ 0 & 0 & 1/k_3 \end{bmatrix}, \qquad k_2 = k_3,$$

where V is an arbitrary orthogonal matrix. It is natural to choose the identity matrix $V = I$. Moreover,

$$A = \begin{bmatrix} 1/k_1 & 0 & 0 \\ 0 & 1/k_2 & 0 \\ 0 & 0 & 1/k_3 \end{bmatrix}, \qquad k_2 = k_3.$$

It is also natural to assume that the velocity field is one-dimensional and is directed along the x_1-axis:

$$\begin{aligned}
&u_1(x_1, x_2, x_3, t) = \frac{x_1}{h(t)} \frac{dh(t)}{dt} = x_1 \frac{\dot{h}(t)}{h(t)}, \\
&u_2(x_1, x_2, x_3, t) = u_3(x_1, x_2, x_3, t) = 0.
\end{aligned}$$

Actually, the velocity is defined by solution of the complete system consisting of conservation laws, as well as equations connecting stresses with deformations. However, if the change of the strain rate $\dot{h}(t)/h(t)$ is small while perturbation waves run through the layer, we can assume that the homogeneity condition are satisfied by deformations of the layer which implies the postulated distribution of velocity.

We write equations for the distortion tensor

$$\frac{\partial a_{ij}}{\partial t} + u_r \frac{\partial a_{ij}}{\partial x_r} + a_{ir} \frac{\partial u_r}{\partial x_j} = -\psi_{ij},$$

where

$$a_{11} = 1/k_1, \quad a_{22} = a_{33} = 1/k_2, \quad a_{ij} = 0 \quad (\text{for } i \neq j),$$

$$-\|\psi_{ij}\| = \frac{1}{\tau}\begin{bmatrix} \frac{1}{k_1}\ln\frac{k_1}{\sqrt[3]{k_1k_2^2}} & 0 & 0 \\ 0 & \frac{1}{k_2}\ln\frac{k_2}{\sqrt[3]{k_1k_2^2}} & 0 \\ 0 & 0 & \frac{1}{k_2}\ln\frac{k_2}{\sqrt[3]{k_1k_2^2}} \end{bmatrix}.$$

They are reduced to two equations

$$\frac{\partial}{\partial t}\left(\frac{1}{k_1}\right) + \frac{1}{k_1}\frac{\dot{h}(t)}{h(t)} - \frac{2}{3\tau k_1}\ln\frac{k_1}{k_2} = 0,$$
$$\frac{\partial}{\partial t}\left(\frac{1}{k_2}\right) - \frac{1}{3\tau k_2}\ln\frac{k_2}{k_1} = 0$$

which do not contain the derivatives with respect to x_1, i.e., we obtain two ordinary equations. It is convenient to write them in the form

$$\begin{aligned} \frac{dk_1}{dt} &= k_1\frac{\dot{h}(t)}{h(t)} - \frac{2}{3\tau}k_1\ln\frac{k_1}{k_2}, \\ \frac{dk_2}{dt} &= -\frac{1}{3\tau}k_2\ln\frac{k_2}{k_1}. \end{aligned} \tag{16.1}$$

The thermal state of a medium is characterized by the entropy S which changes under deformations accompanied by relaxation of shear stresses without the heat influx and without its re-distribution of the heat transfer according to the equations

$$\frac{dS}{dt} = \frac{1}{T}E_{g_{ij}}\varphi_{ij} \tag{16.2}$$

which were derived in Section 15 (for a more general case). In the relaxation model under consideration, we obtain the relation (cf. Sections 9, 10, 15)

$$\begin{aligned} E_{g_{ij}}\varphi_{ij} &= \frac{1}{3\tau}\left[(k_1E_{k_1} - k_2E_{k_2})\ln\frac{k_1}{k_2} + (k_2E_{k_2} - k_3E_{k_3})\ln\frac{k_2}{k_3}\right. \\ &\left. + (k_3E_{k_3} - k_1E_{k_1})\ln\frac{k_3}{k_1}\right] = \frac{2}{3\tau}(k_1E_{k_1} - k_2E_{k_2})\ln\frac{k_1}{k_2} \geqslant 0. \end{aligned} \tag{16.3}$$

Thus, we have the complete system of three equations

$$
\begin{aligned}
\frac{dk_1}{dt} &= k_1 \frac{\dot{h}(t)}{h(t)} - \frac{2}{3\tau} k_1 \ln \frac{k_1}{k_2}, \\
\frac{dk_2}{dt} &= -\frac{1}{3\tau} k_2 \ln \frac{k_2}{k_1}, \\
\frac{dS}{dt} &= \frac{2}{3\tau T} (k_1 E_{k_1} - k_2 E_{k_2}) \ln \frac{k_1}{k_2},
\end{aligned}
\tag{16.4}
$$

$E = E(k_1, k_2, k_3, S)$, $T = E_S(k_1, k_2, k_3, S)$, $\tau = \tau(k_1, k_2, k_3, S)$, $k_2 = k_3$, describing the time-dependence of the parameters k_1, $k_2 = k_3$, S characterizing the stressed and thermal state of the layer under deformation.

The dimension of the ratio

$$\dot{\varepsilon} = \frac{\dot{h}(t)}{h(t)} = \frac{1}{h(t)} \frac{dh(t)}{dt}$$

is the inverse of time and characterizes the strain rate. If we know $\dot{\varepsilon}$, then we can restore the thickness of the layer h as a solution to the equation

$$\frac{dh}{dt} = \dot{\varepsilon} h.$$

The stress s_1 extending the plane and cross-section stresses $s_2 = s_3$ can be found for any t from the equalities

$$s_1 = \frac{\rho_0}{k_2^2} E_{k_1}(k_1, k_2, k_2, S), \quad s_2 = \frac{\rho_0}{k_1 k_2} E_{k_2}(k_1, k_2, k_3, S)\big|_{k_3=k_2}.$$

We proceed by obtaining equations that describe the change of the stressed state of the extending bar. Unlike the above problem, some extending (or compressing) forces act on the lateral surface of the bar, which yields

$$s_2 = s_3 = \frac{\rho_0}{k_1 k_2} E_{k_2}(k_1, k_2, k_3)\big|_{k_3=k_2} = 0.$$

This equation should be used instead of the equation

$$\frac{dk_2}{dt} = -\frac{1}{3\tau} \ln \frac{k_2}{k_1}.$$

We again assume that the velocity u_1 is subject to the law

$$u_1 = x_1 \frac{\dot{h}(t)}{h(t)} = x_1 \dot{\varepsilon}(t),$$

whereas u_2 are u_3 may be nonzero. They provide us with the equalities $E_{k_2} = E_{k_3} = 0$. In the case of a bar, we arrive at the equations

$$\begin{aligned}
&\frac{dh}{dt} = \dot{\varepsilon} h, \\
&\frac{dk_1}{dt} = \dot{\varepsilon} k_1 - \frac{2k_1}{3\tau(k_1, k_2, k_2, S)} \ln \frac{k_1}{k_2}, \\
&E_{k_2}(k_1, k_2, k_3, S)\big|_{k_3=k_2} = 0, \\
&\frac{dS}{dt} = \frac{2}{3\tau E_S(k_1, k_2, k_2, S)} (k_1 E_{k_1} - k_2 E_{k_2}) \ln \frac{k_1}{k_2}.
\end{aligned} \tag{16.5}$$

Instead of deformation without heat influx, as was postulated in obtaining Eqs. (16.4), (16.5), we can consider isothermal deformation with constant temperature ($T = T^*$). Moreover, instead of the equation

$$\frac{dS}{dt} = \frac{2}{3\tau E_S} (k_1 E_{k_1} - k_2 E_{k_2}) \ln \frac{k_1}{k_2},$$

we introduce the equation

$$E_S(k_1, k_2, k_2, S) = T^*. \tag{16.6}$$

By Eq. (16.6), we can assume that the entropy is some definite function $S = S(k_1, k_2)$ of k_1, k_2. By the equality $E_{k_2}(k_1, k_2, k_3, S)\big|_{k_3=k_2} = 0$, we can regard k_2 as a function of k_1. Thus, in an isothermal process, we have $k_2 = k_2(k_1)$, $S = S(k_1, k_2) = S(k_1)$ and the equation for k_1 takes the form

$$\frac{dk_1}{dt} = \dot{\varepsilon}(t) k_1 - \varphi(k_1). \tag{16.7}$$

The last equation differs from the equation for the length $h(t)$ of the bar by only the term $-\varphi(k_1)$. Integrating the system

$$\begin{aligned}
&\frac{dh}{dt} = \dot{\varepsilon}(t) h, \\
&\frac{dk_1}{dt} = \dot{\varepsilon}(t) k_1 - \varphi(k_1)
\end{aligned} \tag{16.8}$$

for some strain rate $\dot{\varepsilon}$, we can investigate how the "effective" extension coefficient (defining the stress $\sigma_{11} = s_1$) depends on the real length $h(t)$ and time t. For many materials, $\varphi(k_1)$ is very small if k_1 is not too large, but has very fast growth if k_1 increases. For a model we take

$$\varphi(k_1) = \dot{\varepsilon}_0 k_1^* \left(\frac{k_1}{k_1^*}\right)^n,$$

where φ is a power function of k_1 with a sufficiently large exponent n.

Let $\dot{\varepsilon}$ be constant and positive ($\dot{\varepsilon} > 0$). The equation

$$\frac{dk_1}{dt} = \dot{\varepsilon} k_1 - \dot{\varepsilon}_0 k_1^* \left(\frac{k_1}{k_1^*}\right)^n \tag{16.9}$$

has integral curves as in Fig. 10. They are obtained each from other by parallel translation along the t-axis. For small k_1 the nonlinear term $-\dot{\varepsilon}_0 (k_1)^n/(k_1^*)^{n-1}$ is small and the solutions are given by the formula $k_1 = \text{const}\, e^{\dot{\varepsilon} t}$ with sufficiently large accuracy.

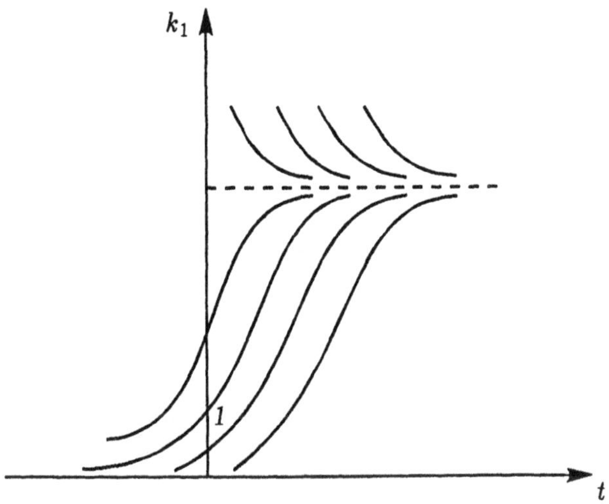

Fig. 10

Although k_1 increases if t increases, the graph never intersects the curve

$$k_1 = k_1^* \left(\frac{\dot{\varepsilon}}{\dot{\varepsilon}_0}\right)^{1/(n-1)}.$$

(It is easy to see that this line is the graph of one of integral curves of our differential equation. It is the asymptote to the other curves. By the uniqueness theorem, different integral curves have no intersection.)

We can assume that the solution (with the initial condition $k_1(0) = 1$) consists of two parts:

$$k_1 = e^{\dot{\varepsilon} t} \qquad \text{if } t < \frac{1}{\dot{\varepsilon}} \ln\left[k_1^* \left(\frac{\dot{\varepsilon}}{\dot{\varepsilon}_0}\right)^{1/(n-1)}\right],$$

$$k_1 = k_1^* \left(\frac{\dot{\varepsilon}}{\dot{\varepsilon}_0}\right)^{1/(n-1)} \qquad \text{if } t > \frac{1}{\dot{\varepsilon}} \ln\left[k_1^* \left(\frac{\dot{\varepsilon}}{\dot{\varepsilon}_0}\right)^{1/(n-1)}\right].$$

Here, we replaced the integral curves tending to the asymptote with the asymptote itself. It is clear that the passage from exponent to a constant is smooth. However, the smoothing of the graph is almost invisible for large n, i.e., if $\varphi(k_1)$ "sharply" depends on the state of the medium, which is a consequence of the "sharp" dependence of τ on the state of a medium. (In the real schemes, τ depends on k_1, k_2, k_3 exponentionally.)

The equation $\dfrac{dh}{dt} = \dot{\varepsilon}h$ for the length of the bar h has the solution $h(t) = h_0e^{\dot{\varepsilon}t}$. Thus, $h(t) = h_0e^{\dot{\varepsilon}t}$ and $k_1(t) = \min\{e^{\dot{\varepsilon}t}, k_1^*(\dot{\varepsilon}/\dot{\varepsilon}_0)^{1/(n-1)}\}$.

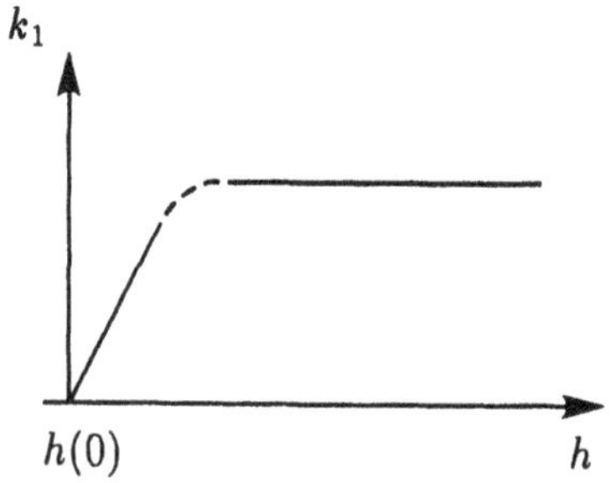

Fig. 11

The graph of the dependence of the "effective" extension coefficient k_1 on the length of the bar h is represented in Fig. 11. It is obvious that the graphs of $k_2(h)$ and $S = S(h)$ are similar. All these quantities first essentially depend on h, but then they become constant because k_2 and S are functions of k_1. Let us clarify what force should be applied to the bar in order that the above deformation process is realized.

The force applied to the bar is equal to the product of the stress s_1 and the area ω of the cross-section. This area also changes under deformation. The total volume occupied by the bar is equal to the product of ωh and the mass of the bar $\rho\omega h$ remains constant: $\rho\omega h = \rho_0\omega_0 h_0$ $(\rho = \rho_0/(k_1k_2^2))$. Hence $\omega = \omega_0 h_0 \dfrac{k_1k_2^2}{h}$ and the force

$$F = s_1\omega = \omega_0 h_0 \,\frac{k_1[k_2(k_1)]^2 s_1(k_1)}{h}$$

is equal to the product of a function of k_1 and the inverse of the real length of the bar h.

Therefore, extending a bar, we first increase F until k_1 attains the maximal value $k_1(\dot{\varepsilon}/\dot{\varepsilon}_0)^{1/(n-1)}$ and becomes constant. After that, F begins to decrease as soon as h increases. It should be noted that it is usual to consider deformations in terms of the change in length of the bar as a percentage. Under such deformations, the presence of h in the denominator in the formulas for F is not essential and the dependence of F on h is approximately the same as in the case of k_1.

We emphasize that the scheme under consideration is rather idealistic because we assume that the relaxation time τ depends only on the stressed state and temperature of the medium. In fact, τ essentially depends on the structural properties of material which can change under deformations. This dependence leads to the hardening or softening when, after exit to some "plateau" and further increase of h, we can observe that s_1 is nonconstant. This stress either rather slowly increases (hardening) or drops. The sharp drop leads to the appearance of the characteristic "yielding peak" at the graphs of $s_1(h)$

We do not discuss the structural properties of materials and clarify how they depend on the deformation process but restrict ourselves to simulating the properties of the medium using the equation of state $E = E(k_1, k_2, k_3, S)$ and the relaxation time τ depending on stresses and temperature. Instead of stresses and temperature in the notation $\tau = \tau(k_1, k_2, k_3, S)$, we indicate the dependence on the parameters k_1, k_2, k_3 and entropy S.

We note that the "height of plateau" in the graph of $k_1 = k_1(h)$ constructed by the solutions to the equations

$$\frac{dk_1}{dt} = \dot{\varepsilon} k_1 - \dot{\varepsilon}_0 k_1^* \left(\frac{k_1}{k_1^*}\right)^n \qquad (k_1(0) = 1),$$

$$\frac{dh}{dt} = \dot{\varepsilon} h \qquad (h(0) = h_0)$$

is determined by the strain rate $\dot{\varepsilon}$:

$$k_{1_{\max}} = k_1^* (\dot{\varepsilon}/\dot{\varepsilon}_0)^{1/(n-1)}.$$

However, this dependence is very weak for large n (or if τ depends on k_1 as an exponential function). The graph of $F(h)$ depends on $\dot{\varepsilon}$ in a similar way. The lifting of "plateau" when $\dot{\varepsilon}$ increases is, in general, small and it should be taken into account only if we consider processes where changes of the strain rate are significant.

We consider the behavior of the bar if we first extend it so that the extending force attains the maximal value and, after that, slightly decrease its length. At the first stage (extension),

$$\frac{dk_1}{dt} = \dot{\varepsilon} k_1 - \dot{\varepsilon}_0 k_1^* \left(\frac{k_1}{k_1^*}\right)^n,$$

$$\frac{dh}{dt} = \dot{\varepsilon} h,$$

whereas at the second stage (shortening):

$$\frac{dk_1}{dt} = -\dot{\varepsilon}k_1 - \dot{\varepsilon}_0 k_1^* \left(\frac{k_1}{k_1^*}\right)^n,$$

$$\frac{dh}{dt} = -\dot{\varepsilon}h.$$

At the second stage, k_1 decreases and the nonlinear term $\dot{\varepsilon}_0 k_1^* \left(\frac{k_1}{k_1^*}\right)^n$ in the equation for k_1 becomes unessential. Then the unloading process is described with high accuracy by the linear equations

$$\frac{dk_1}{dt} = -\dot{\varepsilon}k_1,$$

$$\frac{dh}{dt} = -\dot{\varepsilon}h,$$

which imply the linear dependence of k_1 on h

$$\frac{d}{dt}\left(\frac{k_1}{h}\right) = \frac{-\dot{\varepsilon}k_1 h - (-\dot{\varepsilon}hk_1)}{h^2} = 0, \quad k_1 = \text{const} \cdot h.$$

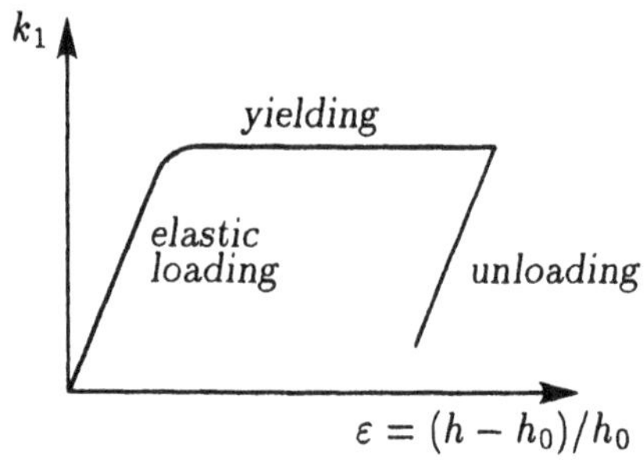

Fig. 12

The complete graph of the dependence of k_1 on the deformation parameter $\varepsilon = (h-h_0)/h_0$ is represented in Fig. 12. The graph of the force F acting on the bar looks similar. The maximal value $F_{\max}$ (more exactly, the corresponding value of s_1) is called the *yield point*. The behavior of materials that are elastic under small deformations and are characterized by yield points under large deformations are studied in plasticity theory. In such models, it is assumed that the passage from elastic to plastic state has a jump-like character.

The Maxwell relaxation model with nonlinear and rather sharp dependence of the relaxation time on the parameters characterizing the state of a medium allows us to study processes appearing from the passage to the yielding state. It is important if we study high rate mechanical phenomena, for example, deformation of metals under explosive loading. As a rule, such deformations occur in different parts of a body at different rates. Therefore, it is necessary to take into account the dependence of the yield point on $\dot{\varepsilon}$. We emphasize that the Maxwell theory accounts to this fact automatically.

To simplify the presentation, we assume the isotropic character of a medium which can be described by the same τ along any direction. In the stressed material, there are some particular planes along which slip is easier than it is along the other directions. Similarly, we ignore the possible dependence of τ on structural and chemical parameters responsible for hardening phenomena.

If, based on the phenomenological approach, we postulate the dependence of the relaxation time on stresses and temperature and the dependence of the internal energy E on elastic deformations and entropy, then we obtain the Maxwell model, which allows us to regard the passage to the "yielding" state as a property of the solutions to some differential equations.

Within framework of this model, we thereby do not intend to "understand" the nature of relaxation processes but only describe them at the most rough level, i.e., in terms of the relaxation time.

§ 17. Equations for One-Dimensional Nonstationary Processes and Structure of Stationary Waves in a Maxwell Medium

Considering the simplest problems in Section 16, we were given a deformation process and computed only a force necessary for the realization of this process. Now, using the momentum conservation law, we connect the acceleration of points of a medium with stresses and density. This leads to a closed system of partial differential equations. We consider equations whose solutions are functions of two independent variables x_1 and t. We assume that the principal axes of the stressed state are directed along the coordinates x_1, x_2, and x_3, whereas the principal stresses s_2 and s_3 are equal. Therefore, the metric effective elastic deformation tensor and effective distortion at each point of the medium take the form

$$G = \begin{bmatrix} 1/k_1^2 & 0 & 0 \\ 0 & 1/k_2^2 & 0 \\ 0 & 0 & 1/k_2^2 \end{bmatrix}, \quad A = \begin{bmatrix} 1/k_1 & 0 & 0 \\ 0 & 1/k_2 & 0 \\ 0 & 0 & 1/k_2 \end{bmatrix}.$$

The effective extension coefficients k_1, k_2 and entropy S are functions of x_1, t, i.e., $k_1 = k_1(x_1, t)$, $k_2 = k_2(x_1, t)$, $S = S(x_1, t)$, the velocity has only one nonzero component $u_1 = u_1(x_1, t)$ $(u_2 = u_3 = 0)$, whereas the density, temperature, and stresses are determined by the formulas

$$\rho = \rho_0/(k_1 k_2^2), \quad T = E_S(k_1, k_2, k_2.S),$$

$$\sigma_{11} = s_1 = \sigma_{11}^{\text{vis}} + \sigma_{11}^{\text{el}} = \left(\zeta + \frac{4}{3}\eta\right)\frac{\partial u_1}{\partial x_1} + \rho k_1 E_{k_1}(k_1, k_2, k_2, S) = \mu\frac{\partial u_1}{\partial x_1} + \sigma_1,$$

$$\sigma_{22} = \sigma_{33} = s_2 = \sigma_{22}^{\text{vis}} + \sigma_{22}^{\text{el}} = \left(\zeta - \frac{2}{3}\eta\right)\frac{\partial u_1}{\partial x_1} + \rho k_2 E_{k_2}(k_1, k_2, k_3)\big|_{k_3=k_2}$$

$$= \left(\zeta - \frac{2}{3}\eta\right)\frac{\partial u_1}{\partial x_1} + \sigma_2,$$

$$\sigma_{ij} = 0 \text{ for } i \neq j,$$

where σ_1 and σ_2 are the principal values of the elastic stress tensor. In this section, we consider only processes without actions of mass forces ($F_i = 0$) and exterior heat sources ($Q = 0$).

The momentum and energy conservation laws lead to the equations

$$\frac{\partial(\rho u_1)}{\partial t} + \frac{\partial}{\partial x_1}\left(\rho u_1^2 - \sigma_1 - \mu\frac{\partial u_1}{\partial x_1}\right) = 0, \tag{17.1}$$

$$\frac{\partial}{\partial t}\left[\rho\left(E + \frac{1}{2}u_1^2\right)\right] + \frac{\partial}{\partial x_1}\left[\rho u_1\left(E + \frac{1}{2}u_1^2\right) - u_1\sigma_1 - \mu u_1\frac{\partial u_1}{\partial x_1} - \varkappa\frac{\partial T}{\partial x_1}\right] = 0. \tag{17.2}$$

Two additional differential equations are obtained from the equations for the effective distortion:

$$\frac{\partial a_{ij}}{\partial t} + u_r\frac{\partial a_{ij}}{\partial x_r} + a_{ir}\frac{\partial u_r}{\partial x_j} = -\psi_{ij},$$

which for

$$A = \begin{bmatrix} 1/k_1 & 0 & 0 \\ 0 & 1/k_2 & 0 \\ 0 & 0 & 1/k_2 \end{bmatrix},$$

$$-\|\psi_{ij}\| = \frac{1}{\tau}\begin{bmatrix} \frac{1}{k_1}\ln\frac{k_1}{\sqrt[3]{k_1k_2^2}} & 0 & 0 \\ 0 & \frac{1}{k_2}\ln\frac{k_2}{\sqrt[3]{k_1k_2^2}} & 0 \\ 0 & 0 & \frac{1}{k_2}\ln\frac{k_2}{\sqrt[3]{k_1k_2^2}} \end{bmatrix},$$

$u_1 = u_1(x_1, t)$, $u_2 = u_3 = 0$ are reduced to the following equations for k_1 and k_2:

$$\frac{\partial k_1}{\partial t} + u_1 \frac{\partial k_1}{\partial x_1} - k_1 \frac{\partial u_1}{\partial x_1} = -\frac{2}{3} \frac{k_1}{\tau} \ln \frac{k_1}{k_2}, \tag{17.3}$$

$$\frac{\partial k_2}{\partial t} + u_1 \frac{\partial k_2}{\partial x_1} = -\frac{1}{3} \frac{k_2}{\tau} \ln \frac{k_2}{k_1}. \tag{17.4}$$

The relaxation terms are taken according to the recommendations of Section 12.

Thus, we arrive at a complete system of four equations with four unknown functions k_1, k_2, S, u_1.

Sometimes, it is preferable to use the consequences of Eqs. (17.3) and (17.4) instead of the equations themselves. The first consequence is the continuity equation

$$\frac{\partial \rho}{\partial t} + \frac{\partial(\rho u_1)}{\partial x_1} = 0 \quad (\rho = \rho_0/(k_1 k_2^2)), \tag{17.5}$$

and the second one is the following equation for $\beta = \ln k_2$:

$$\frac{\partial \beta}{\partial t} + u_1 \frac{\partial \beta}{\partial x_1} + \frac{1}{\tau}(\beta - \ln \sqrt[3]{\rho_0/\rho}) = 0. \tag{17.6}$$

In an isotropic medium, where $E = E(\rho, S)$, the elastic stress tensor (Section 4) is represented in any orthogonal axes as follows: $\sigma_{ij}^{\text{el}} = -p\delta_{ij}$, where $p = p(S, \rho)$ is the pressure $p = \rho^2 E_\rho(\rho, S)$. Gases and fluids can serve as examples of such media. Under large stresses, we can assume that all solid bodies (for example, metals) are characterized by such equations of state.

For media with $E = E(\rho, S)$ from conservation laws we obtain the closed system of differential equations

$$\begin{aligned}
&\frac{\partial \rho}{\partial t} + \frac{\partial(\rho u_1)}{\partial x_1} = 0,\\
&\frac{\partial(\rho u_1)}{\partial t} + \frac{\partial(\rho u_1^2 + p)}{\partial x_1} = \frac{\partial}{\partial x_1}\left(\mu \frac{\partial u_1}{\partial x_1}\right),\\
&\frac{\partial}{\partial t}\rho\left(E + \frac{u_1^2}{2}\right) + \frac{\partial}{\partial x_1}\left[\rho u_1\left(E + \frac{u_1^2}{2}\right) + p u_1 - \mu u_1 \frac{\partial u_1}{\partial x_1}\right]\\
&\quad = \frac{\partial}{\partial x_1}\left(\varkappa \frac{\partial T}{\partial x_1}\right),\\
&E = E(\rho, S), \quad T = E_S(\rho, S), \quad p = \rho^2 E_\rho(\rho, S),\\
&\mu = \mu(\rho, S), \quad \varkappa = \varkappa(\rho, S).
\end{aligned} \tag{17.7}$$

The system (17.7) is known as the *Navier-Stokes equations.*

We consider the simplest example of a Maxwell medium described by one-dimensional equations without internal friction ($\mu = 0$) and without heat conductivity ($\varkappa = 0$). This example demonstrates how the fact that τ drops under large stresses and temperatures brings a Maxwell medium to some state which can be described by the equations of hydrodynamics. We formulate the problem and sketch the justification of its solution illustrating numerical experiments and figures presenting the results of these experiments. All computations were performed on the basis of interpolation formulas for equations of state and for relaxation times for some metals (copper, aluminum, nickel, etc.)

We consider the propagation of plane waves where the graphs of distributions of any quantities are displaced with unchanged shape in the direction of the x_1 -axis. In the case of such a wave, the functions $u_1 = u_1(x_1 - wt)$, $\beta = \beta(x_1 - wt)$, $\rho = \rho(x_1 - wt)$, $S = S(x_1 - wt)$ satisfy the Eqs. (17.1), (17.2), (17.5), (17.6). It is possible to pass to moving coordinates with respect to which the wave is in the rest state. Moreover, $w = 0$ and the solution takes the form $u = u(x)$, $\rho = \rho(x)$, $\beta = \beta(x)$, $S = S(x)$. (Hereinafter, we omit the subscript "1" in the notation u_1, x_1, σ_1, etc.) For $u = u(x)$, $\rho = \rho(x)$, $\beta = \beta(x)$, $S = S(x)$ we obtain the ordinary differential equations

$$\begin{gathered}\frac{\partial(\rho u)}{\partial x} = 0, \quad \frac{d(\rho u^2 - \sigma)}{dx} = 0, \quad \frac{d}{dx}\left[\rho u\left(E + \frac{1}{2}u^2\right) - \sigma u\right] = 0, \\ u\,\frac{d\beta}{dx} = -\frac{1}{\tau}\left(\beta - \ln\sqrt[3]{\rho_0/\rho}\right).\end{gathered} \tag{17.8}$$

We are interested in solutions for which u, ρ, β, and S tend to finite limits and their derivatives vanish as $x \to \pm\infty$. It turns out that such solutions can be used in order to clarify the characteristic features of the structure of shock waves propagated in an elastic-plastic medium.

If the limit values of τ as $x \to \pm\infty$ are finite and differ from zero, then, as $\dfrac{d\beta}{dx} \to 0$, we have

$$\frac{1}{3}\ln\frac{k_2}{k_1} = \ln k_2 - \ln\sqrt[3]{k_1 k_2^2} = \beta - \ln\sqrt[3]{\rho_0/\rho} = 0,$$

i.e., the coefficients k_1 and $k_2 = k_3$ characterizing the effective elastic deformation tend to the same limit $k_1 = k_2 = k_3$ as $x \to \pm\infty$. (The limits can be different at $x = +\infty$ and $x = -\infty$.) Consequently, the limit values σ_1 and $\sigma_2 = \sigma_3$ are also equal ($\sigma_1 = \sigma_2 = \sigma_3 = -p = \rho k_1 E_{k_1}(k_1, k_2, k_3, S)$). Thus, the limit states ($x = \pm\infty$) of the medium should be affected by the action of hydrostatic pressure.

We can integrate the first three equations of the system. The integrals $\rho u =$ const, $\rho u^2 - \sigma =$ const, $\rho u(E + u^2/2) - \sigma u =$ const represent the conservation of mass, momentum, and energy.

The conservation laws should hold not only for smooth (differentiable) distributions $\rho = \rho(x)$, $u = u(x)$, $S = S(x)$, $\sigma = \sigma(x)$, but also for discontinuous distributions which are not treated within the framework of differential equations. Thus, any stationary distributions $\rho = \rho(x)$, $u = u(x)$, $S = S(x)$, $\sigma = \sigma(x)$, smooth or discontinuous, satisfy the equations

$$\rho u = \text{const}, \quad \rho u^2 - \sigma = \text{const}, \quad \rho u(E + u^2/2) - \sigma u = \text{const},$$

and, on the parts, where $\rho(x)$, $u(x), \ldots$ are smooth, the following differential equation holds:

$$\rho u \frac{d\beta}{dx} = -\frac{\rho}{\tau}(\beta - \sqrt[3]{\rho_0/\rho}).$$

Denoting by $m = \rho u = \text{const}$ the mass flux through a cross-section $x = \text{const}$, it is easy to derive from the equalities

$$\rho u = m, \quad \rho u^2 - \sigma = mu - \sigma = \text{const},$$
$$\rho u(E + u^2/2) - \sigma u = m(E + u^2/2) - \sigma u = \text{const}$$

the following relations for all x:

$$u - u_0 = m\left(\frac{1}{\rho} - \frac{1}{\rho_{00}}\right), \quad \sigma + p_0 = m^2\left(\frac{1}{\rho} - \frac{1}{\rho_0}\right) = m(u - u_0),$$
$$E - E_0 + \frac{p_0 - \sigma}{2}\left(\frac{1}{\rho} - \frac{1}{\rho_{00}}\right) = 0.$$

Here, $u_0, p_0 = -\sigma_0 \equiv -\sigma_{10} = -\sigma_{20} = -\sigma_{30}$, ρ_{00} denote the parameters of the state in front of the wave at $x = -\infty$. (We assume that $m > 0$, i.e., the x-axis is directed towards the motion of a medium.)[3] The above relations describing the structure of a wave can be solved according to the following order (u_0, ρ_{00}, ρ_0 and, consequently, S_0, m are given).

First, from the equalities

$$\sigma + p_0 = m^2\left(\frac{1}{\rho} - \frac{1}{\rho_0}\right), \qquad E = E(k_1, k_2, k_2, S),$$
$$\sigma = \rho k_1 E_{k_1}(k_1, k_2, k_2, S),$$
$$E - E_0 + \frac{p_0 - \sigma}{2}\left(\frac{1}{\rho} - \frac{1}{\rho_{00}}\right) = 0, \quad \rho = \rho/(k_1 k_2^2)$$

we determine (in the form of a table) the curve of possible states inside the wave. For this purpose, regarding the density ρ as a parameter and solving the above

---

[3] We describe how to obtain the third equality, which is not obvious:

$$0 = E - E_0 + \frac{u^2 - u_0^2}{2} - \frac{\sigma u - (-p_0)u_0}{m} = E - E_0 + \frac{(u - u_0)(u + u_0)}{2}$$
$$- \frac{[\sigma - (-p_0)](u + u_0)}{2m} - \frac{[\sigma + (-p_0)](u - u_0)}{2m} = E - E_0 + \frac{(\sigma + p_0)(u + u_0)}{2m}$$
$$- \frac{(\sigma + p_0)(u + u_0)}{2m} - \frac{(\sigma - p_0)m\left(\frac{1}{\rho} - \frac{1}{\rho_0}\right)}{2m} = E - E_0 - \frac{\sigma - p_0}{2}\left(\frac{1}{\rho} - \frac{1}{\rho_{00}}\right).$$

equations with respect to k_1, k_2, and S, we construct the tables (or graphs) of the dependence $k_1 = k_1(\rho)$, $k_2 = k_2(\rho)$, $\beta = \ln k_2 = \beta(\rho)$, and $S = S(\rho)$.

By the equality

$$u = u_0 + m\left(\frac{1}{\rho} - \frac{1}{\rho_{00}}\right) = u(\rho),$$

with each value of ρ the corresponding value of the velocity is associated.

We first assume that the structure of the wave is described by smooth functions of x. Then we can use the differential equation

$$m\,\frac{d\beta}{dx} = -\rho\,\frac{\beta - \ln\sqrt[3]{\rho_0/\rho}}{\tau(k_1, k_2, k_3, S)}$$

which can easily be written as the quadrature

$$x = -m\int \frac{\tau d\beta}{\rho(\beta - \ln\sqrt[3]{\rho_0/\rho})} = m\int \frac{3\tau d\beta}{\rho(\ln k_1 - \ln k_2)} = 3m\int \frac{\tau d\beta}{\rho(\alpha - \beta)},$$

where $\alpha = \ln k_1$, $\beta = \ln k_2$. Computing the integral in the above formula, we find the dependence $x = x(\rho)$.

In this section, we consider only those waves behind which ($x = -\infty$) the material (metal) is at the room temperature ($T = 293°\mathrm{K}$) and zero pressure ($p_0 = 0$).

The final hydrostatic state ($\sigma_1 = \sigma_2 = \sigma_3 = -p', \rho', S'$) behind the wave is connected with the state in front of the wave by the equalities

$$p' - p_0 + m^2\left(\frac{1}{\rho'} - \frac{1}{\rho_{00}}\right) = 0, \qquad k_{10} = k_{20} = k_{30} = \sqrt[3]{\rho_0/\rho_{00}},$$

$$E' - E_0 + \frac{p' + p_0}{2}\left(\frac{1}{\rho'} - \frac{1}{\rho_{00}}\right) = 0, \quad k_1' = k_2' = k_3' = \sqrt[3]{\rho_0/\rho'},$$

$$p_0 = -\rho_{00}k_{10}E_{k_{10}}(k_{10}, k_{20}, k_{30}, S_0), \quad E_0 = E(k_{10}, k_{20}, k_{30}, S_0),$$

$$p' = -\rho' k_1' E_{k_1'}(k_1', k_2', k_3', S'), \qquad E' = E(k_1', k_2', k_3', S').$$

One of these equalities, the *Hugoniot curve*,

$$E' - E_0 + \frac{p' + p_0}{2}\left(\frac{1}{\rho'} - \frac{1}{\rho_{00}}\right) = 0$$

represents the dependence $S' = S'(k_{10}', k_{10}, S_0)$ which, in view of the relations

$$\rho' = \rho_0/(k_1')^3, \qquad \rho_{00} = \rho_0/(k_{10})^3,$$

$$p' = -\sigma_1(k_1', k_1', k_1', S'), \quad p_0 = -\sigma_0(k_{10}, k_{10}, k_{10}, S_0),$$

can be expressed as the equation $p' = p'(\rho', \rho_{00}, p_0)$ of some curve in the plane $(p = -\sigma_1, V = 1/\rho)$.

This curve connecting the possible limit states ($x \to +\infty$) in the stationary wave with the given initial state ($x \to -\infty$) takes the form as in Fig. 13.

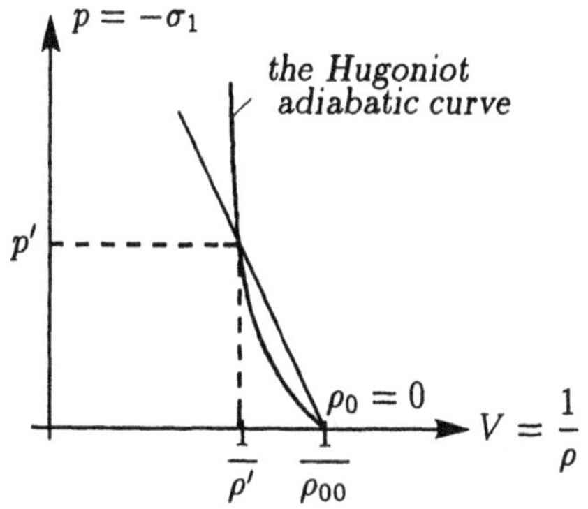

Fig. 13

If the mass flux m is given, the final state ($x \to +\infty$) is determined as the intersection point of the Hugoniot curve and the line

$$p' - p_0 + m^2\left(\frac{1}{\rho'} - \frac{1}{\rho_{00}}\right) = 0,$$

which is called the *Rayleigh line*. The module of its angle coefficient $-m^2$ is equal to squared mass flux of the wave m. We know that

$$-\sigma_1 - p_0 + m^2\left(\frac{1}{\rho} - \frac{1}{\rho_{00}}\right) = 0$$

everywhere inside the wave, i.e., the projections onto the part $(-\sigma_1, 1/\rho)$ of a curve passing through the points σ_1, ρ, S, $\alpha = \ln k_1$, $\beta = \ln k_2$, relative to possible states inside the wave should lie in the Rayleigh line and join the points $(p_0, 1/\rho_{00})$, $(p', 1/\rho')$ at which this line intersects the Hugoniot curve.

At an arbitrary point $(-\sigma_1, 1/\rho)$ of the Rayleigh line, the corresponding state inside the wave can be found because we know the dependence $k_1 = k_1(\rho)$, $k_2 = k_2(\rho)$, $S = S(\rho)$, $u = u(\rho)$. In Fig. 14, we represent a typical form of the projection onto the plane (α, β) of the curve $\alpha = \ln k_1(\rho)$, $\beta = \ln k_2(\rho)$ constructed with the help of this dependence (if M is not too large). The state in front of the wave is associated with the initial point $\alpha_0 = \beta_0 = \ln k_{10} = \ln k_{20} = \ln \sqrt[3]{\rho_0/\rho_{00}}$. The state behind the wave is represented by the point $\alpha' = \beta' = \ln k_1'$ which also lies on the line $\alpha = \beta$. In the figure, a part of the curve is distinguished by a bold line. This part contains the initial point and the final point of the wave, i.e., the interval corresponding to all intermediate states travelling inside the wave.

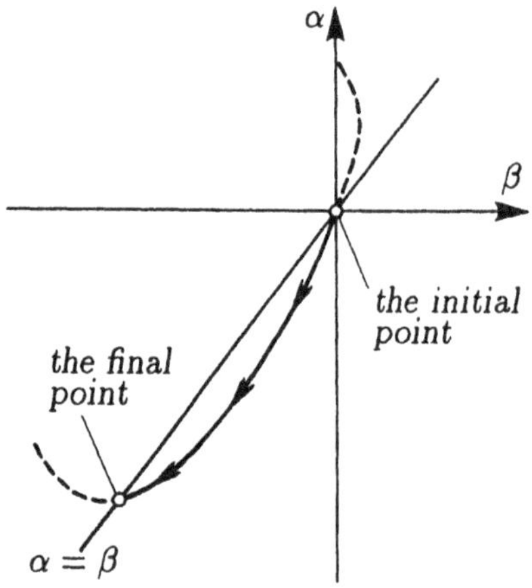

Fig. 14

If we move along this distinguished part from the initial point to the endpoint, then the parameter β decreases, the difference $\alpha - \beta$ is negative everywhere in this part and, consequently, the coordinate x which is restored with the help of the integral

$$x = 3m \int \frac{\tau(k_1, k_2, k_2, S)\, d\beta}{\rho(\alpha - \beta)},$$

will increase. The arrows along the distinguished part of the curve indicate the direction along which x increases.

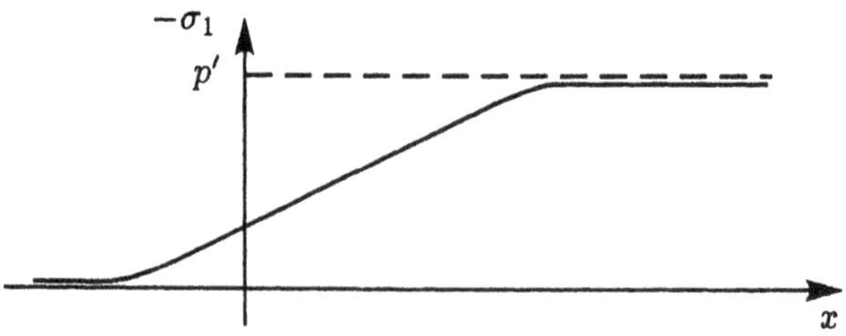

Fig. 15

We note that, as a rule, the integral determining the coordinate x is divergent in a neighborhood of endpoints of the wave where $\alpha - \beta \to 0$. Hence, inside the wave, x changes from $-\infty$ to $+\infty$. The graph of the dependence of stress on x is represented in Fig. 15.

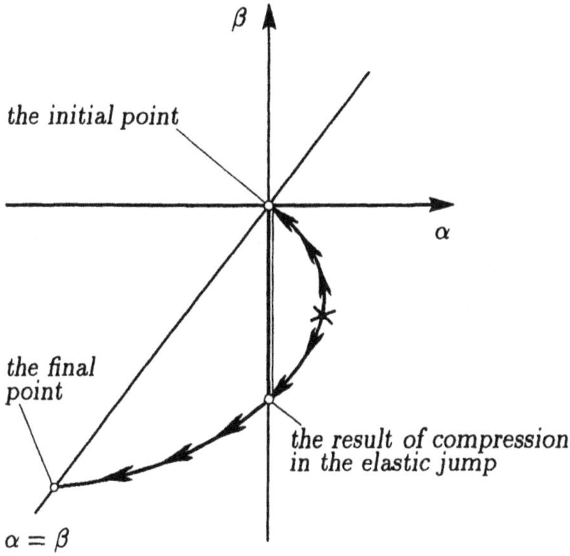

Fig. 16

If the mass flux m increases, the situation changes. For sufficiently large m the possible states in the plane $\alpha = \ln k_1$, $\beta = \ln k_2$ are presented in Fig. 16.

First, this curve goes towards the direction where β increases and, after that, towards the direction where β decreases but remains in the half-plane $\alpha < \beta$. Therefore, along this curve from the initial point to the final point, x is not monotone. The growth directions of x are marked by arrows. Obviously, it is impossible to obtain the continuous structure of the wavefront in this situation.

Results of experimental investigations show that the profiles of density and some other physical parameters can be discontinuous. The values at both sides of the discontinuity are connected by the laws of conservation of mass, momentum, and energy. As we know, the validity of these laws is equivalent to the fact that the corresponding quantities u, ρ, E, S, σ, k_1, k_2 belong to the above-constructed curve of possible states.

Introducing a discontinuity to the solution, we can construct this solution as follows. From the initial point we can reach some point of the curve of possible states by a jump. Starting with this point, the motion along the curve to the final point can be realized with the increase of x. The choice of a point at a path where x increases, starting from which the solution becomes smooth, requires some additional assumptions justified by results of theoretical or experimental research.

By experiment, it is reasonable to assume that k_2 (the effective coefficient of the cross extension) is constant at the jump. Indeed, since the jump is momentary, there is no relaxation of shear stresses because it takes some time and, consequently, it should be described by elasticity laws which imply that k_1, $k_2 = k_3$ can change only under the real change of the shape of elementary volumes of a material. In the case of one-dimensional motion, the velocity vanishes ($u_2 = u_3 = 0$). Therefore, the compression in the jump may occur only along the x-axis ($x \equiv x_1$), which leads to the change of only k_1, whereas k_2 and k_3 remain constant. A jump with unchanged k_2 or, which is the same, with unchanged β is naturally said to be *elastic*. Introducing the elastic jump to the solution, we arrive at the following structure of the wavefront.

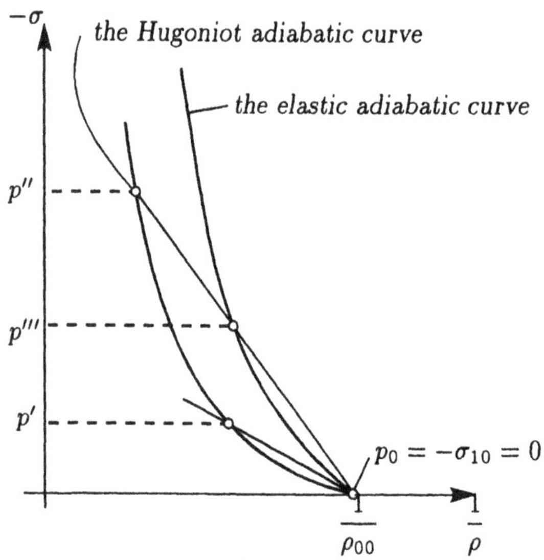

Fig. 17

The wave starts with a jump such that β is constant and $\alpha = \ln k_1$ decreases. In Fig. 16, the part of $\beta = \text{const}$ where α decreases is marked by double line. The state obtained by compression by elastic jump can be joined with the final state by continuous structure. By construction, the relations at the elastic jump are as follows:

$$\rho u = \rho_{00} u_0 = m, \quad \sigma + p_0 = m^2\left(\frac{1}{\rho} - \frac{1}{\rho_{00}}\right) = 0,$$

$$E - E_0 + \frac{p_0 - \sigma}{2}\left(\frac{1}{\rho} - \frac{1}{\rho_{00}}\right), \quad k_2 = k_{20}, \quad \sigma = \sigma_1.$$

The above equalities should be completed by the inequality $S > S_0$ which follows from the thermodynamical arguments. The validity of this inequality was confirmed by computations performed with the use of interpolation state equations for copper, aluminum, and a number of other metals.

Figure 17 presents in the plane $(-\sigma, 1/\rho)$ the Hugoniot curve with the final state satisfying the condition $k_1 = k_2 = k_3$ and the *elastic Hugoniot curve* where $k_2 = k_{20}$ is at the final state.

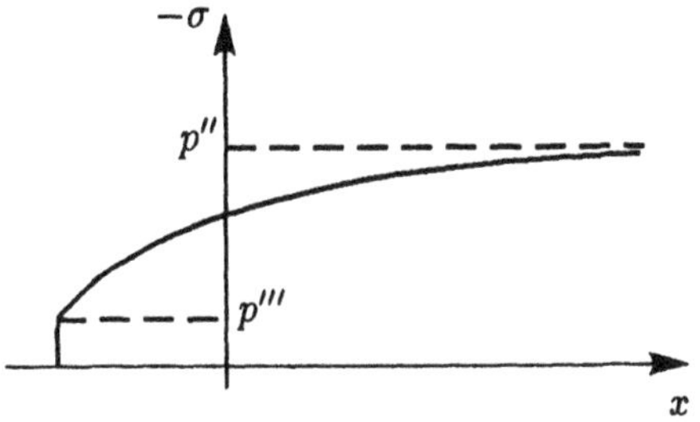

Fig. 18

If the mass flux of the wave m is small, then the line with slope $-m^2$ in the plane $(-\sigma, 1/\rho)$ that passes through the initial point of the Hugoniot curve and joins this point with the final point $(-\sigma = p')$ does not intersect the elastic Hugoniot curve. For large m the line going from the initial point to the endpoint $(\sigma = p'')$ intersects the elastic Hugoniot curve at the point $(-\sigma = p''')$. Moreover, the elastic jump leads a medium to the state with σ relative to this intersection point and, after that, with continuous relaxation, the medium takes to the state with isotropic stress tensor $\sigma_1 = \sigma_2 = \sigma_3 = -p''$. The graph of the change of σ $(\sigma \equiv \sigma_1)$ in such a composite wave is presented in Fig. 18.

As we already know, the "internal structure" of elastic jump cannot be treated within the framework of the Maxwell model. To study this, it is necessary to introduce an additional postulate concerning conservation of k_2. Instead of this, we can assume that, inside the elastic jump, the medium can be described by equations with small but nonzero viscosity coefficient μ and the heat conductivity coefficient $\varkappa$. Moreover, the smearing of the elastic jump by some nonzero thickness allows us to study its structure. However, we do not consider this approach in details here.

In the above consideration of the structure of one-dimensional travelling waves in a Maxwell medium we ignored the strong dependence of the relaxation time τ on stresses and temperature. It turns out that this nonlinearity affecting only the computation of the integral determining the coordinate x can significantly deformate the graphs.

From the formula

$$x = 3m \int \frac{\tau d\beta}{\rho(\alpha - \beta)}$$

we see that the parts corresponding to small τ are short, whereas the parts with rather large τ are long. The structure of the wavefronts in Figs. 17 and 18 was obtained under the assumption that $\tau = \text{const}$.

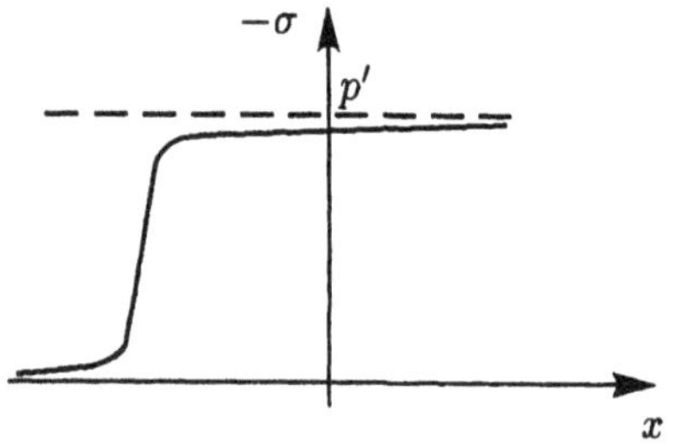

Fig. 19

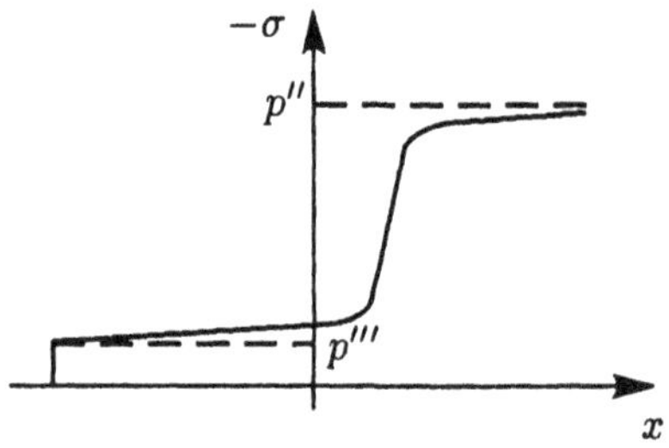

Fig. 20

If we use the interpolation formulas for τ which represent the real dependence of the relaxation time on the parameters of state of the medium, then the structure of wavefronts will be the same as in Figs. 19 and 20. Some time the relaxation goes slowly until the evolving heat leads to a fast decrease of τ. Because of this the relaxation process significantly accelerates.

The situation presented in Fig. 20 expresses the case where the wavefront is, in a sense, decomposed into the *elastic jump*, called the *elastic precursor*, and the *relaxation* or *plastic wave* that follows the precursor at some distance.

However, it should be noted that such a decomposition can be observed only if the amplitude of the wave is not too large and the pressure behind the wave is not larger than 200–300 GPa. In stronger waves, in view of large shear stresses and temperature, τ immediately turns out to be very small and relaxation is realized momentary. To distinguish an elastic jump from the following narrow relaxation layer (the plastic wave) is practically impossible. Therefore, to study strong shock waves in metallic media, we can use the gasodynamical approximation under the assumption that the stresses are isotropic behind the wavefront, i.e., the situation is the same as in hydrodynamics ($\sigma_1 = \sigma_2 = \sigma_3 = -p$).

It should be noted that it is not easy to create experimentally stationary wavefronts with structure as presented in Fig. 20 since for a small pressure the distance between the elastic precursor and plastic wave is large and it is not easy to obtain the stationary wavefront of large length in explosive experiments. But if the pressure increases, the time during passage elastic precursor and plastic wave is very susceptible to the pressure behind the wave, which again creates significant difficulties in experiments. As a rule, decomposition of waves is observed in nonstationary processes, but we do not discuss them here.

Now, we study how to distinguish plastic waves. By the above arguments, it is clear that it is natural to assume that points of the graph of some quantity in the structure of wave are placed inside the plastic wave if $\tau < \tau_0$, where τ_0 is small. For example, we can take $\tau = 1$ microsecond (or $\tau = 100$ microseconds if we study long-time processes).

Given mass flux m of the wave, all states of the medium are described by functions of a single parameter ρ

$$k_1 = k_1(\rho, m), \quad k_2 = k_2(\rho, m) = k_3, \quad S = S(\rho, m).$$

Therefore, the relaxation time $\tau = \tau(k_1, k_2, k_3, S)$ can also be expressed in terms of $m = \sqrt{(\sigma + p_0)/(1/\rho - 1/\rho_{00})}$ and ρ. Therefore, we can, in addition to Hugoniot and elastic Hugoniot curves, draw in the plane $(-\sigma, 1/\rho)$ lines coinciding with τ, for example, the line $\tau = \tau_0$ as is represented in Fig. 21.

In this figure, the Rayleigh lines with different angle coefficients $(-m^2)$ are marked by a dashed line. It is clear that for weak waves (small m) the relaxation layer is rather wide. If m increases, a plastic wave appears inside the layer. For large m it begins to look like the elastic precursor and finally, it and the precursor flow together for further increase of m.

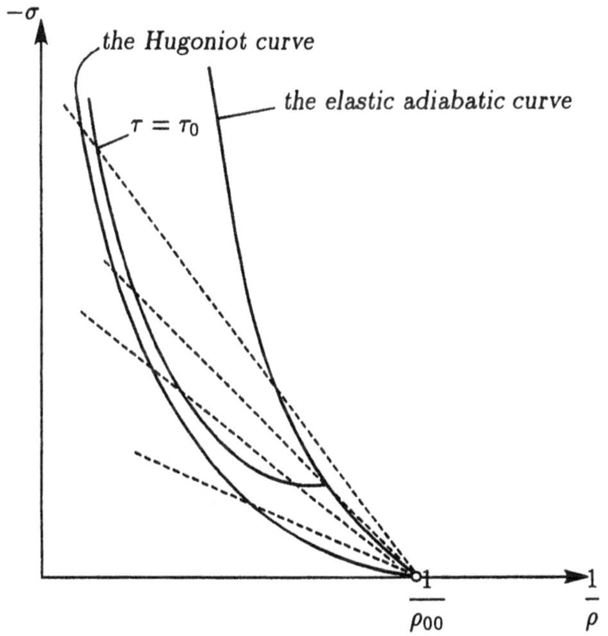

Fig. 21

Chapter IV

Well-Posedness of Differential Equations and Thermodynamics

§ 18. Well-Posedness Conditions for Equations of Elasticity

18.1. In this section, we consider the system of equations describing a Maxwell medium without viscous stresses, heat transfer, external heat sources, and external forces:

$$\frac{\partial a_{ij}}{\partial t} + u_k \frac{\partial a_{ij}}{\partial x_k} + a_{ik} \frac{\partial u_k}{\partial x_j} = -\psi_{ij}. \tag{18.1}$$

We also write equations for the velocity and entropy

$$\frac{\partial u_i}{\partial t} + u_k \frac{\partial u_i}{\partial x_k} - \frac{1}{\rho} \frac{\partial \sigma_{ij}}{\partial x_j} = 0, \tag{18.2}$$

$$\frac{\partial S}{\partial t} + u_k \frac{\partial S}{\partial x_k} = \left(\varphi_{ij} \frac{\partial E}{\partial g_{ij}} \right) \frac{1}{T}, \tag{18.3}$$

and complete the system by the equation of state and interpolation formulas for relaxation terms ψ_{ij} on the right-hand side of (18.1) the relations

$$\begin{aligned} g_{ij} &= a_{ki}a_{kj} \quad (G = A^*A), \\ \varphi_{ij} &= a_{ki}\psi_{kj} + \psi_{ki}a_{kj} \quad (\Phi = A^*\Psi + \Psi^*A), \\ \sigma_{ij} &= -2\rho g_{ik}\frac{\partial E}{\partial g_{kj}}, \quad \rho = \rho_0\sqrt{\det G} = \rho_0 \det A. \end{aligned} \tag{18.4}$$

The system (18.1)–(18.4) can be used to study mechanical processes if the Cauchy problem for this system is well posed.

Petrovsky suggested to call a Kovalevskaya-type system (solved or, at least, solvable with respect to the higher-order t-derivatives) *hyperbolic* if the Cauchy problem for the system is well posed.

As is known, the well-posedness (hyperbolicity) conditions can be expressed as inequalities for the coefficients at the higher-order derivatives of the unknown functions. Therefore, it suffices to obtain the hyperbolicity conditions for the homogeneous system $\psi_{ij} = 0$, $(\varphi_{ij}E_{g_{ij}}) = 0$ (nonlinear elasticity).

We consider solutions of nonlinear elasticity equations with the zero Burgers tensor:

$$B_{ij} = 0 \quad \left(\frac{\partial a_{ik}}{\partial x_j} - \frac{\partial a_{ij}}{\partial x_k} = 0\right). \tag{18.5}$$

In the nonlinear case, the well-posedness is verified only on solutions. Then the well-posedness conditions are valid for all solutions. Since the well-posedness conditions are expressed in terms of the coefficients of the equations but not the derivatives, the additional requirement (18.5) on the Burgers tensor is not essential.

By (18.5), there exists $x_{i0}(x_1, x_2, x_3, t_0)$ such that

$$\frac{\partial x_{i0}}{\partial x_j} = a_{ij}(x_1, x_2, x_3, t_0). \tag{18.6}$$

We find $x_{i0}(x_1, x_2, x_3, t)$ for all t as a solution to the equations

$$\frac{\partial x_{i0}}{\partial t} + u_k\frac{\partial x_{i0}}{\partial x_k} = 0 \tag{18.7}$$

with the initial data

$$x_{i0}(x_1, x_2, x_3, t_0) = x_{i0}. \tag{18.8}$$

As was shown in Section 8, for $x_{i0}(x_1, x_2, x_3, t)$ the distortion

$$\widetilde{a}_{ij}(x_1, x_2, x_3, t) = \frac{\partial x_{i0}(x_1, x_2, x_3, t)}{\partial x_j} \tag{18.9}$$

satisfies the equations

$$\frac{\partial \widetilde{a}_{ij}}{\partial t} + u_k \frac{\partial \widetilde{a}_{ij}}{\partial x_k} + \widetilde{a}_{ik} \frac{\partial u_k}{\partial x_j} = 0. \tag{18.10}$$

It is obvious that $a_{ij}(x_1, x_2, x_3, t_0) = \widetilde{a}_{ij}(x_1, x_2, x_3, t_0)$. Therefore,

$$a_{ij}(x_1, x_2, x_3, t) = \widetilde{a}_{ij}(x_1, x_2, x_3, t) = \frac{\partial x_{i0}(x_1, x_2, x_3, t)}{\partial x_j}. \tag{18.11}$$

We can take $x_{10}(x_1, x_2, x_3, t)$, $x_{20}(x_1, x_2, x_3, t)$, $x_{30}(x_1, x_2, x_3, t)$ for coordinates and write the equations in these new coordinates. The coordinates x_{i0} are connected with the distortion by the equalities (18.11) and are constant along the trajectories of particles of the medium:

$$\frac{dx_{i0}}{dt} = \frac{\partial x_{i0}}{\partial t} + u_k \frac{\partial x_{i0}}{\partial x_k} = 0. \tag{18.12}$$

These Lagrangian coordinates (cf. Section 6) will be referred to as the *canonical Lagrangian coordinates.*

Remark 18.1. Our investigation of the well-posedness conditions becomes much easier if we consider equations in the canonical Lagrangian coordinates because, in this case, we can use the Lagrange variational principle. However, it is not easy to realize the passage from the Eulerian coordinates to the Lagrangian ones.

18.2. Suppose that the motion of a medium is described by displacements of material particles subject to the law $x_i = x_i(x_{10}, x_{20}, x_{30}, t)$. Then we can write the following expressions for the kinetic energy

$$\iiint \rho_0 \frac{\dot{x}_1^2 + \dot{x}_2^2 + \dot{x}_3^2}{2} \, dx_{10} \, dx_{20} \, dx_{30} \tag{18.13}$$

and the potential energy

$$\iiint \rho_0 \mathcal{E}\left(\frac{\partial x_1}{\partial x_{10}}, \frac{\partial x_1}{\partial x_{20}}, \dots, \frac{\partial x_3}{\partial x_{30}}, S\right) dx_{10} \, dx_{20} \, dx_{30}, \tag{18.14}$$

where

$$\mathcal{E}(c_{11}, c_{12}, \dots, c_{33}, S) = E(k_1, k_2, k_3, S)$$

(k_1, k_2, k_3 are roots of the equation $\det \|CC^* - k^2 I\| = 0$). As is known, the Lagrangian $\mathcal{L}$ is equal to the difference between the kinetic and potential energies:

$$\mathcal{L} = \rho_0 \left[\frac{\dot{x}_1^2 + \dot{x}_2^2 + \dot{x}_3^2}{2} - \mathcal{E}\left(\frac{\partial x_1}{\partial x_{10}}, \frac{\partial x_1}{\partial x_{20}}, \dots, \frac{\partial x_3}{\partial x_{30}}, S\right)\right]. \tag{18.15}$$

The corresponding variational equations have the form

$$\frac{\partial}{\partial t}\left[\frac{\partial \mathcal{L}}{\partial \dot{x}_i}\right]+\frac{\partial}{\partial x_{r0}}\left[\frac{\partial \mathcal{L}}{\partial(\partial x_i/\partial x_{r0})}\right]=0 \tag{18.16}$$

and can be written as

$$\ddot{x}_i-\frac{\partial}{\partial x_{r0}}\left[\frac{\partial \mathcal{E}}{\partial(\partial x_i/\partial x_{r0})}\right]=0. \tag{18.17}$$

Since the entropy S is constant along the trajectory of each particle, we also write the equation

$$\dot{S}=0 \quad \left(\frac{dS}{dt}=\frac{\partial S}{\partial t}+u_k\frac{\partial S}{\partial x_k}=0\right). \tag{18.18}$$

Hence we obtain the system

$$\begin{aligned}&\ddot{x}_i-\mathcal{E}_{c_{ir}c_{jm}}\frac{\partial^2 x_j}{\partial x_{r0}\partial x_{m0}}-\mathcal{E}_{c_{ir}S}\frac{\partial S}{\partial x_{r0}}=0,\\&\dot{S}=0\end{aligned} \tag{18.19}$$

with the characteristic matrix

$$\begin{bmatrix}\|\omega^2\delta_{ij}-\xi_r\xi_m\mathcal{E}_{c_{ir}c_{jm}}\| & \|-\xi_r\mathcal{E}_{c_{ir}S}\|\\ 0 & \omega\end{bmatrix}. \tag{18.20}$$

In order for the characteristic values of this matrix be real, the matrix $\|\xi_r\xi_m\mathcal{E}_{c_{ir}c_{jm}}\|$ should be positive definite (more exactly, positive semidefinite) for all admissible vectors (ξ_1,ξ_2,ξ_3). The condition that the characteristic values are real is necessary for the well-posedness of the Cauchy problem. By definition, the *hyperbolicity* of a system implies the *well-posedness* of the Cauchy problem for any system with the same coefficients at the higher-order derivatives. Thus, the hyperbolicity condition is stronger than the well-posedness.

It is not too different from the real situation if we interpret the hyperbolicity as the positive definiteness of the matrix $\|\xi_r\xi_m\mathcal{E}_{c_{ir}c_{jm}}\|=\|M_{ij}\|$. This condition holds if $M_{ij}\eta_i\eta_j>0$ for any nonzero vector (η_1,η_2,η_3), i.e.,

$$\mathcal{E}_{c_{ir}c_{jm}}\eta_i\xi_r\eta_j\xi_m>0. \tag{18.21}$$

18.3. Out next goal is to derive the hyperbolicity condition in the form of inequalities for the derivatives of $E(k_1,k_2,k_3,S)$. Let C be represented as

$$C=U\begin{bmatrix}k_1&0&0\\0&k_2&0\\0&0&k_3\end{bmatrix}V \quad (k_i>0), \tag{18.22}$$

where U and V are orthogonal matrices. Let

$$\mathcal{E}(c_{11},c_{12},\ldots,c_{33},S)=E(k_1,k_2,k_3,S).$$

By definition, $\mathcal{E}$ is invariant under the multiplication of C from the right and from the left by any orthogonal matrix, which is equivalent to the fact that the energy is constant under any orthogonal transformation of the space with Eulerian coordinates (x_1, x_2, x_3), as well as of the space with the canonical Lagrangian coordinates (x_{10}, x_{20}, x_{30}). Under such transformations, the hyperbolicity of a system is preserved.

We choose coordinates such that at some point the matrix C becomes diagonal

$$C = \begin{bmatrix} k_1 & 0 & 0 \\ 0 & k_2 & 0 \\ 0 & 0 & k_3 \end{bmatrix}. \tag{18.23}$$

Consider a close matrix

$$\widetilde{C} = \begin{bmatrix} k_1 + \gamma_{11} & \gamma_{12} & \gamma_{13} \\ \gamma_{21} & k_2 + \gamma_{22} & \gamma_{23} \\ \gamma_{31} & \gamma_{32} & k_3 + \gamma_{33} \end{bmatrix}.$$

The singular values $\widetilde{k}_1$, $\widetilde{k}_2$, $\widetilde{k}_3$ are found from the canonical polar representation

$$\widetilde{C} = U \begin{bmatrix} \widetilde{k}_1 & 0 & 0 \\ 0 & \widetilde{k}_2 & 0 \\ 0 & 0 & \widetilde{k}_3 \end{bmatrix} V$$

and can be found by the formulas

$$\begin{aligned} \widetilde{k}_1 &= k_1 + \gamma_{11} + \frac{1}{4}\frac{(\gamma_{12}+\gamma_{21})^2}{k_1 - k_2} + \frac{1}{4}\frac{(\gamma_{12}-\gamma_{21})^2}{k_1 + k_2} + \frac{1}{4}\frac{(\gamma_{13}+\gamma_{31})^2}{k_1 - k_3} \\ &+ \frac{1}{4}\frac{(\gamma_{13}-\gamma_{31})^2}{k_1 + k_3} + \dots, \\ \widetilde{k}_2 &= k_2 + \gamma_{22} + \frac{1}{4}\frac{(\gamma_{21}+\gamma_{12})^2}{k_2 - k_1} + \frac{1}{4}\frac{(\gamma_{21}-\gamma_{12})^2}{k_2 + k_1} + \frac{1}{4}\frac{(\gamma_{23}+\gamma_{32})^2}{k_2 - k_3} \\ &+ \frac{1}{4}\frac{(\gamma_{23}-\gamma_{32})^2}{k_2 + k_3} + \dots, \\ \widetilde{k}_3 &= k_3 + \gamma_{33} + \frac{1}{4}\frac{(\gamma_{31}+\gamma_{13})^2}{k_3 - k_1} + \frac{1}{4}\frac{(\gamma_{31}-\gamma_{13})^2}{k_3 + k_1} + \frac{1}{4}\frac{(\gamma_{32}+\gamma_{23})^2}{k_3 - k_2} \\ &+ \frac{1}{4}\frac{(\gamma_{32}-\gamma_{23})^2}{k_3 + k_2} + \dots . \end{aligned} \tag{18.24}$$

We prove formulas (18.24) later. Now, we use these formulas to study the dependence of $\mathcal{E}$ on $\widetilde{c}_{ij} = k_i\delta_{ij} + \gamma_{ij}$.

It is easy to see that

$$\mathcal{E}(\widetilde{c}_{11}, \widetilde{c}_{12}, \dots, \widetilde{c}_{33}, S) = E(\widetilde{k}_1, \widetilde{k}_2, \widetilde{k}_3, S) = E(k_1, k_2, k_3, S) + \gamma_{11}E_{k_1} + \gamma_{22}E_{k_2}$$
$$+ \gamma_{33}E_{k_3} + \frac{E_{k_1} - E_{k_2}}{k_1 - k_2}\left(\frac{\gamma_{12}+\gamma_{21}}{2}\right)^2 + \frac{E_{k_1} + E_{k_2}}{k_1 + k_2}\left(\frac{\gamma_{12}-\gamma_{21}}{2}\right)^2$$
$$+ \frac{E_{k_2} - E_{k_3}}{k_2 - k_3}\left(\frac{\gamma_{23}+\gamma_{32}}{2}\right)^2 + \frac{E_{k_2} + E_{k_3}}{k_2 + k_3}\left(\frac{\gamma_{23}-\gamma_{32}}{2}\right)^2$$
$$+ \frac{E_{k_3} - E_{k_1}}{k_3 - k_1}\left(\frac{\gamma_{31}+\gamma_{13}}{2}\right)^2 + \frac{E_{k_3} + E_{k_1}}{k_3 + k_1}\left(\frac{\gamma_{31}-\gamma_{13}}{2}\right)^2 + \frac{1}{2}\sum_{i,j} E_{k_ik_j}\gamma_{ii}\gamma_{jj} + \dots .$$

The quadratic terms form a quadratic form in γ_{ij} equal to $\frac{1}{2}\mathcal{E}_{c_{pq}c_{nm}}\gamma_{pq}\gamma_{nm}$, where $c_{pq} = k_p\delta_{pq}$. Thus,

$$\mathcal{E}_{c_{pq}c_{nm}}\gamma_{pq}\gamma_{nm} = \sum_{i,j} E_{k_ik_j}\gamma_{ii}\gamma_{jj} + \frac{1}{2}\frac{E_{k_1} - E_{k_2}}{k_1 - k_2}(\gamma_{12}+\gamma_{21})^2$$
$$+ \frac{1}{2}\frac{E_{k_1} + E_{k_2}}{k_1 + k_2}(\gamma_{12}-\gamma_{21})^2 + \frac{1}{2}\frac{E_{k_2} - E_{k_3}}{k_2 - k_3}(\gamma_{23}+\gamma_{32})^2$$
$$+ \frac{1}{2}\frac{E_{k_2} + E_{k_3}}{k_2 + k_3}(\gamma_{23}-\gamma_{32})^2 + \frac{1}{2}\frac{E_{k_3} - E_{k_1}}{k_3 - k_1}(\gamma_{31}+\gamma_{13})^2$$
$$+ \frac{1}{2}\frac{E_{k_3} + E_{k_1}}{k_3 + k_1}(\gamma_{31}-\gamma_{13})^2. \tag{18.25}$$

We recall that the hyperbolicity condition (18.21) of the system (18.17) means that $\mathcal{E}_{c_{pq}c_{nm}}\xi_q\eta_p\xi_m\eta_n$ is positive for any nonzero real vectors (ξ_1, ξ_2, ξ_3) and (η_1, η_2, η_3). For example, if $\xi_1 = 1$, $\xi_2 = 0$, $\xi_3 = 0$, $\eta_1 = 0$, $\eta_2 = 1$, $\eta_3 = 0$, then $\gamma_{ii} = \xi_i\eta_i = 0$, $\gamma_{12} + \gamma_{21} = \gamma_{12} - \gamma_{21} = 1$, $\gamma_{13} + \gamma_{31} = \gamma_{13} - \gamma_{31} = 0$, $\gamma_{23} + \gamma_{32} = \gamma_{23} - \gamma_{32} = 0$, which implies

$$\mathcal{E}_{c_{pq}c_{nm}}\xi_q\eta_p\xi_m\eta_n = \mathcal{E}_{c_{pq}c_{nm}}\gamma_{pq}\gamma_{nm} = \frac{1}{2}\frac{E_{k_1} - E_{k_2}}{k_1 - k_2} + \frac{1}{2}\frac{E_{k_1} + E_{k_2}}{k_1 + k_2} > 0. \tag{18.26}$$

Thus, the following inequality is a necessary condition for hyperbolicity:

$$\frac{1}{2}\frac{E_{k_1} - E_{k_2}}{k_1 - k_2} + \frac{1}{2}\frac{E_{k_1} + E_{k_2}}{k_1 + k_2} > 0. \tag{18.27}$$

Similarly, we establish the necessity of the inequalities

$$\frac{1}{2}\frac{E_{k_2} - E_{k_3}}{k_2 - k_3} + \frac{1}{2}\frac{E_{k_2} + E_{k_3}}{k_2 + k_3} > 0, \tag{18.28}$$
$$\frac{1}{2}\frac{E_{k_3} - E_{k_1}}{k_3 - k_1} + \frac{1}{2}\frac{E_{k_3} + E_{k_1}}{k_3 + k_1} > 0. \tag{18.29}$$

• *In Section 9, the inequalities* (18.27)–(18.29) *were postulated in order to provide the entropy condition for a medium with relaxation of shear stresses. Here, these conditions appear as a necessary condition for the system of nonlinear elasticity to be hyperbolic.*

18.4. We justify the decomposition of the singular values $\widetilde{k}_i$ of the matrix

$$\widetilde{C} = \begin{bmatrix} \widetilde{c}_{11} & \widetilde{c}_{12} & \widetilde{c}_{13} \\ \widetilde{c}_{21} & \widetilde{c}_{22} & \widetilde{c}_{23} \\ \widetilde{c}_{31} & \widetilde{c}_{32} & \widetilde{c}_{33} \end{bmatrix} = \begin{bmatrix} k_1 + \gamma_{11} & \gamma_{12} & \gamma_{13} \\ \gamma_{21} & k_2 + \gamma_{22} & \gamma_{23} \\ \gamma_{31} & \gamma_{32} & k_3 + \gamma_{33} \end{bmatrix}. \tag{18.30}$$

The squared singular values are roots of the characteristic equation $\det \|\widetilde{C}\widetilde{C}^* - k^2 I\| = \det \|H - \widetilde{k}^2 I\| = 0$. The entries h_{ik} of the matrix $\widetilde{C}\widetilde{C}^*$ are found by the formulas

$$\begin{aligned} h_{11} &= k_1^2 + 2\gamma_{11}k_1 + \gamma_{11}^2 + \gamma_{12}^2 + \gamma_{13}^2, \\ h_{22} &= k_2^2 + 2\gamma_{22}k_2 + \gamma_{21}^2 + \gamma_{22}^2 + \gamma_{23}^2, \\ h_{33} &= k_3^2 + 2\gamma_{33}k_3 + \gamma_{31}^2 + \gamma_{32}^2 + \gamma_{33}^2, \\ h_{32} &= h_{23} = \gamma_{32}k_2 + \gamma_{23}k_3 + \dots, \\ h_{31} &= h_{13} = \gamma_{31}k_1 + \gamma_{13}k_3 + \dots, \\ h_{12} &= h_{21} = \gamma_{12}k_2 + \gamma_{21}k_1 + \dots, \end{aligned}$$

where the dots denote terms containing the second-order terms with respect to γ_{pq}. As we will see below, these terms are not required in the expansion of $\widetilde{k}_i$ with respect to γ_{pq} since we will neglect third-order terms.

We consider an auxiliary problem about eigenvalues of a symmetric matrix with small nondiagonal entries ξ, η, ζ and distinct nonzero diagonal entries p, q, r

$$\begin{aligned} 0 &= \begin{vmatrix} p-\lambda & \zeta & \eta \\ \zeta & q-\lambda & \xi \\ \eta & \xi & r-\lambda \end{vmatrix} \\ &= (p-\lambda)(q-\lambda)(r-\lambda) - [(p-\lambda)\xi^2 + (q-\lambda)\eta^2 + (r-\lambda)\zeta^2] + 2\xi\eta\zeta. \end{aligned}$$

It is easy to see that the roots of this characteristic equation satisfy the approximate formulas

$$\begin{aligned} \lambda_1 &= p + \frac{\eta^2}{p-r} + \frac{\zeta^2}{p-q} + O([\xi^2+\eta^2+\zeta^2]^{3/2}), \\ \lambda_2 &= q + \frac{\xi^2}{q-r} + \frac{\zeta^2}{q-p} + O([\xi^2+\eta^2+\zeta^2]^{3/2}), \\ \lambda_3 &= r + \frac{\xi^2}{r-q} + \frac{\eta^2}{r-p} + O([\xi^2+\eta^2+\zeta^2]^{3/2}). \end{aligned}$$

We will use these formulas in order to compute the eigenvalues $\widetilde{k}_i^2$ of the matrix $H = ||h_{ik}||$, by setting

$$p = h_{11} = k_1^2 + 2\gamma_{11}k_1 + \gamma_{11}^2 + \gamma_{12}^2 + \gamma_{13}^2 = (k_1 + \gamma_{11})^2 + \gamma_{12}^2 + \gamma_{13}^2,$$
$$q = h_{22} = k_2^2 + 2\gamma_{22}k_2 + \gamma_{21}^2 + \gamma_{22}^2 + \gamma_{23}^2 = (k_2 + \gamma_{22})^2 + \gamma_{21}^2 + \gamma_{23}^2,$$
$$r = h_{33} = k_3^2 + 2\gamma_{33}k_3 + \gamma_{31}^2 + \gamma_{32}^2 + \gamma_{33}^2 = (k_3 + \gamma_{33})^2 + \gamma_{31}^2 + \gamma_{32}^2,$$

$$\xi = h_{32} = h_{23} = \gamma_{32}k_2 + \gamma_{23}k_3 + \dots,$$
$$\eta = h_{31} = h_{13} = \gamma_{31}k_1 + \gamma_{13}k_3 + \dots,$$
$$\zeta = h_{12} = h_{21} = \gamma_{12}k_2 + \gamma_{21}k_1 + \dots.$$

Then

$$\lambda_1 = \widetilde{k}_1^2 = (k_1 + \gamma_{11})^2 + \gamma_{12}^2 + \gamma_{13}^2 + \frac{(\gamma_{21}k_1 + \gamma_{12}k_2)^2}{k_1^2 - k_2^2} + \frac{(\gamma_{31}k_1 + \gamma_{13}k_3)^2}{k_1^2 - k_3^2} + \dots,$$
$$\lambda_2 = \widetilde{k}_2^2 = (k_2 + \gamma_{22})^2 + \gamma_{21}^2 + \gamma_{23}^2 + \frac{(\gamma_{32}k_2 + \gamma_{23}k_3)^2}{k_2^2 - k_3^2} + \frac{(\gamma_{21}k_1 + \gamma_{12}k_2)^2}{k_2^2 - k_1^2} + \dots,$$
$$\lambda_3 = \widetilde{k}_3^2 = (k_3 + \gamma_{33})^2 + \gamma_{31}^2 + \gamma_{32}^2 + \frac{(\gamma_{31}k_1 + \gamma_{13}k_3)^2}{k_3^2 - k_1^2} + \frac{(\gamma_{32}k_2 + \gamma_{23}k_3)^2}{k_3^2 - k_2^2} + \dots.$$

We also use the approximate formula for taking the root

$$\sqrt{(k + \varepsilon_1)^2 + \varepsilon_2} = k + \varepsilon_1 + \frac{\varepsilon_2}{2k} + \dots,$$

where ε_1 is of order 1 and ε_2 is of order 2. Then

$$\begin{aligned}
\widetilde{k}_1 &= k_1 + \gamma_{11} + \frac{\gamma_{12}^2 + \gamma_{13}^2}{2k_1} + \frac{(\gamma_{21}k_1 + \gamma_{12}k_2)^2}{2k_1(k_1^2 - k_2^2)} + \frac{(\gamma_{31}k_1 + \gamma_{13}k_3)^2}{2k_1(k_1^2 - k_3^2)} + \dots \\
&= k_1 + \gamma_{11} + \frac{(k_1^2 - k_2^2)\gamma_{12}^2 + (\gamma_{21}k_1 + \gamma_{12}k_2)^2}{2k_1(k_1^2 - k_2^2)} \\
&\quad + \frac{(k_1^2 - k_3^2)\gamma_{13}^2 + (\gamma_{31}k_1 + \gamma_{13}k_3)^2}{2k_1(k_1^2 - k_3^2)} + \dots \\
&= k_1 + \gamma_{11} + \frac{k_1^2\gamma_{21}^2 + 2k_1k_2\gamma_{12}\gamma_{21} + k_1^2\gamma_{12}^2}{2k_1(k_1^2 - k_2^2)} + \frac{k_1^2\gamma_{31}^2 + 2k_1k_3\gamma_{13}\gamma_{31} + k_1^2\gamma_{13}^2}{2k_1(k_1^2 - k_3^2)} + \dots \\
&= k_1 + \gamma_{11} + \frac{k_1^2\gamma_{21}^2 + 2k_1k_2\gamma_{12}\gamma_{21} + k_1^2\gamma_{12}^2}{2k_1(k_1 + k_2)(k_1 - k_2)} + \frac{k_1^2\gamma_{31}^2 + 2k_1k_3\gamma_{13}\gamma_{31} + k_1^2\gamma_{13}^2}{2k_1(k_1 + k_3)(k_1 - k_3)} + \dots \\
&= k_1 + \gamma_{11} + \frac{(k_1 + k_2)(\gamma_{21} + \gamma_{12})^2 + (k_1 - k_2)(\gamma_{21} - \gamma_{12})^2}{4(k_1 + k_2)(k_1 - k_2)} \\
&\quad + \frac{(k_1 + k_3)(\gamma_{31} + \gamma_{13})^2 + (k_1 - k_3)(\gamma_{31} - \gamma_{13})^2}{4(k_1 + k_3)(k_1 - k_3)} + \dots \\
&= k_1 + \gamma_{11} + \frac{1}{4}\frac{(\gamma_{12} + \gamma_{21})^2}{k_1 - k_2} + \frac{1}{4}\frac{(\gamma_{12} - \gamma_{21})^2}{k_1 + k_2} + \frac{1}{4}\frac{(\gamma_{13} + \gamma_{31})^2}{k_1 - k_3} + \frac{1}{4}\frac{(\gamma_{13} - \gamma_{31})^2}{k_1 + k_3} + \dots
\end{aligned}$$

Thus, we have justified the decomposition

$$\widetilde{k}_1 = k_1 + \gamma_{11} + \frac{1}{4}\frac{(\gamma_{12}+\gamma_{21})^2}{k_1-k_2} + \frac{1}{4}\frac{(\gamma_{12}-\gamma_{21})^2}{k_1+k_2} + \frac{1}{4}\frac{(\gamma_{13}+\gamma_{31})^2}{k_1-k_3} + \frac{1}{4}\frac{(\gamma_{13}-\gamma_{31})^2}{k_1+k_3} + \ldots .$$

The arguments for $\widetilde{k}_2$ and $\widetilde{k}_3$ are the same.

§ 19. Equations of Linear Elasticity as Symmetric Hyperbolic Equations

We continue to study the well-posedness of systems of equations of the theory of elasticity. In this section, we focus on the simplest case of linear equations corresponding to the case of small deformations.

19.1. We will consider only isentropic (or isothermal) processes where the stress tensor $\|\sigma_{ij}\|$ is connected with the strain tensor $\|\varepsilon_{ij}\|$ by Hook's law (cf. Section 7):

$$\sigma_{ij} = \lambda\delta_{ij}(\varepsilon_{11}+\varepsilon_{22}+\varepsilon_{33}) + 2\mu\varepsilon_{ij}. \tag{19.1}$$

The equations (cf. Section 8)

$$\frac{\partial g_{ij}}{\partial t} + u_k\frac{\partial g_{ij}}{\partial x_k} + g_{ir}\frac{\partial u_r}{\partial x_j} + g_{rj}\frac{\partial u_r}{\partial x_i} = 0 \tag{19.2}$$

for the metric deformation tensor $\|g_{ij}\| = \|\delta_{ij} - 2\varepsilon_{ij}\|$ in terms of ε_{ij} are written as follows:

$$\frac{\partial \varepsilon_{ij}}{\partial t} + \underline{u_k}\frac{\partial \varepsilon_{ij}}{\partial x_k} - \frac{1}{2}(\delta_{ir} - 2\underline{\varepsilon_{ir}})\frac{\partial u_r}{\partial x_j} - \frac{1}{2}(\delta_{rj} - 2\underline{\varepsilon_{rj}})\frac{\partial u_r}{\partial x_i} = 0. \tag{19.3}$$

Since u_k and ε_{ir} are small, we can neglect the underlined small terms. Then we obtain the equations of linear elasticity

$$\frac{\partial \varepsilon_{ij}}{\partial t} - \frac{1}{2}\left(\frac{\partial u_i}{\partial x_j} + \frac{\partial u_j}{\partial x_i}\right) = 0. \tag{19.4}$$

In the same way, instead of the equations (cf. Section 15)

$$[\rho_0 + \underline{(\rho-\rho_0)}]\left(\frac{\partial u_i}{\partial t} + \underline{u_k}\frac{\partial u_i}{\partial x_k}\right) - \frac{\partial \sigma_{ij}}{\partial x_j} = 0, \tag{19.5}$$

we will consider the equations

$$\rho_0\frac{\partial u_i}{\partial t} - \frac{\partial \sigma_{ij}}{\partial x_j} = 0 \tag{19.6}$$

eliminating the underlined small terms in (19.5).

19.2. Thus, we study the system

$$\begin{aligned} &\frac{\partial \varepsilon_{ij}}{\partial t} - \frac{1}{2}\left(\frac{\partial u_i}{\partial x_j} + \frac{\partial u_j}{\partial x_i}\right) = 0, \\ &\rho_0 \frac{\partial u_i}{\partial t} - \frac{\partial \sigma_{ij}}{\partial x_j} = 0, \\ &\sigma_{ij} = \lambda\delta_{ij}(\varepsilon_{11} + \varepsilon_{22} + \varepsilon_{33}) + 2\mu\varepsilon_{ij}. \end{aligned} \tag{19.7}$$

Our goal is to find conditions on λ and μ under which the system (19.7) is hyperbolic. We show that the hyperbolilcity conditions are expressed as the inequalities $3\lambda+2\mu > 0$, $\mu > 0$ and consider a method for reducing this system to a symmetric hyperbolic system. Simultaneously, we develop the symmetrization technique which will be also used for equations of gas dynamics in Section 20.

19.3. As is known from the theory of partial differential equations, from symmetric systems we can obtain an equation expressing a conservation law, the energy integral. In the case of linear isentropic elasticity, we obtain the energy conservation law (in the isothermal case, the conservation of free energy).

For an elastic medium the specific energy is the sum of the specific kinetic energy

$$\rho_0 \frac{u_1^2 + u_2^2 + u_3^2}{2} = \rho_0 \frac{u_k u_k}{2}$$

and the specific internal energy

$$\rho_0 E = \frac{\lambda}{2}(\varepsilon_{11} + \varepsilon_{22} + \varepsilon_{33})^2 + \mu\varepsilon_{ij}\varepsilon_{ji}.$$

If λ and μ are isothermal moduli (cf. Section 7), then the energy is said to be *free*.

It is natural to expect that the conservation law has the differential form

$$\frac{\partial}{\partial t}\left[\rho_0 E + \rho_0 \frac{u_1^2 + u_2^2 + u_3^2}{2}\right] + \frac{\partial \Pi_1}{\partial x_1} + \frac{\partial \Pi_2}{\partial x_2} + \frac{\partial \Pi_3}{\partial x_3} = 0. \tag{19.8}$$

We can derive (19.8) from the above system. In particular, to find an expression for Π_i, we multiply (19.6) by a factor q_i and (19.4) by a factor r_{ij} and add the products. But we should clarify if the resulting relation corresponds to the required conservation law.

Recall that, in the case of linear elasticity (cf. Section 7),

$$\sigma_{jk} = \rho_0 \frac{\partial E}{\partial \varepsilon_{jk}} = \frac{\partial}{\partial \varepsilon_{jk}}(\rho_0 E). \tag{19.9}$$

Therefore, $d(\rho_0 E) = \sigma_{jk} d\varepsilon_{jk}$. Furthermore,

$$d\left(\rho_0 \frac{u_1^2 + u_2^2 + u_3^2}{2}\right) = \rho_0 u_i \, du_i. \tag{19.10}$$

Hence

$$\frac{\partial}{\partial t}\left[\rho_0 E + \rho_0 \frac{u_i u_i}{2}\right] = u_i \frac{\partial(\rho_0 u_i)}{\partial t} + \sigma_{jk} \frac{\partial \varepsilon_{jk}}{\partial t}. \tag{19.11}$$

This equality shows that q_i and r_{jk} should have the form $q_i = u_i$ and $r_{jk} = \sigma_{jk}$. Consider the sum

$$\begin{aligned} & q_i \left[\frac{\partial(\rho_0 u_i)}{\partial t} - \frac{\partial \sigma_{jk}}{\partial x_k}\right] + r_{jl}\left[\frac{\partial \varepsilon_{jl}}{\partial t} - \frac{1}{2}\left(\frac{\partial u_j}{\partial x_l} + \frac{\partial u_l}{\partial x_j}\right)\right] \\ &= u_i \left[\frac{\partial(\rho_0 u_i)}{\partial t} - \frac{\partial \sigma_{ik}}{\partial x_k}\right] + \sigma_{jl}\left[\frac{\partial \varepsilon_{jl}}{\partial t} - \frac{1}{2}\left(\frac{\partial u_j}{\partial x_l} + \frac{\partial u_l}{\partial x_j}\right)\right] \\ &= \frac{\partial[\rho_0 E + \rho_0 u_i u_i/2]}{\partial t} - u_i \frac{\partial \sigma_{ik}}{\partial x_k} - \frac{1}{2}\sigma_{jl} \frac{\partial u_j}{\partial x_l} - \frac{1}{2}\sigma_{jl} \frac{\partial u_l}{\partial x_j} \\ &= \frac{\partial[\rho_0 E + \rho_0 u_i u_i/2]}{\partial t} - u_i \frac{\partial \sigma_{ik}}{\partial x_k} - \sigma_{ik} \frac{\partial u_i}{\partial x_k} \\ &= \frac{\partial[\rho_0 E + \rho_0 u_i u_i/2]}{\partial t} - \frac{\partial(u_i \sigma_{ik})}{\partial x_k} = 0. \end{aligned}$$

Setting

$$\Pi_k = -u_i \sigma_{ik} = -u_1 \sigma_{1k} - u_2 \sigma_{2k} - u_3 \sigma_{3k}, \tag{19.12}$$

we obtain the conservation law.

19.4. We show how to reduce the system (19.7) to a symmetric system. Starting with the equations

$$\begin{aligned} &\frac{\partial(\rho_0 u_i)}{\partial t} + \frac{\partial F_{ik}}{\partial x_k} = 0 \quad (F_{ik} = -\sigma_{ik}), \\ &\frac{\partial \varepsilon_{ij}}{\partial t} + \frac{\partial H_{ijk}}{\partial x_k} = 0 \quad \left(H_{ijk} = -\frac{1}{2}\delta_{ik} u_j - \frac{1}{2}\delta_{jk} u_i\right), \end{aligned} \tag{19.13}$$

we obtain the conservation law

$$\frac{\partial \Pi_0}{\partial t} + \frac{\partial \Pi_k}{\partial x_k} = 0, \tag{19.14}$$

where

$$d\Pi_0 = q_i d(\rho_0 u_i) + r_{ij} d\varepsilon_{ij}, \quad d\Pi_k = q_i dF_{ik} + r_{ij} dH_{ijk}. \tag{19.15}$$

Making a simple transformation, we can write

$$\begin{aligned} \rho_0 u_i dq_i + \varepsilon_{ij} dr_{ij} &= d(q_i u_i + r_{ij}\varepsilon_{ij} - \Pi_0) = dL^0, \\ F_{ik} dq_i + H_{ijk} dr_{ij} &= d(q_i F_{ik} + r_{ij} H_{ijk} - \Pi_k) = dL^k. \end{aligned} \tag{19.16}$$

From (19.16) we obtain the equalities

$$\rho_0 u_i = \frac{\partial L^0}{\partial q_i}, \quad F_{ik} = \frac{\partial L^k}{\partial q_i}, \quad \varepsilon_{ij} = \frac{\partial L^0}{\partial r_{ij}}, \quad H_{ijk} = \frac{\partial L^k}{\partial r_{ij}}. \tag{19.17}$$

By (19.17), we have

$$\left.\begin{aligned} \frac{\partial(\rho_0 u_i)}{\partial t} + \frac{\partial F_{ik}}{\partial x_k} = 0, \\ \frac{\partial \varepsilon_{ij}}{\partial t} + \frac{\partial H_{ijk}}{\partial x_k} = 0, \end{aligned}\right\} \to \left\{\begin{aligned} \frac{\partial L^0_{q_i}}{\partial t} + \frac{\partial L^k_{q_i}}{\partial x_k} = 0, \\ \frac{\partial L^0_{r_{ij}}}{\partial t} + \frac{\partial L^k_{r_{ij}}}{\partial x_k} = 0. \end{aligned}\right. \tag{19.18}$$

Multiplying the equations in (19.18) by q_i and r_{ij} respectively and adding the products, we arrive at the energy conservation law

$$\frac{\partial\left(q_i L^0_{q_i} + r_{ij} L^0_{r_{ij}} - L^0\right)}{\partial t} + \frac{\partial\left(q_i L^k_{q_i} + r_{ij} L^k_{r_{ij}} - L^k\right)}{\partial x_k} = 0. \tag{19.19}$$

Here, we used the identities ($L = L^0, L^1, L^2, L^3$)

$$\begin{aligned} q_i dL_{q_i} + r_{ij} dL_{r_{ij}} &= q_i dL_{q_i} + L_{q_i} dq_i + r_{ij} dL_{r_{ij}} + L_{r_{ij}} dr_{ij} - (L_{q_i} dq_i + L_{r_{ij}} dr_{ij}) \\ &= d(q_i L_{q_i}) + d(r_{ij} L_{r_{ij}}) - dL = d(q_i L_{q_i} + r_{ij} dL_{r_{ij}} - L). \end{aligned}$$

Here, F_{ij}, H_{ijk}, ε_{ij} are linear forms in u_i and σ_{ij}, i.e., in q_i and r_{ij}. Since Π_0 and Π_k are quadratic forms, we conclude that L^0 and L^k are quadratic forms in q_i and r_{ij}. The expressions $L^0_{q_i}$, $L^0_{r_{ij}}$, $L^k_{q_i}$, $L^k_{r_{ij}}$ are linear forms in the same variables. Let us show that the matrices of these linear forms are symmetric. For the sake of simplicity, we denote q_i and r_{ij} by q_i but increase the number of admissible values of i. We verify that the linear system of equations

$$\frac{\partial L^0_{q_i}}{\partial t} + \frac{\partial L^k_{q_i}}{\partial x_k} = 0 \tag{19.20}$$

is a symmetric hyperbolic system in the sense of Friedrichs. Indeed, (19.20) can be written in the form

$$L^0_{q_i q_j} \frac{\partial q_j}{\partial t} + L^k_{q_i q_j} \frac{\partial q_j}{\partial x_k} = 0. \tag{19.21}$$

The matrices of coefficients $L^0_{q_i q_j}$, $L^k_{q_i q_j}$ are constant because they are matrices of the second-order derivatives of quadratic forms. The symmetry property is obviously valid. To show hyperbolisity, we additionally require the positive definiteness of $L^0_{q_i q_j}$, i.e., as it follows from the Euler identity for homogeneous functions $q_i q_j L^0_{q_i q_j} = 2L^0$, L^0 is nonnegative and vanishes only if all the q_i's vanish.

We recall that L^0 is a homogeneous second-order function of q_i. Therefore, $q_i L^0_{q_i} = 2L^0$ and $q_i L^0_{q_i} - L^0 = L^0$. On the other hand,

$$q_i L^0_{q_i} - L^0 = \Pi_0 = \rho_0 E + \rho_0 \frac{u_1^2 + u_2^2 + u_3^2}{2}.$$

(In the previous notation, $q_i L^0_{q_i} + r_{ij} L^0_{r_{ij}} - L^0 = \rho_0 E + \rho_0 \dfrac{u_1^2+u_2^2+u_3^2}{2}$.) Thus, the positive definiteness of $\|L^0_{q_i q_j}\|$ follows from the positive definiteness of the energy regarded as a function of q_i. We recall that the notation q_i includes u_1, u_2, u_3 and all the components of the stress tensor σ_{ij}. The total energy is equal to the sum of the kinetic energy $\rho_0 \dfrac{u_1^2+u_2^2+u_3^2}{2}$ and the internal energy

$$\rho_0 E = \frac{\lambda}{2}(\varepsilon_{11}+\varepsilon_{22}+\varepsilon_{33})^2 + \mu(\varepsilon_{11}^2+\varepsilon_{22}^2+\varepsilon_{33}^2+2\varepsilon_{12}\varepsilon_{21}+2\varepsilon_{13}\varepsilon_{31}+2\varepsilon_{23}\varepsilon_{32}).$$

Using (19.1) and the expressions

$$\sigma_{11}+\sigma_{22}+\sigma_{33} = (3\lambda+2\mu)(\varepsilon_{11}+\varepsilon_{22}+\varepsilon_{33}),$$

$$\varepsilon_{11}+\varepsilon_{22}+\varepsilon_{33} = \frac{1}{3\lambda+2\mu}(\sigma_{11}+\sigma_{22}+\sigma_{33}),$$

$$\varepsilon_{ij} = \frac{1}{2\mu}\left[\sigma_{ij} - \frac{\lambda}{3\lambda+2\mu}\delta_{ij}(\sigma_{11}+\sigma_{22}+\sigma_{33})\right],$$

$$\varepsilon_{ij} - \frac{1}{3}\delta_{ij}(\varepsilon_{11}+\varepsilon_{22}+\varepsilon_{33}) = \frac{1}{2\mu}\left[\sigma_{ij} - \frac{1}{3}\delta_{ij}(\sigma_{11}+\sigma_{22}+\sigma_{33})\right],$$

we write $\rho_0 E$ in the form

$$\begin{aligned}
\rho_0 E &= \frac{\lambda}{2}(\varepsilon_{11}+\varepsilon_{22}+\varepsilon_{33})^2 + \mu(\varepsilon_{11}^2+\varepsilon_{22}^2+\varepsilon_{33}^2+2\varepsilon_{12}\varepsilon_{21}\\
&+ 2\varepsilon_{13}\varepsilon_{31}+2\varepsilon_{23}\varepsilon_{32}) = \left(\frac{\lambda}{2}+\frac{\mu}{3}\right)(\varepsilon_{11}+\varepsilon_{22}+\varepsilon_{33})^2\\
&+ \mu\left[\left(\varepsilon_{11} - \frac{\varepsilon_{11}+\varepsilon_{22}+\varepsilon_{33}}{3}\right)^2 + \left(\varepsilon_{22} - \frac{\varepsilon_{11}+\varepsilon_{22}+\varepsilon_{33}}{3}\right)^2\right.\\
&+ \left.\left(\varepsilon_{33} - \frac{\varepsilon_{11}+\varepsilon_{22}+\varepsilon_{33}}{3}\right)^2 + 2\varepsilon_{12}\varepsilon_{21}+2\varepsilon_{13}\varepsilon_{31}+2\varepsilon_{23}\varepsilon_{32}\right]\\
&= \frac{1}{6(3\lambda+2\mu)}(\sigma_{11}+\sigma_{22}+\sigma_{33})^2 + \frac{1}{4\mu}\left[\left(\sigma_{11} - \frac{\sigma_{11}+\sigma_{22}+\sigma_{33}}{3}\right)^2\right.\\
&+ \left(\sigma_{22} - \frac{\sigma_{11}+\sigma_{22}+\sigma_{33}}{3}\right)^2 + \left(\sigma_{33} - \frac{\sigma_{11}+\sigma_{22}+\sigma_{33}}{3}\right)^2\\
&\left. + 2\sigma_{12}\sigma_{21}+2\sigma_{13}\sigma_{31}+2\sigma_{23}\sigma_{32}\right].
\end{aligned}$$

Since $\sigma_{ik} = \sigma_{ki}$, the quadratic form $\rho_0 E$ is positive definite for $\lambda/2+\mu/3 > 0$, $\mu > 0$.

Since σ_{ik} should be equal to σ_{ki}, we need to make some modifications. We recall that the sum

$$q_i\left[\frac{\partial(\rho_0 u_i)}{\partial t}-\frac{\partial\sigma_{ik}}{\partial x_k}\right]+r_{jl}\left[\frac{\partial\varepsilon_{jl}}{\partial t}-\frac{1}{2}\left(\frac{\partial u_j}{\partial x_l}+\frac{\partial u_l}{\partial x_j}\right)\right]$$

contains the term

$$r_{12}\left[\frac{\partial\varepsilon_{12}}{\partial t}-\frac{1}{2}\left(\frac{\partial u_1}{\partial x_2}+\frac{\partial u_2}{\partial x_1}\right)\right]=\sigma_{12}\left[\frac{\partial\varepsilon_{12}}{\partial t}-\frac{1}{2}\left(\frac{\partial u_1}{\partial x_2}+\frac{\partial u_2}{\partial x_1}\right)\right],$$

as well as the equal term

$$r_{21}\left[\frac{\partial\varepsilon_{21}}{\partial t}-\frac{1}{2}\left(\frac{\partial u_1}{\partial x_2}+\frac{\partial u_2}{\partial x_1}\right)\right]=\sigma_{21}\left[\frac{\partial\varepsilon_{21}}{\partial t}-\frac{1}{2}\left(\frac{\partial u_1}{\partial x_2}+\frac{\partial u_2}{\partial x_1}\right)\right].$$

Since we identify σ_{12} with σ_{21} and consider σ_{ik} for $i\neq k$, we will deal with the single term

$$2\sigma_{21}\left[\frac{\partial\varepsilon_{21}}{\partial t}-\frac{1}{2}\left(\frac{\partial u_1}{\partial x_2}+\frac{\partial u_2}{\partial x_1}\right)\right]$$

obtained by the multiplication of the left-hand side of some of the equivalent equations

$$\frac{\partial\varepsilon_{12}}{\partial t}-\frac{1}{2}\left(\frac{\partial u_1}{\partial x_2}+\frac{\partial u_2}{\partial x_1}\right)=0,\quad \frac{\partial\varepsilon_{21}}{\partial t}-\frac{1}{2}\left(\frac{\partial u_2}{\partial x_1}+\frac{\partial u_1}{\partial x_2}\right)=0$$

by $\widetilde{r}_{12}=2r_{21}=2\sigma_{21}$. Now,

$$q_1=u_1,\quad q_2=u_2,\quad q_3=u_3,\quad q_4=\sigma_{11},\quad q_5=\sigma_{22},$$
$$q_6=\sigma_{33},\quad q_7=2\sigma_{21},\quad q_8=2\sigma_{31},\quad q_9=2\sigma_{32}.$$

The total energy L^0 is expressed in terms of q_i as follows:

$$\begin{aligned}L^0&=\rho_0E+\rho_0\frac{u_1^2+u_2^2+u_3^2}{2}=\rho_0\frac{q_1^2+q_2^2+q_3^2}{2}\\&+\frac{1}{6(3\lambda+2\mu)}(q_4+q_5+q_6)^2+\frac{1}{4\mu}\left[\left(q_4-\frac{q_4+q_5+q_6}{3}\right)^2\right.\\&\left.+\left(q_5-\frac{q_4+q_5+q_6}{3}\right)^2+\left(q_6-\frac{q_4+q_5+q_6}{3}\right)^2\right]+\frac{1}{8\mu}(q_7^2+q_8^2+q_9^2).\end{aligned}$$

It is obvious that this quadratic form is positive definite for $\rho_0>0$, $\lambda/2+\mu/3>0$, $\mu>0$. It is easy to see that if at least one sign in these inequalities changed, then L^0 takes negative values. For example, let $\lambda/2+\mu/3<0$. Setting $q_1=q_2=q_3=0$, $q_7=q_8=q_9=0$, $q_4=q_5=q_6=(3\lambda+2\mu)/3$, we see that the value of L^0 is equal to $\lambda/2+\mu/3$ and, consequently, is negative. The necessity of the inequalities $\rho_0>0$ and $\mu>0$ is obvious.

Thus, the system of linear elasticity becomes a symmetric hyperbolic system with respect to q_i if ρ_0, λ, μ satisfy certain inequalities. We will see below that such a choice of unknown functions is often used in the symmetrization procedure.

• The functions $L^0(q_1, q_2, \ldots)$, $L^1(q_1, q_2, \ldots)$, $L^2(q_1, q_2, \ldots)$, $L^3(q_1, q_2, \ldots)$ are called the *generators of the system* (19.20) (or (19.21)).

We note that the generators L^1, L^2, L^3 are the quadratic forms

$$\begin{aligned} L^1 &= \frac{1}{2}(2q_1q_4 + q_2q_7 + q_3q_8), \\ L^2 &= \frac{1}{2}(q_1q_7 + 2q_2q_5 + q_3q_9), \\ L^3 &= \frac{1}{2}(q_1q_5 + q_2q_9 + 2q_3q_6). \end{aligned} \tag{19.22}$$

Remark 19.1. Since we know how to choose variables in order to reduce the system (19.7) to the symmetric form, we can directly perform the symmetrization by using the formula

$$\varepsilon_{ij} = -\frac{\lambda}{2\mu(3\lambda + 2\mu)}\delta_{ij}(\sigma_{11} + \sigma_{22} + \sigma_{33}) + \frac{1}{2\mu}\sigma_{ij}$$

and passing to the variables u_1, u_2, u_3, σ_{11}, σ_{22}, σ_{33}, $2\sigma_{12}$, $2\sigma_{13}$, $2\sigma_{23}$. A verifiaction of this in detail is left to the reader as an exercise.

§ 20. Symmetrization of Equations of Gas Dynamics

Using the technique developed in Section 19, we show how to write the equations of gas dynamics in the form of a symmetric hyperbolic system.

20.1. We begin with the equations describing barotropic gas flow with the equation of state $E = E(\rho)$, $p = \rho^2 E_\rho(\rho)$:

$$\begin{aligned} &\frac{\partial \rho}{\partial t} + \frac{\partial(\rho u_j)}{\partial x_j} = 0, \\ &\frac{\partial(\rho u_i)}{\partial t} + \frac{\partial[\rho u_i u_j + \delta_{ij}\rho^2 E_\rho]}{\partial x_j} = 0. \end{aligned} \tag{20.1}$$

Here, we do not take into account body forces, viscous friction, heat transfer, and heat influx. We write the energy conservation law

$$\frac{\partial \rho(E + u_k u_k/2)}{\partial t} + \frac{\partial \rho(E + p/\rho + u_k u_k/2)u_j}{\partial x_j} = 0. \tag{20.2}$$

In the case of a barotropic gas, the energy conservation law holds only if the density and velocity are continuous functions, i.e., no shock waves occur.

We show that the smooth functions ρ and u_i satisfy the energy conservation law which, in our case, is obtained as a combination of the basic equations (20.1):

$$\frac{\partial\rho(E+u_k u_k/2)}{\partial t}+\frac{\partial\rho(E+p/\rho+u_k u_k/2)u_j}{\partial x_j}$$
$$\equiv q_0\left[\frac{\partial\rho}{\partial t}+\frac{\partial(\rho u_j)}{\partial x_j}\right]+q_i\left[\frac{\partial(\rho u_i)}{\partial t}+\frac{\partial(\rho u_i u_j+\delta_{ij}\rho^2E_\rho)}{\partial x_j}\right]. \tag{20.3}$$

To do this, we choose q_0, q_1, q_2, q_3 from the conditions

$$q_0 d\rho+q_i d(\rho u_i)=d\left[\rho\left(E+\frac{u_k u_k}{2}\right)\right], \tag{20.4}$$

$$q_0 d(\rho u_j)+q_i d(\rho u_i u_j+\delta_{ij}\rho^2E_\rho)=d\left[\rho u_j\left(E+\frac{p}{\rho}+\frac{u_k u_k}{2}\right)\right]. \tag{20.5}$$

In fact, to determine these factors, it suffices to consider only (20.4) and verify that they satisfy the remaining required conditions.

We write (20.4) in the form

$$d\left[\rho E+\frac{(\rho u_1)^2+(\rho u_2)^2+(\rho u_3)^2}{2\rho}\right]$$
$$=q_0 d\rho+q_1 d(\rho u_1)+q_2 d(\rho u_2)+q_3 d(\rho u_3). \tag{20.6}$$

It is easy to see that

$$q_0=\frac{\partial}{\partial\rho}\left[\rho E+\frac{(\rho u_1)^2+(\rho u_2)^2+(\rho u_3)^2}{2\rho}\right]\Bigg|_{\rho u_j=\text{const}}$$
$$=\frac{\partial(\rho E)}{\partial\rho}-\frac{(\rho u_1)^2+(\rho u_2)^2+(\rho u_3)^2}{2\rho^2}$$
$$=(\rho E_\rho+E)-\frac{u_1^2+u_2^2+u_3^2}{2}=E+\frac{p}{\rho}-\frac{u_1^2+u_2^2+u_3^2}{2},$$

$$q_1=\frac{\partial}{\partial(\rho u_1)}\left[\rho E+\frac{(\rho u_1)^2+(\rho u_2)^2+(\rho u_3)^2}{2\rho}\right]\Bigg|_{\rho,\rho u_2,\rho u_3=\text{const}}$$
$$=\frac{1}{2\rho}\cdot 2(\rho u_1)=u_1,$$

$$q_2=u_2,$$
$$q_3=u_3.$$

A direct elementary calculation shows that the remaining relations are also valid.

The equality (20.4) implies

$$-d\left[\rho\left(E+\frac{u_k u_k}{2}\right)-q_0\rho-q_i\rho u_i\right]=\rho dq_0+\rho u_i dq_i.$$

In other words,

$$-d\left[\rho\left(E+\frac{u_k u_k}{2}\right)-\left(E+\rho E_\rho-\frac{u_k u_k}{2}\right)\rho-u_i\rho u_i\right]=\rho dq_0+\rho u_i dq_i,$$

$$d[\rho^2 E_\rho]=\rho dq_0+\rho u_i dq_i.$$

We set $p\equiv\rho^2 E_\rho=L^0(q_0,q_1,q_2,q_3)$. Then

$$\begin{aligned} dL^0 &= L^0_{q_0}dq_0+L^0_{q_1}dq_1+L^0_{q_2}dq_2+L^0_{q_3}dq_3\\ &=\rho dq_0+\rho u_1 dq_1+\rho u_2 dq_2+\rho u_3 dq_3. \end{aligned} \tag{20.7}$$

Similarly, (20.5) implies

$$-d\left[\rho u_j\left(E+\rho E_\rho+\frac{u_k u_k}{2}\right)-q_0\rho u_j-q_i(\rho u_i u_j+\delta_{ij}\rho^2 E_\rho)\right]$$
$$=\rho u_j dq_0+(\rho u_i u_j+\delta_{ij}\rho^2 E_\rho)dq_i,$$

which can be written as

$$d(u_j p)=d(u_j\rho^2 E_\rho)=\rho u_j dq_0+(\rho u_i u_j+\delta_{ij}\rho^2 E_\rho)dq_i.$$

20.2. We introduce $L^j=u_j\rho^2 E_\rho=u_j p$, $j=1,2,3$, and consider the relations

$$dL^j=\rho u_j dq_0+(\rho u_i u_j+\delta_{ij}\rho^2 E_\rho)dq_i. \tag{20.8}$$

• The functions L^0, L^1, L^2, L^3 of q_0, q_1, q_2, q_3 defined by the parametric equalities

$$\begin{aligned} &q_0=E+\rho E_\rho-(u_1^2+u_2^2+u_3^2)/2, \quad q_1=u_1,\\ &q_2=u_2, \quad q_3=u_3,\\ &L^0=\rho^2 E_\rho, \quad L^1=u_1\rho^2 E_\rho,\\ &L^2=u_2\rho^2 E_\rho, \quad L^3=u_3\rho^2 E_\rho. \end{aligned} \tag{20.9}$$

are called the *generating functions*.

We have

$$L^0_{q_0}=\rho, \quad L^j_{q_0}=\rho u_j, \qquad L^0_{q_i}=\rho u_i, \quad L^j_{q_i}=\rho u_j u_i+\delta_{ij}\rho^2 E_\rho, \quad j,i\geqslant 1. \tag{20.10}$$

Hence we can write the equations of barotropic gas in the form

$$\frac{\partial L^0_{q_\alpha}}{\partial t}+\frac{\partial L^j_{q_\alpha}}{\partial x_j}=0, \quad \alpha=0,1,2,3. \tag{20.11}$$

To obtain the energy conservation law, we multiply each equation by q_α and add the products:

$$q_\alpha\left(\frac{\partial L^0_{q_\alpha}}{\partial t}+\frac{\partial L^j_{q_\alpha}}{\partial x_j}\right)=0.$$

By the identities $q_\alpha dL^\beta_{q_\alpha}=d(q_\alpha L^\beta_{q_\alpha}-L^\beta)$, the energy conservation law takes the form

$$\frac{\partial(q_\alpha L^0_{q_\alpha})}{\partial t}+\frac{\partial(q_\alpha L^j_{q_\alpha}-L^j)}{\partial x_j}=0. \tag{20.12}$$

So, everything is the same as in the theory of linear elasticity. In particular, we can write the equations in the symmetric form

$$L^0_{q_\alpha q_\beta}\frac{\partial q_\beta}{\partial t}+L^j_{q_\alpha q_\beta}\frac{\partial q_\beta}{\partial x_j}=0, \tag{20.13}$$

but, in this case, the matrices of coefficients are not constant but depend on the solution. The obtained system is quasilinear.

20.3. Similarly, we can symmetrize the system of gas dynamics in the general case $E=E(\rho,S)$. In this case, the complete system of equations (body forces and heat influx are not taken into account) can be written as follows:

$$\frac{\partial\rho}{\partial t}+\frac{\partial(\rho u_j)}{\partial x_j}=0, \tag{20.14}$$

$$\frac{\partial\rho u_j}{\partial t}+\frac{\partial(\rho u_i u_j+\delta_{ij}\rho^2E_\rho)}{\partial x_j}=0,\quad i=1,2,3, \tag{20.15}$$

$$\frac{\partial[\rho(E+u_\alpha u_\alpha/2)]}{\partial t}+\frac{\partial[\rho u_j(E+\rho E_\rho+u_\alpha u_\alpha/2)]}{\partial x_j}=0. \tag{20.16}$$

In this case, the entropy conservation law

$$\frac{\partial(\rho S)}{\partial t}+\frac{\partial(\rho S u_j)}{\partial x_j}=0. \tag{20.17}$$

is obtained as a consequence of the basic equations (20.14)–(20.16). Indeed, multiplying (20.14)–(20.16) by the corresponding factors, we obtain an equation expressing the conservation of entropy.[1)] For this purpose, we use the differential

[1)] Recall that the equation for entropy was derived from the equations of continuity for momentum and energy in Section 15. We repeat the arguments of Section 15 with slight modifications.

relations

$$d(\rho S) = \frac{1}{E_S} d\left[\rho\left(E + \frac{u_\alpha u_\alpha}{2}\right)\right] - \frac{E + \rho E_\rho - S E_S - u_\alpha u_\alpha/2}{E_S} d\rho - \frac{u_i}{E_S} d(\rho u_i),$$

$$d(\rho S u_j) = \frac{1}{E_S} d\left[\rho u_j\left(E + \rho E_\rho + \frac{u_\alpha u_\alpha}{2}\right)\right] - \frac{E + \rho E_\rho - S E_S - u_\alpha u_\alpha/2}{E_S} d(\rho u_j)$$
$$- \frac{u_i}{E_S} d(\rho u_i u_j + \rho^2 E_\rho \delta_{ij}).$$

To verify them, we express both sides in terms of the differentials $d\rho$, dS, du_α and collect like terms. Introducing

$$q_0 = \frac{1}{E_S}, \quad q_1 = -\frac{u_1}{E_S}, \quad q_2 = -\frac{u_2}{E_S},$$
$$q_3 = -\frac{u_3}{E_S}, \quad q_4 = -\frac{E + \rho E_\rho - S E_S - u_\alpha u_\alpha/2}{E_S},$$

we find

$$q_0\left\{\frac{\partial[\rho(E + u_\alpha u_\alpha/2)]}{\partial t} + \frac{\partial[\rho u_j(E + \rho E_\rho + u_\alpha u_\alpha/2)]}{\partial x_j}\right\}$$
$$+ \sum_{i=1}^{3} q_i\left\{\frac{\partial(\rho u_i)}{\partial t} + \frac{\partial(\rho u_i u_j + \delta_{ij}\rho^2 E_\rho + u_\alpha u_\alpha/2)}{\partial x_j}\right\}$$
$$+ q_4\left\{\frac{\partial \rho}{\partial t} + \frac{\partial(\rho u_j)}{\partial x_j}\right\} \equiv \frac{\partial(\rho S)}{\partial t} + \frac{\partial(\rho S u_j)}{\partial x_j}.$$

As above, we set

$$L^0 = q_0\left[\rho\left(E + \frac{u_\alpha u_\alpha}{2}\right)\right] + \sum_{i=1}^{3} q_i \rho u_i + q_4 \rho - \rho S, \tag{20.18}$$

$$L^j = q_0\left[\rho u_j\left(E + \rho E_\rho + \frac{u_\alpha u_\alpha}{2}\right)\right]$$
$$+ \sum_{i=1}^{3} q_i\left(\rho u_i u_j + \delta_{ij}\rho^2 E_\rho\right) + q_4 \rho u_j - \rho S u_j. \tag{20.19}$$

Then

$$dL^0 = \rho\left(E + \frac{u_\alpha u_\alpha}{2}\right) dq_0 + \sum_{i=1}^{3} \rho u_i dq_i + \rho dq_4, \tag{20.20}$$

$$dL^j = \rho u_j\left(E + \rho E_\rho + \frac{u_\alpha u_\alpha}{2}\right) dq_0$$
$$+ \sum_{i=1}^{3}\left(\rho u_i u_j + \delta_{ij}\rho^2 E_\rho\right) dq_i + \rho u_j dq_4. \tag{20.21}$$

We see that the system admits the same form as (20.11):

$$\frac{\partial L^0_{q_\alpha}}{\partial t} + \frac{\partial L^j_{q_\alpha}}{\partial x_j} = 0. \tag{20.22}$$

We can simplify formulas for L^j by substituting for q_0, q_1, q_2, q_3, q_4 their expressions in terms of ρ, u_α, S and collecting like terms:

$$\begin{aligned} L^0 &= -\frac{\rho^2 E_\rho}{E_S} = -\frac{p}{T}, \\ L^j &= -u_j\,\frac{\rho^2 E_\rho}{E_S} = -u_j\,\frac{p}{T}. \end{aligned} \tag{20.23}$$

Remark 20.1. The symmetry of $L^0_{q_i q_k}$, $L^j_{q_i q_k}$ and the coefficients of the system (20.22) do not imply the hyperbolicity of the system. It is necessary to require that the matrix of coefficients $L^0_{q_i q_k}$ to be positive definite. In other words, the function $L^0 = L^0(q_1, q_2, \ldots, q_n)$ should be convex.

20.4. To obtain a hyperbolicity criterion for the equations of gas dynamics we need to establish some important properties of convex functions.

Let $L(q_1, q_2, \ldots, q_n)$ be a convex function of $q_1, q_2, \ldots, q_n$. Consider the function $x_i q_i - L(q_1, q_2, \ldots, q_n)$ depending on $x_1, x_2, \ldots, x_n$. Assume that this function attains the maximal value at some point $q_{10}, q_{20}, \ldots, q_{n0}$. This assumption is valid if $\frac{1}{t}\, L(q_1 + t\varkappa_1, q_2 + t\varkappa_2, \ldots, q_n + t\varkappa_n) \to \infty$ as $|t| \to \infty$ for any q_i, $\varkappa_i$. Indeed, $x_i q_i - L(q_1, q_2, \ldots, q_n)$ is a convex function tending to $-\infty$ along any direction. Hence it attains the maximum and it is a unique singular point of level curves.

As a rule, in applications, the variables q_i cannot run over the entire space. Therefore, the above arguments should be modified to the case where the variables are allowed to run over a bounded domain, which is possible in many practical cases.

Thus, we assume that

$$\max_{q_j} [x_i q_i - L(q_1, q_2, \ldots, q_n)] = H(x_1, x_2, \ldots, x_n).$$

We prove that $H(x_1, x_2, \ldots, x_n)$ is a convex function. Let $\alpha \geqslant 0$ and $\beta \geqslant 0$ be such that $\alpha + \beta = 1$. Then

$$\begin{aligned} \alpha H(x_1, x_2, \ldots, x_n) &= \max_{q_1, q_2, \ldots, q_n} [(\alpha x_i) q_i - \alpha L(q_1, q_2, \ldots, q_n]), \\ \beta H(y_1, y_2, \ldots, y_n) &= \max_{r_1, r_2, \ldots, r_n} [(\beta y_i) r_i - \beta L(r_1, r_2, \ldots, r_n)], \end{aligned}$$

$$\begin{aligned}
&\alpha H(x_1, x_2, \dots, x_n) + \beta H(y_1, y_2, \dots, y_n) \\
&\quad = \max_{q_j}[(\alpha x_i) q_i - \alpha L(q_1, q_2, \dots, q_n)] + \max_{r_j}[(\beta y_i) r_i - \beta L(r_1, r_2, \dots, r_n)] \\
&\quad \geqslant \max_{q_j}\{[(\alpha x_i) q_i - \alpha L(q_1, q_2, \dots, q_n)] + [(\beta y_i) q_i - \beta L(q_1, q_2, \dots, q_n)]\} \\
&\quad = \max_{q_j}[(\alpha x_i + \beta y_i) q_i - L(q_1, q_2, \dots, q_n)] \\
&\quad = H(\alpha x_1 + \beta y_1, \alpha x_2 + \beta y_2, \dots, \alpha x_n + \beta y_n).
\end{aligned}$$

Hence H is convex. We note that the strict convexity of L implies the strict convexity of H.

It is obvious that $x_i q_i - L(q_1, q_2, \dots, q_n) \leqslant H(x_1, x_2, \dots, x_n)$ for any $x_1, x_2, \dots, x_n$ and $q_1, q_2, \dots, q_n$.

- The inequality

$$x_i q_i \leqslant L(q_1, q_2, \dots, q_n) + H(x_1, x_2, \dots, x_n) \tag{20.24}$$

is called the *Young inequality*.

20.5. Let us show how to compute $H(x_1, x_2, \dots, x_n)$ from $L(q_1, q_2, \dots, q_n)$. For this purpose, we find q_j at which $x_i q_i - L(q_1, q_2, \dots, q_n)$ attains the maximal value. We have $\dfrac{\partial}{\partial q_j}[x_i q_i - L(q_1, q_2, \dots, q_n)] = 0$, i.e., $x_j = L_{q_j}$ and, consequently, $H(x_1, x_2, \dots, x_n) = q_i L_{q_i} - L(q_1, q_2, \dots, q_n)$. The Jacobian $\det \left\| \dfrac{\partial x_j}{\partial q_i} \right\| = \det \|L_{q_i q_j}\|$ does not vanish because L is convex. Consequently, $q_i \to x_j$ is a one-to-one transformation, called the *Legendre transform*.

We note that

$$dH(x_1, x_2, \dots, x_n) = d(q_i L_{q_i} - L) = q_i dL_{q_i} + L_{q_i} dq_i - L_{q_i} dq_i = q_i dL_{q_i} = q_i dx_i,$$

i.e., $H_{x_i} = q_i$. Therefore, the Legendre transform constructed from the function H is given by

$$x_i H_{x_i} - H = L_{q_i} q_i - (q_i L_{q_i} - L) = L(q_1, q_2, \dots, q_n), \quad H_{x_i} = q_i.$$

The Legendre transform of the Legendre transform yields the initial functions and variables. We have

$$\|L_{q_i q_j}\| = \left\| \frac{\partial x_i}{\partial q_j} \right\|, \quad \|H_{x_i x_j}\| = \left\| \frac{\partial q_i}{\partial x_j} \right\| = \left\| \frac{\partial x_j}{\partial q_i} \right\|^{-1} = \|L_{q_i q_j}\|^{-1}. \tag{20.25}$$

This formula shows that the convexity of $L(q_1, q_2, \dots, q_n)$ implies the convexity of $H(x_1, x_2, \dots, x_n)$.

Remark 20.2. Formulas (20.25) can be used locally if q_i run over a bounded domain. For example, let $E(V, S)$ be a convex function of V, S. It is clear that $V > 0$ since $V = 1/\rho$ is the specific volume in physical applications.

It is obvious that if the thermodynamical potential $E(V,S)$ is convex, then $E(V,S)+u_iu_i/2$ is convex with respect to the variables V, S, u_1, u_2, u_3. By the Legendre transform, $VE_V+SE_S+(u_iu_i)/2-E$ is a convex function of E_V, E_S, u_1, u_2, u_3. As a rule, $E_V=-p<0$ and $E_S=T>0$ in gas dynamics.

20.6. We continue to consider transformations of convex functions.

Let $L(q_1,q_2,\dots,q_n)$ be a convex function, and let $L_{q_0}<0$. We write the convexity condition

$$\begin{aligned}&L(\alpha\overline{q}_0+\beta\overline{\overline{q}}_0,\alpha\overline{q}_1+\beta\overline{\overline{q}}_1,\dots,\alpha\overline{q}_n+\beta\overline{\overline{q}}_n)\\&\leqslant\alpha L(\overline{q}_0,\dots,\overline{q}_n)+\beta L(\overline{\overline{q}}_0,\dots,\overline{\overline{q}}_n)=\alpha\overline{L}+\beta\overline{\overline{L}}.\end{aligned}\tag{20.26}$$

We find q_0 from the equation

$$L(q_0,\alpha\overline{q}_1+\beta\overline{\overline{q}}_1,\dots,\alpha\overline{q}_n+\beta\overline{\overline{q}}_n)=\alpha\overline{L}+\beta\overline{\overline{L}}.$$

Since $L_{q_0}<0$, we have

$$\begin{aligned}&q_0\leqslant\alpha\overline{q}_0+\beta\overline{\overline{q}}_0,\quad q_0(\alpha\overline{L}+\beta\overline{\overline{L}},\alpha\overline{q}_1+\beta\overline{\overline{q}}_1,\dots,\alpha\overline{q}_n+\beta\overline{\overline{q}}_n)\\&\leqslant\alpha q_0(\overline{L},\overline{q}_1,\dots,\overline{q}_n)+\beta q_0(\overline{\overline{L}},\overline{\overline{q}}_1,\dots,\overline{\overline{q}}_n).\end{aligned}$$

Thus, we have proved the following assertion.

• *If $q_0(L,q_1,q_2,\dots,q_n)$ satisfies the equation $L(q_0,q_1,q_2,\dots,q_n)=L$, then q_0 is convex.*

For example, from the convexity of $L=VE_V+SE_S+\dfrac{u_iu_i}{2}-E$ with respect to $q_0=-E_V$, $q_1=u_1$, $q_2=u_2$, $q_3=u_3$, $q_4=E_S$ and the inequality $L_{q_0}=-V<0$ it follows that $-E_V=p$ is a convex function of $E-VE_V-SE_S-(u_iu_i)/2$, u_1, u_2, u_3, $E_S=T$.

Let $L(q_1,q_2,\dots,q_n)$ be convex with respect to $q_0>0$ and $q_2,q_3,\dots,q_n$. Then $\mathcal{L}=\dfrac{1}{q_0}L(q_0,q_1,\dots,q_n)$ is convex with respect to $r_0=1/q_0$, $r_1=q_1/q_0,\dots,r_n=q_n/q_0$.

We use the inequality

$$\begin{aligned}&\alpha L(\overline{q}_0,\overline{q}_1,\dots,\overline{q}_n)+\beta L(\overline{\overline{q}}_0,\overline{\overline{q}}_1,\dots,\overline{\overline{q}}_n)\\&\qquad\geqslant L(\alpha\overline{q}_0+\beta\overline{\overline{q}}_0,\alpha\overline{q}_1+\beta\overline{\overline{q}}_1,\dots,\alpha\overline{q}_n+\beta\overline{\overline{q}}_n)\end{aligned}$$

which can be written in the form

$$\begin{aligned}&\frac{\alpha\overline{q}_0}{\alpha\overline{q}_0+\beta\overline{\overline{q}}_0}\,\frac{1}{\overline{q}_0}\,L(\overline{q}_0,\overline{q}_1,\dots,\overline{q}_n)+\frac{\beta\overline{\overline{q}}_0}{\alpha\overline{q}_0+\beta\overline{\overline{q}}_0}\,\frac{1}{\overline{\overline{q}}_0}\,L(\overline{\overline{q}}_0,\overline{\overline{q}}_1,\dots,\overline{\overline{q}}_n)\\&\geqslant\frac{1}{\alpha\overline{q}_0+\beta\overline{\overline{q}}_0}\,L(\alpha\overline{q}_0+\beta\overline{\overline{q}}_0,\alpha\overline{q}_1+\beta\overline{\overline{q}}_1,\dots,\alpha\overline{q}_n+\beta\overline{\overline{q}}_n).\end{aligned}$$

We note that

$$\frac{1}{\alpha\bar{q}_0+\beta\bar{\bar{q}}_0}=\frac{\alpha\bar{q}_0}{\alpha\bar{q}_0+\beta\bar{\bar{q}}_0}\frac{1}{\bar{q}_0}+\frac{\beta\bar{\bar{q}}_0}{\alpha\bar{q}_0+\beta\bar{\bar{q}}_0}\frac{1}{\bar{\bar{q}}_0},$$

$$\frac{\alpha\bar{q}_j+\beta\bar{\bar{q}}_j}{\alpha\bar{q}_0+\beta\bar{\bar{q}}_0}=\frac{\alpha\bar{q}_0}{\alpha\bar{q}_0+\beta\bar{\bar{q}}_0}\frac{\bar{q}_j}{\bar{q}_0}+\frac{\beta\bar{\bar{q}}_0}{\alpha\bar{q}_0+\beta\bar{\bar{q}}_0}\frac{\bar{\bar{q}}_j}{\bar{\bar{q}}_0}.$$

Introducing the notation

$$\frac{\alpha\bar{q}_0}{\alpha\bar{q}_0+\beta\bar{\bar{q}}_0}=\gamma,\qquad \frac{\beta\bar{\bar{q}}_0}{\alpha\bar{q}_0+\beta\bar{\bar{q}}_0}=\delta,$$

we can write

$$\gamma\mathcal{L}(\bar{r}_0,\bar{r}_1,\dots,\bar{r}_n)+\delta\mathcal{L}(\bar{\bar{r}}_0,\bar{\bar{r}}_1,\dots,\bar{\bar{r}}_n)\geqslant\mathcal{L}(\gamma\bar{r}_0+\delta\bar{\bar{r}}_0,\dots,\gamma\bar{r}_n+\delta\bar{\bar{r}}_n),$$

which means the convexity of $\mathcal{L}(r_0,r_1,\dots,r_n)$. It suffices to note that $\gamma\geqslant 0$, $\delta\geqslant 0$, $\gamma+\delta=1$ and, if α and β take all admissible value, γ and δ also run over the admissible domain.

We give an example of such a transformation. We know that $p=-E_V$ is a convex function of $E_S=T>0$, u_1, u_2, u_3, $E-VE_V-SE_S-u_iu_i/2$. Consequently, $p/T=-E_V/E_S$ is a convex function of the variables

$$\frac{1}{E_S}=\frac{1}{T},\frac{u_1}{T},\frac{u_2}{T},\frac{u_3}{T},\frac{E-VE_V-SE_S-(u_iu_i)/2}{E_S}=\frac{E+\rho E_\rho-TS-(u_iu_i)/2}{T}.$$

Each transformation is *involutive*, i.e., applying it twice, we return to the initial functions and variables. This allows us to make the following conclusion.

• *Let p/T be a convex function of $1/T$, u_i/T, $(E-\rho E_\rho-TS-u_iu_i/2)/T$. Transforming the functions and variables in the inverse order, we see that $E(V,S)+u_iu_i/2$ is convex with respect to V, S, u_1, u_2, u_3, i.e., the matrix*

$$\begin{bmatrix} E_{VV} & E_{VS} & 0 & 0 & 0\\ E_{SV} & E_{SS} & 0 & 0 & 0\\ 0 & 0 & 1 & 0 & 0\\ 0 & 0 & 0 & 1 & 0\\ 0 & 0 & 0 & 0 & 1\end{bmatrix}$$

is positive definite, which is equivalent to the inequalities $E_{VV}>0$ and $E_{VV}E_{SS}-E_{VS}^2>0$.

20.7. We recall that the equations of gas dynamics are written as

$$\frac{\partial L^0_{q_i}}{\partial t}+\frac{\partial L^j_{q_i}}{\partial x_j}=0,$$

where $L^0=-p/T$, $L^j=-(u_jp)/T$, $q_0=1/T$, $q_1=-u_1/T$, $q_2=-u_2/T$, $q_3=-u_3/T$, $q_4=-(E+\rho E_\rho-SE_S-u_iu_i/2)/T$. This system is symmetric and hyperbolic if $L^0(q_0, q_1, \dots, q_n)$ is convex, which happens if $E_{VV} > 0$ and $E_{VV}E_{SS} - E_{VS}^2 > 0$. It should be noted that these inequalities are sufficient for the well-posedness of the Cauchy problem, but they are not necessary. The necessary condition for the well-posedness is the inequality $E_{VV} > 0$. Our conditions are necessary and sufficient for the system obtained by the above symmetrization to be symmetric and hyperbolic.

§ 21. Influence of Small Viscosity on the Behavior of Solutions

In this section, we consider systems obtained from the hyperbolic systems of gas dynamics by adding some terms describing friction forces and heat transfer. It is natural to expect that the perturbations generated by small additional terms of the equations will be also small. But this is not so in general. We demonstrate this by simple model problems.

21.1. We consider the equations

$$\frac{\partial L^0_{q_i}}{\partial t} + \frac{\partial L^j_{q_i}}{\partial x_j} = \frac{\partial}{\partial x_j}\left(\varepsilon b^{jl}_{ik}\frac{\partial q_k}{\partial x_l}\right), \tag{21.1}$$

obtained by inserting dissipative terms with a small parameter ε into the right-hand sides of the hyperbolic system

$$\frac{\partial L^0_{q_i}}{\partial t} + \frac{\partial L^j_{q_i}}{\partial x_j} = 0. \tag{21.2}$$

In actual media, the matrix of dissipative coefficients is, as a rule, singular or positive semidefinite. In spite of this fact, we will assume that the matrix is strictly dissipative, i.e.,

$$\sum_{i,k,j,l} b^{jl}_{ik}\eta_{ij}\eta_{kl} \geqslant a^2 \sum_{m,n} \eta^2_{mn} \quad (a > 0). \tag{21.3}$$

This assumption allows us to suggest a simple method of analyzing properties of solutions and make important conclusions of qualitative character.

We consider a solution to (21.1) in the form of a plane wave that propagates along the x_1-axis. The gradients vanish along all the remaining directions, and we will not account them in our consideration. Introducing the notation $x_1 = x$, $L^0 = L$, and $L^1 = M$, we write (21.1) in the form

$$\frac{\partial L_{q_i}}{\partial t} + \frac{\partial M_{q_i}}{\partial x} = \frac{\partial}{\partial x}\left(\varepsilon b_{ik}\frac{\partial q_k}{\partial x}\right). \tag{21.4}$$

The solution can be written as

$$q_i = q_i\left(w^2 \frac{t - x/w}{\varepsilon}\right),$$

where w is the velocity and ε plays the role of a microscope, i.e., it stretches the wave zone up to size of order 1, which can be accessible to observation.

Introduce the variable

$$\zeta = \frac{w^2}{\varepsilon}\left(t - \frac{x}{w}\right)$$

and pass to the ordinary differential equations

$$\frac{w^2}{\varepsilon}(L_{q_i})' - \frac{w^2}{w\varepsilon}(M_{q_i})' = \frac{w^2}{w\varepsilon}\left(\varepsilon b_{ik}\frac{w^2}{w\varepsilon}q_k'\right)'$$

which can be written as

$$\left(L_{q_i} - \frac{1}{w}M_{q_i}\right)' = (b_{ik}q_k')'.$$

Integrating this equation, we find

$$L_{q_i} - \frac{1}{w}M_{q_i} - A_i = b_{ik}q_k',$$

where A_i are constants. Introducing the notation

$$\Lambda(q_1, q_2, \dots, q_n) = L - (1/w)M - A_i q_i, \quad D = (1/2)b_{ik}\beta_i\beta_k, \tag{21.5}$$

we write the ordinary differential equations in the following elegant form:

$$\Lambda_{q_i} = D_{\beta_i}, \quad \frac{dq_i}{d\zeta} = \beta_i, \tag{21.6}$$

which admits a geometrical interpretation. Namely, we can imagine that the space $(q_1, q_2, \dots, q_n)$ is divided into the level surfaces

$$\Lambda(q_1, q_2, \dots, q_n) = \text{const}. \tag{21.7}$$

Let $(\alpha_1, \alpha_2, \dots, \alpha_n)$ be a tangent vector at the point $(q_1, q_2, \dots, q_n)$ to the surface $\Lambda = \text{const}$ passing through this point. The vector $(\alpha_1, \alpha_2, \dots, \alpha_n)$ is orthogonal to the gradient of Λ, i.e., $\alpha_i\Lambda_{q_i} = 0$. Let $\beta_i = \dfrac{dq_i}{d\zeta}$ be the components of a tangent vector to the trajectory of the differential equation in (21.6). Since $\Lambda_{q_i} = D_{\beta_i}$ and $0 = \alpha_i D_{\beta_i} = b_{ik}\alpha_i\beta_k$, we see that the trajectories of (21.6) are orthogonal to the level surfaces $\Lambda(q_1, q_2, \dots, q_n) = \text{const}$ relative to the inner product $(\alpha, \beta) = b_{ik}\alpha_i\beta_k$. Moving along the trajectory toward to the direction in which the parameter ζ increases, the function Λ increases:

$$\frac{d\Lambda}{d\zeta} = \Lambda_{q_i}\frac{dq_i}{d\zeta} = D_{\beta_i}\beta_i = 2D > 0. \tag{21.8}$$

This follows from the positive definiteness of the matrix $||b_{ik}||$. We consider the solutions $q_i = q_i(\zeta)$ that tend to some limits $q_i(\pm\infty)$ as $\zeta \to \pm\infty$ such that $q_i'(\zeta) = \dfrac{dq_i}{d\zeta} = \Lambda_{q_i} \to 0$ as $\zeta \to \pm\infty$.

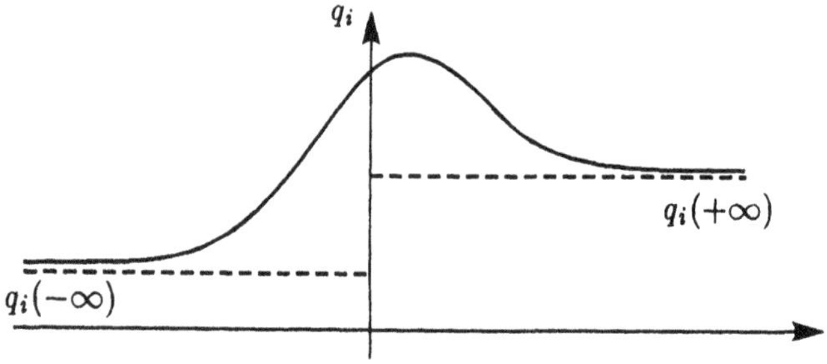

Fig. 22

The graph of such a solution is presented in Fig. 22. It is clear that $q_i(\pm\infty)$ satisfy the equalities $\Lambda_{q_j}(q_1, q_2, \dots, q_n) = 0$. Thus, $q_i(\pm\infty)$ are zeros of the gradient of Λ, i.e., they are singular points of the level surface $\Lambda = \text{const}$. These points are said to be *critical*.

Consider the "two-dimensional" case where there are only two variables q_i and the level surfaces $\Lambda = \text{const}$ have the form presented in Fig. 23. In this picture, the arrows indicate the direction along which Λ grows.

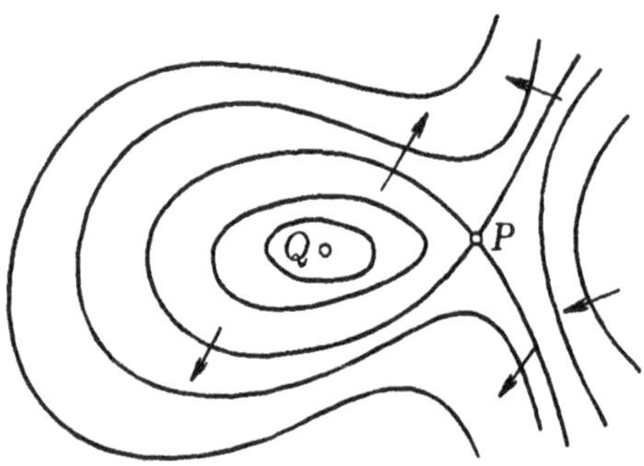

Fig. 23

The level curves and orthogonal trajectories in a neighborhood of a critical point P are presented in Fig. 24. The pictures show that only one curve entering the point P and starting in a domain surrounded by the closed level curve passing through P and containing other critical point Q. On the other hand, all trajectories that leave the domain, start at the point Q which is the minimum point of Λ (cf. Fig. 25). It is clear that there exists a unique trajectory of the required form: the trajectory joining P and Q. Moreover, $q_i(-\infty)$ are the coordinates of Q and $q_i(+\infty)$ are the coordinates of P.

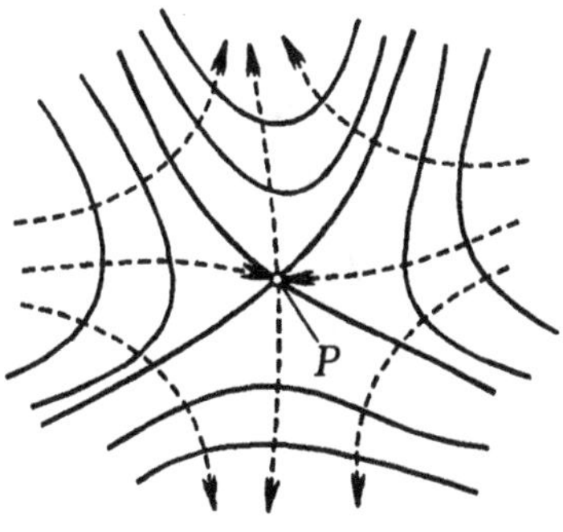

Fig. 24

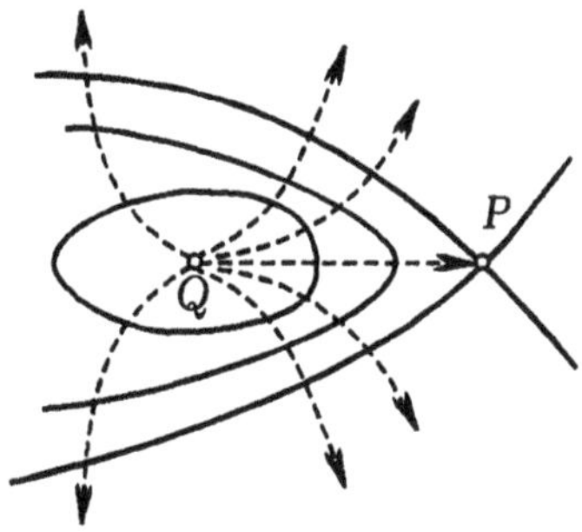

Fig. 25

To justify the above arguments, it suffices to study the behavior of solutions to the ordinary equations in a neighborhood of critical points. We do not discuss this in detail.

21.2. We return to the general case and consider a system of n equations. It is natural to assume that for sufficiently small ε solutions to the equations

$$\frac{\partial L_{q_i}}{\partial t} + \frac{\partial M_{q_i}}{\partial x} = \frac{\partial}{\partial x}\left(\varepsilon b_{ik}\,\frac{\partial q_k}{\partial x}\right) \tag{21.9}$$

are close to solutions to the "inviscous equations"

$$\frac{\partial L_{q_i}}{\partial t} + \frac{\partial M_{q_i}}{\partial x} = 0. \tag{21.10}$$

As is known, inviscous equations of the form (21.10) can have discontinuous solutions (for example, shock waves in gas dynamics). Therefore, we need to introduce the notion of a weak solution.

21.3. To define a weak solution, integral equations instead of the differential equations are usually used. For example,

$$\oint L_{q_i}\,dx - M_{q_i}\,dt = 0,$$

where the integral is taken along any contour in the domain under consideration. Such integral relations lead to certain conditions on discontinuities. Assume that the solution is discontinuous along the line AB (cf. Fig. 26). We choose the contour $(a_1b_1b_2a_2)$ bounding a narrow strip containing a part of the line along which we have the discontinuity. We neglect small parts of the contour (a_2a_1)

and $(b_1 b_2)$ and, for definiteness, assume that the integration is taken along the left or right side of the discontinuity line.[2] We have

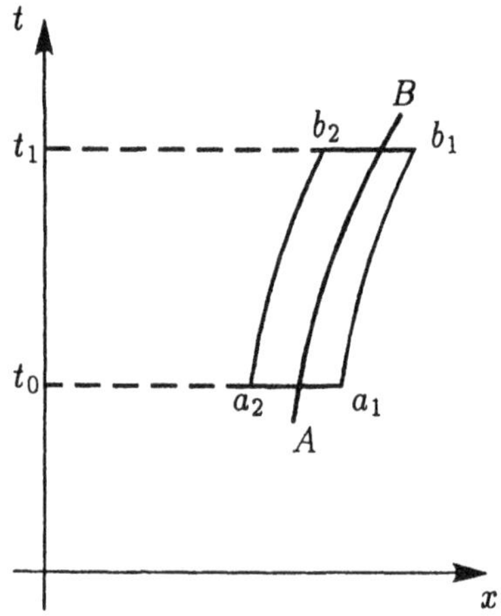

Fig. 26

$$\int_{t_0}^{t_1} \left\{ \left[(L_{q_i})_{\text{right}} - (L_{q_i})_{\text{left}} \right] \frac{dx}{dt} - \left[(M_{q_i})_{\text{right}} - (M_{q_i})_{\text{left}} \right] \right\} dt = 0.$$

Since t_0 and t_1 are arbitrary, on the discontinuity line for all t we have

$$\left[(L_{q_i})_{\text{right}} - (L_{q_i})_{\text{left}} \right] \frac{dx}{dt} - \left[(M_{q_i})_{\text{right}} - (M_{q_i})_{\text{left}} \right] = 0.$$

Introducing the notation $\dfrac{dx}{dt} = w$, we can write the last equality as follows:

$$(L_{q_i})_{\text{right}} - (L_{q_i})_{\text{left}} = \frac{1}{w} (M_{q_i})_{\text{right}} - \frac{1}{w} (M_{q_i})_{\text{left}}.$$

Thus, $(L - (1/w)M)_{q_i}$ has the same value on both sides of the discontinuity line. We denote this value by A_i. Thus, the relations on the discontinuity line take the form

$$(\Lambda_{q_i})_{\text{left}} = \left\{ \left(L - \frac{1}{w} M - A_j q_j \right)_{q_i} \right\}_{\text{left}} = 0,$$

$$(\Lambda_{q_i})_{\text{right}} = \left\{ \left(L - \frac{1}{w} M - A_j q_j \right)_{q_i} \right\}_{\text{right}} = 0$$

[2] Neglecting small parts of the contour, we, actually, assume that the integrand is a bounded function.

or, shortly,

$$w[L_{q_i}] - [M_{q_i}] = 0, \tag{21.11}$$

where $[f]$ denotes the jump of f on the discontinuity line: $[f] = f_{\text{right}} - f_{\text{left}}$.

As was already noted, the gradient Λ_{q_i} vanishes at the points $q_{i\,\text{left}}$ and $q_{i\,\text{right}}$, i.e., these points are singular points of the family of level surfaces $\Lambda(q_1, q_2, \dots, q_n) = \text{const}$. It turns out that the character of these singular points is connected with the behavior of the characteristics of the system

$$\frac{\partial L_{q_i}}{\partial t} + \frac{\partial M_{q_i}}{\partial x} \equiv L_{q_i q_k} \frac{\partial q_k}{\partial t} + M_{q_i q_k} \frac{\partial q_k}{\partial x} = 0 \tag{21.12}$$

in a neighborhood of the discontinuity.

If $||L_{q_i q_k}||$ is a positive definite matrix, then (21.12) is a hyperbolic system of order n and has n real characteristics. Let the slopes $\dfrac{dx}{dt}$ of these characteristics be as follows:

$$k_1^+ \geqslant k_2^+ \geqslant k_3^+ \geqslant \dots \geqslant k_n^+ \text{ to the right of the discontinuity,}$$
$$k_1^- \geqslant k_2^- \geqslant k_3^- \geqslant \dots \geqslant k_n^- \text{ to the left of the discontinuity.}$$

We find the number of incoming and outgoing characteristics (from the left and from the right to the discontinuity line). Let

$$k_1^- \geqslant k_2^- \geqslant \dots \geqslant k_r^- > w > k_{r+1}^- \geqslant \dots \geqslant k_n^-,$$
$$k_1^+ \geqslant k_2^+ \geqslant \dots \geqslant k_s^+ > w > k_{s+1}^+ \geqslant \dots \geqslant k_n^+.$$

Then r characteristics come from the left and $(n-s)$ characteristics come from the right to the discontinuity line. It is easy to see that the slopes of the characteristics k_i are obtained as characteristic roots of the equation

$$\det ||kL_{q_i q_j} - M_{q_i q_j}|| = 0 \tag{21.13}$$

which can be written in the form

$$\det \left\| \left(L_{q_i q_j} - \frac{1}{w} M_{q_i q_j} \right) - \left(1 - \frac{k}{w} \right) L_{q_i q_j} \right\| = 0. \tag{21.14}$$

We assume that $w > 0$. The case $w < 0$ is left to the reader as an easy but useful exercise. It is clear that the number of roots $k_i \geqslant w$ is equal to the number of negative eigenvalues of the matrix $\left\| L_{q_i q_j} - \dfrac{1}{w} M_{q_i q_j} \right\|$ (the matrix $L_{q_i q_j}$ is positive definite). The number of roots $k_i \leqslant w$ is equal to the number of positive eigenvalues of the matrix $\left\| L_{q_i q_j} - \dfrac{1}{w} M_{q_i q_j} \right\|$. We note that

$$\Lambda_{q_i q_j} = \left(L - \frac{1}{w} M - A_m q_m \right)_{q_i q_j} = L_{q_i q_j} - \frac{1}{w} M_{q_i q_j}. \tag{21.15}$$

Thus, r is equal to the number of negative eigenvalues of the matrix $\|\Lambda^{\text{left}}_{q_iq_j}\|$ and $n-s$ is the number of positive eigenvalues of the matrix $\|\Lambda^{\text{right}}_{q_iq_j}\|$.

We know that $\Lambda_{q_i}=0$ at points of the regions adjoining to the wave. In a neighborhood of such a point, we have

$$\Lambda = \frac{1}{2}\,\Lambda_{q_iq_j}\Delta_{q_i}\Delta_{q_j} + \dots . \tag{21.16}$$

Consequently, the level curves $\Lambda=$ const passing through these (critical) points are of the same character as the level curves of the quadratic form $\Lambda_{q_iq_j}\Delta_{q_i}\Delta_{q_j}$. In turn, the topology of these curves is completely determined by the number of positive and negative eigenvalues of the quadratic form. These numbers are connected with the number of incoming characteristics.

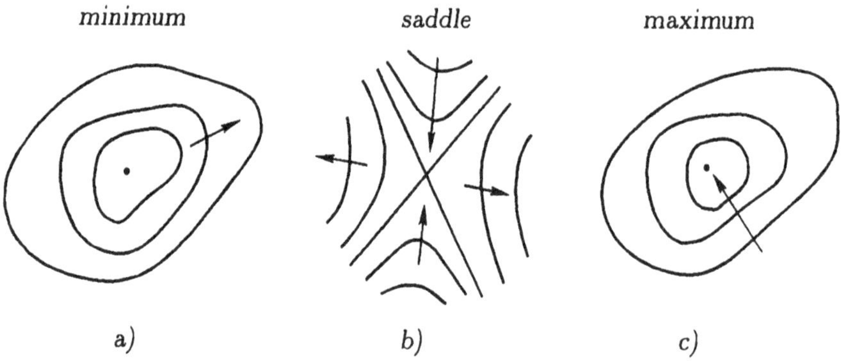

Fig. 27

Example 21.1. Let $n=2$. If $\|\Lambda_{q_iq_j}\|$ has two positive roots, then the level curves are the same as in Fig. 27 (a). If the number of positive roots is equal to 1 or to 0, the level curves are presented in Fig. 27(b) and 27(c) respectively. The arrows in Fig. 27 indicate the direction of the growth of Λ.

21.4. We show that if $\Lambda(q_1,q_2)$ is known and in the plane q_1, q_2 there is a trajectory that joins two critical points and does not touch the level curve of Λ, then we can choose a viscosity, i.e., a positive definite matrix $\|b_{ik}(q_1,q_2)\|$, such that the trajectory is orthogonal to the level curve relative to the inner product $(\alpha,\beta)=b_{ik}\alpha_i\beta_k$.

Let a trajectory be given by the differential equations

$$\frac{dq_1}{ds}=\varphi_1(s),\quad \frac{dq_2}{ds}=\varphi_2(s). \tag{21.17}$$

Along this trajectory, we consider the matrix

$$\Phi = \begin{bmatrix} \varphi_{11} & \varphi_{12} \\ \varphi_{21} & \varphi_{22} \end{bmatrix} = \begin{bmatrix} \varphi_1(s) & \Lambda_{q_2}(q_1(s), q_2(s)) \\ \varphi_2(s) & -\Lambda_{q_1}(q_1(s), q_2(s)) \end{bmatrix}$$

and introduce the inner product relative to the matrix

$$\begin{bmatrix} b_{11} & b_{12} \\ b_{21} & b_{22} \end{bmatrix} = (\Phi^{-1})^* \Phi^{-1}.$$

The tangent vector

$$\begin{pmatrix} f_1 \\ f_2 \end{pmatrix} = \lambda \begin{pmatrix} \varphi_1 \\ \varphi_2 \end{pmatrix} = \lambda \begin{bmatrix} \varphi_1 & \Lambda_{q_2} \\ \varphi_2 & -\Lambda_{q_1} \end{bmatrix} \begin{pmatrix} 1 \\ 0 \end{pmatrix} = \lambda \Phi \begin{pmatrix} 1 \\ 0 \end{pmatrix}$$

satisfies the equality

$$\Phi^{-1} \begin{pmatrix} f_1 \\ f_2 \end{pmatrix} = \lambda \begin{pmatrix} 1 \\ 0 \end{pmatrix},$$

and the tangent vector

$$\begin{pmatrix} g_1 \\ g_2 \end{pmatrix} = \mu \begin{pmatrix} \Lambda_{q_2} \\ -\Lambda_{q_1} \end{pmatrix} = \mu \begin{bmatrix} \varphi_1 & \Lambda_{q_2} \\ \varphi_2 & -\Lambda_{q_1} \end{bmatrix} \begin{pmatrix} 1 \\ 0 \end{pmatrix} = \mu \Phi \begin{pmatrix} 0 \\ 1 \end{pmatrix}$$

satisfies the equality

$$\Phi^{-1} \begin{pmatrix} g_1 \\ g_2 \end{pmatrix} = \mu \begin{pmatrix} 0 \\ 1 \end{pmatrix}.$$

Now, it is obvious that

$$\begin{aligned} b_{ik} f_i g_k &= \left(\|b_{ik}\| \begin{pmatrix} f_1 \\ f_2 \end{pmatrix}, \begin{pmatrix} g_1 \\ g_2 \end{pmatrix} \right) = \left((\Phi^{-1})^* \Phi^{-1} \begin{pmatrix} f_1 \\ f_2 \end{pmatrix}, \begin{pmatrix} g_1 \\ g_2 \end{pmatrix} \right) \\ &= \left(\Phi^{-1} \begin{pmatrix} f_1 \\ f_2 \end{pmatrix}, \Phi^{-1} \begin{pmatrix} g_1 \\ g_2 \end{pmatrix} \right) = \left(\begin{pmatrix} \lambda \\ 0 \end{pmatrix}, \begin{pmatrix} 0 \\ \mu \end{pmatrix} \right) = \lambda \cdot 0 + 0 \cdot \mu = 0, \end{aligned}$$

i.e., the vectors f and g are orthogonal.

The inner product exists if Φ is nonsingular, i.e.,

$$\det \Phi = -\varphi_1 \Lambda_{q_1} - \varphi_2 \Lambda_{q_2} = -\lambda_{q_1} \frac{dq_1}{ds} - \Lambda_{q_2} \frac{dq_2}{ds} = -\frac{d\Lambda}{ds} \neq 0$$

since the trajectory does not touch the level curves of Λ.

Thus, on any part of any smooth trajectory that does not touch the level curves of Λ and does not pass through any critical point λ, it is possible to define a smooth matrix $\|b_{ik}\|$ such that the trajectory is orthogonal to the level curves. A modification of this construction in a neighborhood of a critical point[3)] is used in the proof of the following important assertion.

• *Suppose that for some velocity w and constants A_1, A_2 from $L(q_1, q_2)$ and $M(q_1, q_2)$ we obtain a function $\Lambda(q_1, q_2)$ with certain topological structure of*

[3)] We do not discuss this in detail.

level surfaces and a number of critical points. If there exists a trajectory joining two points that do not touch the level curves, then we can choose an inner product (i.e., viscosity) such that the above junction is realized by a trajectory that is orthogonal to the level curves.

It turns out that it is possible to write the equation of state $E = E(\rho)$ of a barotropic gas such that for one-dimensional equations of gas dynamics (for some velocity w) the level curves of $\Lambda(q_1, q_2)$ are the same as in Fig. 28.

Fig. 28

In this case, there are two different viscosities $||b_{ik}^{(1)}||$ and $||b_{ik}^{(2)}||$ such that the first viscosity "allows the shock passage" from state 1 to state 3 but prohibits any passage from 2 to 3, whereas the second viscosity allows only the passage from 2 to 3. The words "allows the shock passage" mean the existence of a solution to the equations with viscosity

$$q_i = q_i\left[\frac{w^2}{\varepsilon}\left(t - \frac{x}{w}\right)\right],$$

whose asymptotic values as $\zeta = \dfrac{w^2}{\varepsilon}\left(t - \dfrac{x}{w}\right) \to \pm\infty$ coincide with a fixed pair of critical points of Λ.

This example shows that properties of "discontinuous" solutions to equations of continuous media should be studied by not only "inviscous" equations. In each special case, it is necessary to study the kinetics of processes in narrow transition layers. In our case, the kinetics was described by solutions of some "viscous" equations. The above examples show that the "microscopic" kinetics can lead to qualitatively different "macroscopic" situations.

We do not discuss these interesting phenomena in detail but only emphasize some difficulties connected with introducing the notion of a "weak solution" to general quasilinear equations.

21.5. To conclude the study of systems

$$\frac{\partial L_{q_i}}{\partial t} + \frac{\partial M_{q_i}}{\partial x} = 0,$$

we consider inequalities which describe the entropy condition inside narrow zones of shock transition in actual gas dynamical processes. We know that "along the discontinuity" the following equations hold:

$$\Lambda_{q_i} = D_{\beta_i}, \ \frac{dq_i}{d\zeta} = \beta_i, \ \zeta = \frac{w^2}{\varepsilon}\left(t - \frac{x}{w}\right), \ \Lambda = L - \frac{1}{w}M - A_j q_j, \ \frac{d\Lambda}{d\zeta} > 0.$$

Integrating them, we find

$$\int_{-\infty}^{+\infty} \frac{d(q_i\Lambda_{q_i} - \Lambda)}{d\zeta} d\zeta = [q_i\Lambda_{q_i}]_{\zeta=-\infty}^{\zeta=+\infty} - \int_{-\infty}^{+\infty} \frac{d\Lambda}{d\zeta} d\zeta = 0 - [\Lambda]_{\zeta=-\infty}^{\zeta=+\infty} < 0.$$

Since $\Lambda = L - (1/w)M - A_i q_i$, we can write the last relation in the form

$$[q_i L_{q_i} - L] - \frac{1}{w}[q_i M_{q_i} - M] < 0. \tag{21.18}$$

In gas dynamics, the inequality (21.18) is the well-known entropy condition on shock waves.

It can be formulated as follows.

- *The equations*

$$\frac{\partial L_{q_i}}{\partial t} + \frac{\partial M_{q_i}}{\partial x} = 0 \tag{21.19}$$

and the conservation of entropy (on smooth solutions)

$$\frac{\partial (q_i L_{q_i} - L)}{\partial t} + \frac{\partial (q_i M_{q_i} - M)}{\partial x} = 0 \tag{21.20}$$

should be written on discontinuous solutions in the integral form

$$\begin{gathered} \oint L_{q_i}\, dx - M_{q_i}\, dt = 0, \\ \oint (q_i L_{q_i} - L)\, dx - (q_i M_{q_i} - M)\, dt \geqslant 0. \end{gathered} \tag{21.21}$$

Thus, studying discontinuous solutions, it is necessary to distinguish n differential equations that are replaced with integral conservation laws so that the $n+1$th relation is replaced with an integral inequality which can be interpreted as a prohibition inequality for waves to the decrease of entropy. This is possible because of the relation

$$q_i\left(\frac{\partial L_{q_i}}{\partial t} + \frac{\partial M_{q_i}}{\partial x}\right) - 1 \cdot \left[\frac{\partial (q_i L_{q_i} - L)}{\partial t} + \frac{\partial (q_i M_{q_i} - M)}{\partial x}\right] = 0. \tag{21.22}$$

Owing to (21.22) which is valid only on smooth solutions, it is reasonable to take the condition that the contour integrals

$$\oint (q_i L_{q_i} - L)\, dx + (q_i M_{q_i} - M)\, dt \tag{21.23}$$

do not vanish for the characteristic of the presence of discontinuities and verify the entropy condition depending on the sign of these integrals.

We note that the relation

$$q_0\left(\frac{\partial L_{q_0}}{\partial t} + \frac{\partial M_{q_0}}{\partial x}\right) + q_i\left(\frac{\partial L_{q_i}}{\partial x} + \frac{\partial M_{q_i}}{\partial x}\right)$$
$$-1\cdot\left[\frac{\partial(q_0 L_{q_0} + q_i L_{q_i} - L)}{\partial t} + \frac{\partial(q_0 M_{q_0} + q_i M_{q_i} - M)}{\partial x}\right] = 0$$

can be written as follows:

$$\frac{1}{q_0}\left[\frac{\partial(q_0 L_{q_0} + q_i L_{q_i} - L)}{\partial t} + \frac{\partial(q_0 M_{q_0} + q_i M_{q_i} - M)}{\partial x}\right]$$
$$+\left(-\frac{q_i}{q_0}\right)\left(\frac{\partial L_{q_j}}{\partial t} + \frac{\partial M_{q_j}}{\partial x}\right) - 1\cdot\left(\frac{\partial L_{q_0}}{\partial t} + \frac{\partial M_{q_0}}{\partial x}\right) = 0.$$

This means that if we regard $r_0 = 1/q_0$ and $r_j = -q_j/q_0$ as new unknown functions, then we can find new generating functions $\mathcal{L}$ and $\mathcal{M}$ of the system and write the system in the canonical form

$$\frac{\partial \mathcal{L}_{r_0}}{\partial t} + \frac{\partial \mathcal{M}_{r_0}}{\partial x} = 0,$$
$$\frac{\partial \mathcal{L}_{r_i}}{\partial t} + \frac{\partial \mathcal{M}_{r_i}}{\partial x} = 0$$

with the conservation law

$$\frac{\partial(r_0\mathcal{L}_{r_0} + r_i\mathcal{L}_{r_i} - \mathcal{L})}{\partial t} + \frac{\partial(r_0\mathcal{M}_{r_0} + r_i\mathcal{M}_{r_i} - \mathcal{M})}{\partial x} = 0.$$

Furthermore, $\mathcal{L} = (1/q_0)L$, $\mathcal{M} = (1/q_0)M$, $\mathcal{L}_{r_0} = q_0 L_{q_0} + q_i L_{q_i} - L$, $\mathcal{M}_{r_0} = q_0 M_{q_0} + q_i M_{q_i} - M$, $\mathcal{L}_{r_j} = L_{q_j}$, $\mathcal{M}_{q_j} = M_{q_j}$, $r_0\mathcal{L}_{r_0} + r_i\mathcal{L}_{r_i} - \mathcal{L} = L_{q_0}$, $r_0\mathcal{M}_{r_0} + r_i\mathcal{M}_{r_i} - \mathcal{M} = M_{q_0}$. As we know, all the conservation laws are formally equal in rights and the choice of one of them in the form of an integral inequality requires more delicate physical arguments. Here, one should take into account kinetics of processes in shock waves.

It is of interest to note that shock waves were discovered by Riemann theoretically but not in an experimental way. Riemann proposed relations on waves based on the following wrong system of conservation laws:

mass momentum entropy

whereas the relations should be derived from the conservation laws

mass momentum energy

Thus, the passage from q_i to r_i and replacements of roles of conservation laws can be called a formalization of "Riemann's mistake."

We recall that the passage from q_0 and q_i to $r_0 = 1/q_0$ and $r_j = -q_j/q_0$ is one of the transformations used in the study of hyperbolic equations of gas dynamics (cf. Section 20).

§ 22. Well-Posedness, Stability, and Postulates of Phenomenological Thermodynamics

22.1. Investigating equations in continuum mechanics, we can see many times that properties of these equations essentially depend on inequalities satisfied by the derivatives of the equation of state with respect to parameters of the medium. Properties of the equation of state and inequalities for it are studied in thermodynamics. As we have seen, there exists a deep analogy between the phenomenological thermodynamics of a medium and the behavior of solutions to equations describing macroscopic physical processes occurring in the medium. Thermodynamics suggests two different approaches to the study of physical phenomena: *phenomenological* and *statistical*. The statistical approach is based on laws stated from the study of microstructure of a medium, whereas the phenomenological approach deals with properties of systems derived from axiomatic statements about the character of the behavior of solutions. For example, axiomatizing the impossibility of the existence of the perpetual motion machine of the second kind, the phenomenological thermodynamics proves that all possible equations of state $E = E(V,T)$, $p = p(V,T)$ for gases are connected by the existence of a universal (the same for all gases) function of temperature $\mu(T)$ such that the expression $\mu(T)(dE + d\,dV)$ is the exact differential.

In the theory of differential equations, the Hadamard postulate is accepted. This postulate says that the problems for differential equations describing the evolution of physical processes should be well posed. This postulate is of phenomenological character. The modern theory of differential equations mainly deals with conditions caused by the requirement of the well-posedness of systems of differential equations. Thus, in 1937, the founder of this direction I.G. Petrovsky discovered that the well-posedness of the Cauchy problem depends on the structure of the characteristic conoid of the system. He distinguished a class of hyperbolic equations for which the Cauchy problem is well posed.

These arguments confirm that some features of an analogy between the thermodynamical inequalities and the well-posedness of equations of continuum mechanics are not accidental.

To illustrate all what we said above, we give some examples. We begin with one-dimensional equations of hydrodynamics with heat conductivity

$$\begin{aligned}
&\frac{\partial \rho}{\partial t}+\frac{\partial(\rho u)}{\partial x}=0, \\
&\frac{\partial(\rho u)}{\partial t}+\frac{\partial(p+\rho u^2)}{\partial x}=0, \\
&\frac{\partial}{\partial t}\left[\rho\left(E+\frac{u^2}{2}\right)\right]+\frac{\partial}{\partial x}\left[\rho u\left(E+\frac{p}{\rho}+\frac{u^2}{2}\right)-K\frac{\partial T}{\partial x}\right]=0,
\end{aligned} \tag{22.1}$$

which are equivalent to the conservation laws in the integral form

$$\begin{aligned}
&\oint \rho\,dx+\rho u\,dt=0, \\
&\oint \rho u\,dx-(p+\rho u^2)\,dt=0, \\
&\oint \rho\left(E+\frac{u^2}{2}\right)dx-\left[\rho u\left(E+\frac{p}{\rho}+\frac{u^2}{2}\right)-K\frac{\partial T}{\partial x}\right]dt=0.
\end{aligned} \tag{22.2}$$

According to the mass conservation law, the integral

$$\int \rho\,dx-\rho u\,dt$$

is independent of a path joining any two points. This allows us to introduce a new "coordinate" m by the formula $dm=\rho\,dx-\rho u\,dt=\rho(dx-u\,dt)$. We have $dx-u\,dt=0$ along the trajectory of any fixed particle. Consequently, m is constant. This means that m is the Lagrangian coordinate connected with particles themselves but not with the location of these particles in the space. In the variables m, t, the conservation laws are written as follows:

$$\begin{aligned}
&\oint u\,dm-p\,dt\equiv\oint u(\rho\,dx-u\,dt)-p\,dt\equiv\oint \rho u\,dx-(p+\rho u^2)\,dt=0, \\
&\oint\left(E+\frac{u^2}{2}\right)dm-\left(pu-K\rho\frac{\partial T}{\partial m}\right)dt\equiv\oint\rho\left(E+\frac{u^2}{2}\right)dx \\
&\qquad-\left[\rho u\left(E+\frac{p}{\rho}+\frac{u^2}{2}\right)-K\frac{\partial T}{\partial x}\right]dt=0.
\end{aligned} \tag{22.3}$$

In addition, a conservation law follows from the fact that $dx=\dfrac{1}{\rho}\,dm+u\,dt$ is the exact differential:

$$\oint\frac{1}{\rho}\,dm+u\,dt=0.$$

For smooth functions the integral conservation laws

$$
\begin{aligned}
&\oint \frac{1}{\rho}\,dm + u\,dt = 0,\\
&\oint u\,dm - p\,dt = 0,\\
&\oint \left(E + \frac{u^2}{2}\right) dm - \left(pu - K\rho\,\frac{\partial T}{\partial m}\right) dt = 0
\end{aligned}
\tag{22.4}
$$

are equivalent to the differential equations

$$
\begin{aligned}
&\frac{\partial(1/\rho)}{\partial t} - \frac{\partial u}{\partial m} = 0,\\
&\frac{\partial u}{\partial t} + \frac{\partial p}{\partial m} = 0,\\
&\frac{\partial(E + u^2/2)}{\partial t} + \frac{\partial(pu)}{\partial m} = \frac{\partial}{\partial m}\left(K\rho\,\frac{\partial T}{\partial m}\right)
\end{aligned}
\tag{22.5}
$$

which describe processes in a heat-conducting gas in the Lagrangian coordinates. Here, instead of the density ρ we use the specific volume $V = 1/\rho$ and denote $\varkappa = K\rho$. We note that the pressure $p = p(V,S)$ is connected with the equation of state $E = E(V,S)$ by the equalities

$$
p = -E_V(V,S), \quad T = E_S(V,S). \tag{22.6}
$$

Therefore, we can write the equations in the form

$$
\begin{aligned}
&\frac{\partial V}{\partial t} - \frac{\partial u}{\partial m} = 0,\\
&\frac{\partial u}{\partial t} - \frac{\partial E_V}{\partial m} = 0,\\
&\frac{\partial(E + u^2/2)}{\partial t} - \frac{\partial(uE_V)}{\partial m} = \frac{\partial}{\partial m}\left(\varkappa\,\frac{\partial E_S(V,S)}{\partial m}\right),
\end{aligned}
\tag{22.7}
$$

which implies that the equation of state $E = E(V,S)$ and the thermal conductivity coefficient $\varkappa(V,S)$ completely determine this system and properties of solutions. We consider only the case where $\varkappa$ is constant and nonnegative.

22.2. We consider the propagation of small perturbations in a motionless homogeneous medium

$$
\begin{aligned}
&u = \varepsilon\overline{u}, \quad S = S_0 + \varepsilon\overline{S}, \quad V = V_0 + \varepsilon\overline{V},\\
&E_V = E_V(V_0,S_0) + \varepsilon\overline{V}E_{VV}(V_0,S_0) + \varepsilon\overline{S}E_{VS}(V_0,S_0) + \dots,\\
&E_S = E_S(V_0,S_0) + \varepsilon\overline{V}E_{SV}(V_0,S_0) + \varepsilon\overline{S}E_{SS}(V_0,S_0) + \dots,\\
&E = E(V_0,S_0) + \varepsilon\overline{V}E_V(V_0,S_0) + \varepsilon\overline{S}E_S(V_0,S_0) + \dots.
\end{aligned}
\tag{22.8}
$$

If we insert these representations into the initial system and neglect the quantities of order ε^2, then we obtain a linear system with constant coefficients

$$\begin{aligned} &\frac{\partial \overline{u}}{\partial t} - E_{VV}\frac{\partial \overline{V}}{\partial m} - E_{VS}\frac{\partial \overline{S}}{\partial m} = 0, \\ &\frac{\partial \overline{V}}{\partial t} - \frac{\partial \overline{u}}{\partial m} = 0, \\ &E_V\frac{\partial \overline{V}}{\partial t} + E_S\frac{\partial \overline{S}}{\partial t} - E_V\frac{\partial \overline{u}}{\partial m} = \varkappa E_{SS}\frac{\partial^2 \overline{S}}{\partial m^2} + \varkappa E_{SV}\frac{\partial^2 \overline{V}}{\partial m^2}. \end{aligned} \tag{22.9}$$

We simplify this system. Subtracting the second equation multiplied by E_V from the last equation, we find

$$\begin{aligned} &\frac{\partial \overline{u}}{\partial t} - E_{VV}\frac{\partial \overline{V}}{\partial m} - E_{VS}\frac{\partial \overline{S}}{\partial m} = 0, \\ &\frac{\partial \overline{V}}{\partial t} - \frac{\partial \overline{u}}{\partial m} = 0, \\ &E_S\frac{\partial \overline{S}}{\partial t} = \varkappa E_{SS}\frac{\partial^2 \overline{S}}{\partial m^2} + \varkappa E_{SV}\frac{\partial^2 \overline{V}}{\partial m^2}. \end{aligned} \tag{22.9}$$

We are interested in the well-posedness of the initial-value problem for this system with initial data $\overline{u}$, $\overline{V}$, $\overline{S}$ at time $t = 0$. As is known, to study the well-posedness of a problem, it is necessary to consider special solutions of the form $e^{i(\alpha x+\lambda t)}\begin{pmatrix}\widehat{u}\\ \widehat{V}\\ \widehat{S}\end{pmatrix}$. If $\operatorname{Im}(-\lambda)$ remains bounded from above as $\alpha \to \infty$, then the problem is well posed. However, if $\operatorname{Im}(-\lambda)$ tends to $+\infty$ faster than some positive power of α, the problem is ill posed.

To find special solutions, it is necessary to solve the homogeneous system of linear equations

$$\begin{aligned} &i\lambda\widehat{u} - i\alpha E_{VV}\widehat{V} - i\alpha E_{VS}\widehat{S} = 0, \\ &-i\alpha\widehat{u} + i\lambda\widehat{V} = 0, \\ &\alpha^2\varkappa E_{SV}\widehat{V} + (\alpha^2\varkappa E_{SS} + i\alpha E_S)\widehat{S} = 0. \end{aligned} \tag{22.10}$$

The system (22.10) is solvable if

$$\begin{vmatrix} i\lambda & -i\alpha E_{VV} & -i\alpha E_{VS} \\ -i\alpha & i\lambda & 0 \\ 0 & \alpha^2\varkappa E_{SV} & i\lambda E_S + \alpha^2\varkappa E_{SS} \end{vmatrix}$$

$$= -\varkappa\alpha^4 \begin{vmatrix} \dfrac{\lambda}{\alpha} & -E_{VV} & -E_{VS} \\ -1 & \dfrac{\lambda}{\alpha} & 0 \\ 0 & E_{SV} & i\dfrac{\lambda}{\varkappa\alpha^2}E_S + E_{SS} \end{vmatrix} = 0$$

or

$$\left[\left(\frac{\lambda}{\alpha}\right)^2 - E_{VV}\right]\left(i\frac{\lambda E_S}{\varkappa\alpha^2} + E_{SS}\right) + E_{VS}^2 = 0. \tag{22.11}$$

It is natural to expect that $\lambda/\alpha \to \text{const}\,(\lambda/\alpha^2 \to 0)$ for some roots, and $\lambda/\alpha^2 \to \text{const}$ (and $\lambda/\alpha \to \infty$) for other roots as $\alpha \to \infty$. The asymptotic values are found from the equations

$$\left[\left(\frac{\lambda}{\alpha}\right)_\infty^2 - E_{VV}\right]E_{SS} + E_{VS}^2 = 0, \quad i\left(\frac{\lambda}{\alpha^2}\right)_\infty \frac{E_S}{\varkappa} + E_{SS} = 0. \tag{22.12}$$

Solving equations (22.12), we find the approximate roots of the characteristic equations

$$\begin{aligned} \lambda_1 &\approx +\sqrt{\frac{E_{VV}E_{SS} - E_{VS}^2}{E_{SS}}}\,\alpha, \\ \lambda_2 &\approx -\sqrt{\frac{E_{VV}E_{SS} - E_{VS}^2}{E_{SS}}}\,\alpha, \\ \lambda_3 &\approx i\,\frac{\varkappa E_{SS}\alpha^2}{E_S}. \end{aligned} \tag{22.13}$$

Moreover, it is easy to show that for large α $(\alpha \to \infty)$

$$\begin{aligned} i\lambda_1 &= i\sqrt{\frac{E_{VV}E_{SS} - E_{VS}^2}{E_{SS}}}\,\alpha + O(1), \\ i\lambda_2 &= -i\sqrt{\frac{E_{VV}E_{SS} - E_{VS}^2}{E_{SS}}}\,\alpha + O(1), \\ i\lambda_3 &= -\frac{\varkappa E_{SS}\alpha^2}{E_S} + O(1). \end{aligned} \tag{22.14}$$

For the well-posedness of the problem it is necessary that the real parts of the exponents $i\lambda_i$ are bounded from above as $\alpha \to \infty$. Since $i\lambda_3$ is bounded from above and E_S, $\varkappa$ are positive, the inequality $E_{SS} > 0$ is necessary. Since E_{SS} is positive and $i\lambda_1$, $i\lambda_2$ are bounded, the root

$$\sqrt{\frac{E_{VV}E_{SS} - E_{VS}^2}{E_{SS}}} \tag{22.15}$$

is real. Consequently,

$$E_{VV}E_{SS} - E_{VS}^2 > 0. \tag{22.16}$$

Thus, we have proved the following assertion.

• *If the problem about propagation of sound waves in a inviscous heat-conductive gas is well posed, then the following inequalities hold:*

$$E_{SS} > 0, \quad E_{VV}E_{SS} - E_{VS}^2 > 0, \tag{22.17}$$

i.e., the matrix $\begin{bmatrix} E_{VV} & E_{VS} \\ E_{SV} & E_{SS} \end{bmatrix}$ *is positive definite.*

In other words, the well-posedness of the problem implies the convexity of the function $E(V, S)$. Conversely, if $E(V, S)$ is convex, it is easy to justify the well-posedness of the linear problem. To do this, one can prove that the real part of each root $i\lambda$ of the characteristic equation is bounded from above. Let us show that all $i\lambda$ lie in the left half-plane and the imaginary part of each $i\lambda$ is negative (respectively the real part of λ is positive). The characteristic equation can be written in the form

$$\frac{E_S}{\varkappa\alpha^4}(i\lambda)^3 + \frac{E_{SS}}{\alpha^2}(i\lambda)^2 + \frac{E_{VV}E_S}{\varkappa\alpha^2}(i\lambda) + (E_{VV}E_{SS} - E_{VS}^2) = 0. \tag{22.18}$$

Introducing the notation

$$i\lambda = z, \; a_3 = \frac{E_S}{\varkappa\alpha^4}, \quad a_2 = \frac{E_{SS}}{\alpha^2} > 0,$$
$$a_1 = \frac{E_{VV}E_S}{\varkappa\alpha^2} > 0, \quad a_0 = E_{VV}E_{SS} - E_{VS}^2 > 0,$$

we see that z satisfies the cubic equation $a_3z^3 + a_2z^2 + a_1z + a_0 = 0$ with positive coefficients. As is known from algebra, all the roots of a cubic equation with positive coefficients are located in the left half-plane of the complex plane z if and only if $a_3a_0 - a_1a_2 < 0$. In our case, this means that

$$a_3a_0 - a_1a_2 = \frac{E_S}{\varkappa\alpha^4}(E_{VV}E_{SS} - E_{VS}^2) - \frac{E_{VV}E_S}{\varkappa\alpha^2}\frac{E_{SS}}{\alpha^2} =$$
$$- \frac{E_S}{\varkappa\alpha^4}E_{SV}^2 < 0. \tag{22.19}$$

Thus, we have proved that the linear system is well posed.

22.3. In classical phenomenological thermodynamics, the inequalities leading to the convexity of $E(V, S)$ are usually considered as the conditions for the stability of thermodynamical state. Perhaps, the term "well-posedness" is more suitable in this case than the term "stability."

The notion of well-posedness is concerned with systems with infinitesimal number of degrees of freedom and can be regarded as unstability with unbounded decrement of perturbation growth. The shorter the length of the perturbation wave, the faster is the growth of this perturbation.

To trace an analogy between unstability and well-posedness, we derive the convexity of $E(V, S)$ from the stability conditions for the simplest mechanical construction containing a gas. Our arguments are similar to those used in the classical approach of phenomenological thermodynamics.

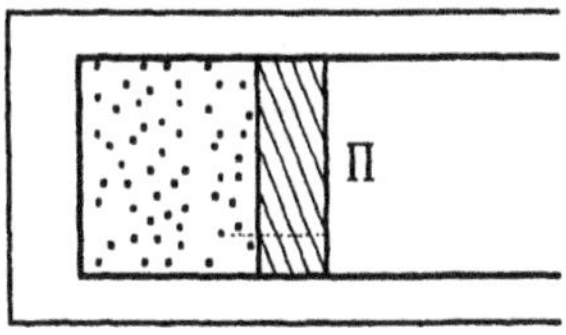

Fig. 29

22.4. We consider the problem about oscillations of a piston Π in a cylinder of unit cross-section containing the unit mass of a gas. The system under consideration is represented in Fig. 29. We assume that the pressure p_0 and the temperature T_0 are constant from the right to the piston Π. From the left to the piston, there is a gas of volume V_0 in the equilibrium state with the entropy S_0, i.e., $p_0 = -E_V(V_0, S_0)$ and $T_0 = E_S(V_0, S_0)$. We study small oscillations in a neighborhood of the equilibrium state under the assumptions that the piston is heat-conductive. It is easy to see that the equations describing oscillations are as follows:

$$
\begin{aligned}
& m\frac{\partial u}{\partial t} = p(V, S) - p_0, \\
& \frac{dE}{dt} + p\frac{dV}{dt} = -k(T - T_0), \\
& \frac{dV}{dt} = u,
\end{aligned}
\tag{22.20}
$$

where m is the mass, u is the velocity, and k is the (positive) heat-conductivity coefficient of the piston.

Considering only small oscillations, i.e., setting $u = \varepsilon\overline{u}$, $V = V_0 + \varepsilon\overline{V}$, $S = S_0 + \varepsilon\overline{S}$, we arrive at the following linear system:

$$
\begin{aligned}
& m\frac{d\overline{u}}{dt} + E_{VV}\overline{V} + E_{VS}\overline{S} = 0, \\
& \varkappa\frac{d\overline{S}}{dt} + E_{VS}\overline{V} + E_{SS}\overline{S} = 0, \\
& \frac{d\overline{V}}{dt} - \overline{u} = 0,
\end{aligned}
\tag{22.21}
$$

where $\varkappa = T_0/k$. For the stability of the equilibrium state of the piston it is necessary that the real parts of the roots of the characteristic equation of the linear system to be nonpositive. The characteristic equation has the form

$$\varkappa m\lambda^3 + mE_{SS}\lambda^2 + \varkappa E_{VV}\lambda + (E_{VV}E_{SS} - E_{VS}^2) = 0.$$

The cubic equation $a_3\lambda^3 + a_2\lambda^2 + a_1\lambda + a_0 = 0$ has no roots in the right half-plane if and only if all the coefficients a_j have the same sign and the following inequality holds: $a_3a_0 - a_1a_2 < 0$. In the case under consideration, we have $a_3 = \varkappa m$, $a_2 = mE_{SS}$, $a_1 x E_{VV}$, $a_0 = E_{VV}E_{SS} - E_{VS}^2$. We can assume that $a_3 = \varkappa m$ is positive because m is the mass of the piston. Therefore, the stability conditions can be written as the inequalities

$$\begin{aligned} &a_2 = mE_{SS} > 0, \quad a_1 = \varkappa E_{VV} > 0, \\ &a_0 = E_{VV}E_{SS} - E_{VS}^2, \\ &a_3a_0 - a_1a_2 = \varkappa m(E_{VV}E_{SS} - E_{VS}^2) - \varkappa mE_{SS}E_{VV} \\ &\quad = -\varkappa mE_{VS}^2 < 0 \end{aligned} \tag{22.22}$$

which are valid (for positive $\varkappa$) if and only if the matrix $\begin{bmatrix} E_{VV} & E_{VS} \\ E_{SV} & E_{SS} \end{bmatrix}$ is positive definite, i.e., the function $E(V, S)$ is convex.

22.5. We consider differential equations describing processes in constructions with a finite number of freedom degrees. We show that the stability conditions imply the existence of a universal integrating factor $\mu(T)$ for the differentials $dE + p\,dV$ constructed from the equations of state for gases in the construction.

We define the notions used in our further consideration.

• A *gas* is a material that acts on the boundary surface of a container occupied by this material with the pressure depending only on the volume of the container per unit mass and on the temperature. This property of a gas is expressed as the *equation of thermal state* $p = p(V, T)$.

In addition, it is necessary to introduce the so-called *equation of caloric state* $E = E(V, T)$. Here, E is the heat obtained if we cool a gas element of unit mass from the temperature T to some standard temperature T^* so that the volume is preserved.

By the energy conservation law, the amount of heat dQ received by a gas element of unit mass is connected with the change of the state of the gas by the equation $dQ = dE + p\,dV$. If, for example, we are given the change of volume dV, we can determine the change of temperature.

Up to now, we used an intuitive notion of temperature. Now, we discuss it in more detail. As is known from experiments, all bodies can be ordered in such a way that the heat can transfer only from bodies with larger numbers to bodies with smaller numbers. Namely, if there are three bodies A, B, C and we know that the heat can transfer from A to B and from B to C, then, if the bodies A and C touch, the heat transfers from A to C. In this case, we say that the "number" of the body A (the temperature of A) is larger than the "number" of the body B (the temperature of B) and write this fact as $T_A > T_B$. The inequalities $T_B > T_C$ and $T_A > T_C$ are understood in the same way.

We say that two bodies have *equal temperature* if no heat transfer occurs when the bodies touch. We can introduce the number characteristic for temperature measured, for example, by the volume occupied by some standard gas of prescribed mass with some standard pressure.

If two bodies touch through some obstacle, the amount of heat transferred from the body with temperature T_2 to the body with temperature T_1 during the time dt is expressed as the following heat conductivity law: $dQ = k(T_2 - T_1)\,dt$, $k = k(T_1, T_2)$. For the heat transfers from a hotter body to a colder body, it is necessary for k to be positive. The form of the function $k(T_2, T_1)$ depends on the properties of the obstacle. If $T_2 \approx T_1 \approx T_0$, which will be assumed, then $k(T_2, T_1) \approx k(T_0, T_0)$ and we can neglect the dependence of k on temperature. Therefore, we simply write $dQ = k(T_2 - T_1)\,dt$, $k \geqslant 0$. We say that an obstacle is not heat-conductive if $k = 0$. Another limit case corresponds to an infinitely heat-conducting obstacle under the condition that both bodies separated by the obstacle have the same temperature at each time moment.

We complete a brief survey of the basic notions of the elementary physics. These notions are used in the proof of the Carathéodory theorem asserting the existence of a universal integrating factor. In fact, all further arguments are a modification of the proof of this theorem in the simplest situation. We present them in order to illustrate once more an analogy between thermodynamic properties of a medium and differential equations describing physical processes in the medium.

22.6. We proceed by considering the simplest construction (cf. Fig. 30). We assume that the area of the cross-section of a cylinder is equal to 1, Π' is a weightless and infinitely heat-conductive piston such that the temperature and pressure are the same from both sides of it. The values of the temperature and pressure are denoted by T and p respectively. The pressure p_0 and the temperature T_0 are constant from the right to the piston. Two different gases of mass x_1 and x_2 occupy boxes 1 and 2. The quantities relative to box 1 or 2 are marked by subscript 1 or 2 respectively. We will neglect the kinetic energy of the motion of gases in boxes 1 and 2, i.e., we assume that $x_1 \ll m$ and $x_2 \ll m$.

We denote by u the velocity of the piston Π and write the equation of motion

$$m\,\frac{du}{dt} = p - p_0. \tag{22.23}$$

The volume occupied by the gas in boxes 1 and 2 is expressed in terms the specific volume and mass as follows:

$$V = x_1 V_1 + x_2 V_2.$$

Since we consider a cylinder with unit cross-section, it is obvious that $\frac{dV}{dt} = u$ which implies the differential equation

$$x_1 \frac{dV_1}{dt} + x_2 \frac{dV_2}{dt} = u, \quad V_1 = V_1(T,p), \quad V_2 = V_2(T,p).$$

The total internal energy E of the gas in boxes 1 and 2 can be expressed in terms of the internal energy and mass as follows: $E = x_1E_1 + x_2E_2$. We know that $dE + p\,dV = dQ = k(T_0 - T)\,dt$. This assertion can be written as the differential equation

$$x_1\left[\frac{dE_1(p,T)}{dt} + p\,\frac{dV_1(p,T)}{dt}\right] + x_2\left[\frac{dE_2(p,T)}{dt} + p\,\frac{dV_2(p,T)}{dt}\right] = -k(T - T_0).$$

Thus, we obtain a system of three equations for three unknown functions u, p, T of time:

$$\begin{aligned}
&m\,\frac{du}{dt} - (p - p_0) = 0,\\
&x_1 \frac{dV_1(p,T)}{dt} + x_2 \frac{dV_2(p,T)}{dt} = u,\\
&x_1\left[\frac{dE_1(p,T)}{dt} + p\,\frac{dV_1(p,T)}{dt}\right] + x_2\left[\frac{dE_2(p,T)}{dt} + p\,\frac{dV_2(p,T)}{dt}\right] = -k(T - T_0).
\end{aligned}$$

To derive linear equations for small oscillations in a neighborhood of the state of equilibrium, we represent the solution in the form

$$u = \varepsilon\overline{u}, \quad p = p_0 + \varepsilon\overline{p}, \quad T = T_0 + \varepsilon\overline{T}.$$

Eliminating second-order terms with respect to ε, we obtain the linear system

$$\begin{aligned}
&m\,\frac{d\overline{u}}{dt} - \overline{p} = 0,\\
&(x_1A_1 + x_2A_2)\,\frac{d\overline{p}}{dt} + (x_1B_1 + x_2B_2)\,\frac{d\overline{T}}{dt} + \overline{u} = 0,\\
&(x_1C_1 + x_2C_2)\,\frac{d\overline{p}}{dt} + (x_1D_1 + x_2D_2)\,\frac{d\overline{T}}{dt} + k\overline{T} = 0,
\end{aligned}$$

where

$$\begin{aligned}
A_1 &= -\frac{\partial V_1(p,T)}{\partial p}, \quad A_2 = -\frac{\partial V_2(p,T)}{\partial p},\\
B_1 &= -\frac{\partial V_1(p,T)}{\partial T}, \quad B_2 = -\frac{\partial V_2(p,T)}{\partial T},\\
C_1 &= \frac{\partial E_1(p,T)}{\partial p} + p_0\,\frac{\partial V_1(p,T)}{\partial p}, \quad C_2 = \frac{\partial E_2(p,T)}{\partial p} + p_0\,\frac{\partial V_2(p,T)}{\partial p},\\
D_1 &= \frac{\partial E_1(p,T)}{\partial T} + p_0\,\frac{\partial V_1(p,T)}{\partial T}, \quad D_2 = \frac{\partial E_2(p,T)}{\partial T} + p_0\,\frac{\partial V_2(p,T)}{\partial T}.
\end{aligned}$$

The characteristic equation of the linear system has the form

$$\begin{vmatrix} m\lambda & -1 & 0 \\ 1 & (x_1A_1 + x_2A_2)\lambda & (x_1B_1 + x_2B_2)\lambda \\ 0 & (x_1C_1 + x_2C_2)\lambda & (x_1D_1 + x_2D_2)\lambda \end{vmatrix} = 0$$

or, in detail,

$$m[(x_1A_1 + x_2A_2)(x_1D_1 + x_2D_2) - (x_1B_1 + x_2B_2)(x_1C_1 + x_2C_2)]\lambda^3 + km(x_1A_1 + x_2A_2)\lambda^2 + (x_1D_1 + x_2D_2)\lambda + k = 0.$$

All the roots λ lie in the left half-plane (which corresponds to stability) if

$$(x_1A_1 + x_2A_2)(x_1D_1 + x_2D_2) - (x_1B_1 + x_2B_2)(x_1C_1 + x_2C_2) > 0,$$
$$x_1A_1 + x_2A_2 > 0, \quad x_1D_1 + x_2D_2 > 0,$$
$$(x_1B_1 + x_2B_2)(x_1C_1 + x_2C_2) > 0.$$

Here we used the above-mentioned conditions on the coefficients of the cubic equation under which the roots lie in the left half-plane.

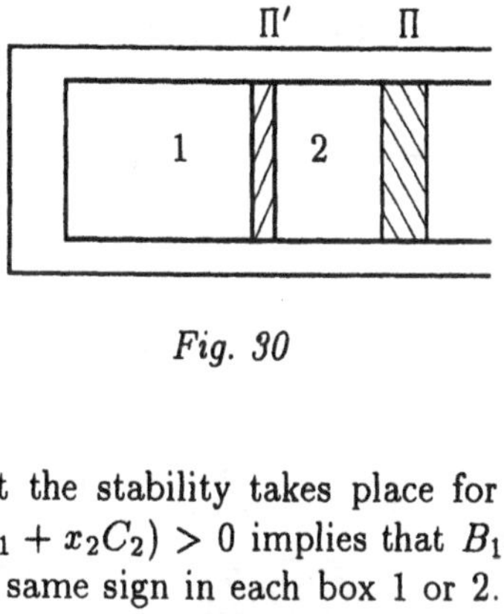

Fig. 30

We assume that the stability takes place for any fraction x_1/x_2. Then $(x_1B_1 + x_2B_2)(x_1C_1 + x_2C_2) > 0$ implies that $B_1C_1 > 0$ and $B_2C_2 > 0$, i.e., B and C are of the same sign in each box 1 or 2. It is easy to show that the above inequality leads to the following alternate.

either B_1 and B_2 (consequently, C_1 and C_2) have the same sign or $\dfrac{B_1}{B_2} = \dfrac{C_1}{C_2}$.

Indeed, let B_1 and B_2 have different signs. Then $-B_1/B_2 > 0$ and $-C_1/C_2 > 0$. Let $B_1/B_2 \neq C_1/C_2$. Then it is possible to choose a positive number μ such that $(\mu + B_1/B_2)(\mu + C_1/C_2) < 0$. Taking $x_2/x_1 = \mu$, we arrive at the inequality $(x_1B_1 + x_2B_2)(x_1C_1 + x_2C_2) < 0$ which contradicts the stability. Thus, the alternate is established.

22.7. We temporarily stop to study properties of the solution and consider the case of a construction where the state of equilibrium is stable and the equation describing small oscillations of this construction is as follows:

$$m\,\frac{d\overline{u}}{dt} = \overline{p},$$
$$(y_1A_1 + y_2A_2)\,\frac{d\overline{p}}{dt} + (y_1B_1 - y_2B_2)\,\frac{d\overline{T}}{dt} = -\overline{u},$$
$$(y_1C_1 - y_2C_2)\,\frac{d\overline{p}}{dt} + (y_1D_1 + y_2D_2)\,\frac{d\overline{T}}{dt} = -k\overline{T}.$$

The parameters $y_1 > 0$ and $y_2 > 0$ are arbitrary masses of gases. It is easy to see that one of the stability conditions for this system is the following inequality:

$$(y_1B_1 - y_2B_2)(y_1C_1 - y_2C_2) > 0.$$

If the stability is established experimentally, the following alternate takes place:

either B_1 and B_2 have different signs or $\dfrac{B_1}{B_2} = \dfrac{C_1}{C_2}$.

Assuming the stability of both constructions, we can combine the alternates and conclude that

$$\frac{B_1}{B_2} = \frac{C_1}{C_2} \quad \left(\frac{C_1}{B_1} = \frac{C_2}{B_2}\right).$$

Hence we have the following assertion.

• *Let both constructions be stable in some domain of parameters T_0, p_0 for any masses x_1, x_2, y_1, y_2. Then, in this domain,*

$$\frac{\dfrac{\partial E_1(p,T)}{\partial p} + p\,\dfrac{\partial V_1(p,T)}{\partial p}}{\dfrac{\partial V_1(p,T)}{\partial T}} = \frac{\dfrac{\partial E_2(p,T)}{\partial p} + p\,\dfrac{\partial V_2(p,T)}{\partial p}}{\dfrac{\partial V_2(p,T)}{\partial T}} \quad \textit{for } p = p_0,\ T = T_0.$$

In other words, for gases occupied both constructions the stability takes place, the fraction

$$\frac{\dfrac{\partial E(p,T)}{\partial p} + p\,\dfrac{\partial V(p,T)}{\partial p}}{\dfrac{\partial V(p,T)}{\partial T}}$$

is a universal function $\alpha(p,T)$ of p and T, i.e., the equations of state for all such gases are connected by the equality

$$\frac{\partial E(p,T)}{\partial p}+p\frac{\partial V(p,T)}{\partial T}=\alpha(p,T)\frac{\partial V(p,T)}{\partial T}.$$

Considering a more complicated construction, we can assume that the gases occupied different boxes are in equilibrium and have the same temperature but different pressure and the equations describing their oscillations are the same as in the above cases if

$$A_i=-\frac{\partial V_i(p,T)}{\partial p},\quad B_i=-\frac{\partial V_i(p,T)}{\partial T},$$
$$C_i=\frac{\partial E_i(p,T)}{\partial p}+p_{i0}\frac{\partial V_i(p,T)}{\partial p},\quad D_i=\frac{\partial E_i(p,T)}{\partial T}+p_{i0}\frac{\partial V_i(p,T)}{\partial T},$$

where the derivatives are computed at the point (p_{10},T_0) if $i=1$ and at the point (p_{20},T_0) if $i=2$. If such a construction is stable, then the fractions

$$\frac{\dfrac{\partial E(p,T)}{\partial p}+p\dfrac{\partial V(p,T)}{\partial p}}{\dfrac{\partial V(p,T)}{\partial T}}$$

are equal for all admissible gases at the same temperature but different, in general, pressures. Thereby the relation

$$\frac{\dfrac{\partial E(p,T)}{\partial p}+p\dfrac{\partial V(p,T)}{\partial p}}{\dfrac{\partial V(p,T)}{\partial T}}=\alpha(T)$$

depends only on T. It turns out that the universal function $\alpha(p,T)$ is independent of the pressure $\alpha(p,T)=\alpha(T)$.

22.8. We will describe construction with desired properties later. Now, we show that the relation

$$\frac{\partial E(p,T)}{\partial p}+p\frac{\partial V(p,T)}{\partial p}=\alpha(T)\frac{\partial V(p,T)}{\partial T}$$

implies the existence of the integrating factor $\mu(T)$ using which we transform $\mu(dE+p\,dV)$ an exact differential. Setting $\mu(T)=e^{\int\frac{dT}{\alpha(T)}}$, we have $\dfrac{\mu}{\mu_T}=\alpha=\dfrac{E_p+pV_p}{V_T}$. The last equality is equivalent to the relation $\mu V_T=\mu_T(E_p+pV_p)$, which means that $\mu(dE+p\,dV)=\mu(E_T+p\,V_T)\,dT+\mu(E_p+pV_p)\,dp$ is the exact differential.

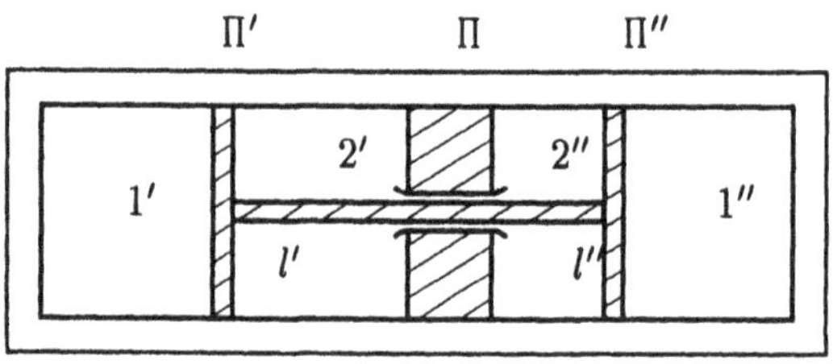

Fig. 31

22.9. To complete the proof of the assertion about the integrating factor, it remains to describe constructions where oscillations are described by the above equations. There are two such constructions. The first one is represented in Fig. 31. It consists of a cylinder of unit cross-section divided into four boxes by three moving pistons Π', Π, Π''. The piston Π has the mass $2m$ and the heat-conductivity coefficient $k/2$. The pistons Π' and Π'' are assumed to be weightless and infinitely heat-conductive. These pistons are rigidly joined by a rod $l'l''$. (This rod passes through the hole in Π and is not joined with Π.) We assume that the gas cannot pass from box $2''$ to box $2'$ through the gap between the rod and the walls of the hole in Π. We neglect the mass of the rod. Box $1'$ is occupied by gas of mass x_1 described by the equation of state $V_1 = V_1(p,T)$, $E_1 = E_1(p,T)$. Box $1''$ is occupied by the same gas of the same mass x_1. Each of the boxes $2'$ and $2''$ is occupied by a gas of mass x_2 with the equation of state $V_2 = V_2(p,T)$, $E_2 = E_2(p,T)$.

We denote by p_1', p_2', p_1'', p_2'' the pressures in boxes $1', 2', 1'', 2''$. Respectively, T' is the total temperature of boxes $1'$, $2'$, and T'' is the total temperature of boxes $1''$, $2''$. The subscript "0" indicates that the corresponding quantity characterizes the state of equilibrium.

If a system is in equilibrium, then $T_0' = T_0''$, $p_{20}' = p_{20}''$, $p_{10}' + p_{20}' = p_{10}'' + p_{20}''$. The last two equalities imply $p_{10}' = p_{10}''$. Introduce the notation $T_0' = T_0'' = T_0$, $p_{10}' = p_{10}'' = p_{10}$, and $p_{20}' = p_{20}'' = p_{20}$. It is obvious that the state of equilibrium exists for any T_0, p_{10}, p_{20}. To study small oscillations, in a neighborhood of the state of equilibrium, we set $u = \varepsilon\overline{u}$, $T' = T_0 + \varepsilon\overline{T}'$, $T'' = T_0 + \varepsilon\overline{T}''$, $p_1' = p_{10} + \varepsilon\overline{p}_1'$, $p_2' = p_{20} + \varepsilon\overline{p}_2'$, $p_1'' = p_{10} + \varepsilon\overline{p}_1''$, $p_2'' = p_{20} + \varepsilon\overline{p}_2''$. We obtain linear equations for $\overline{u}$, $\overline{p}_1'$, $\overline{p}_2'$, $\overline{p}_1''$, $\overline{p}_2''$, $\overline{T}'$, $\overline{T}''$ by a standard procedure.

Since the problem under consideration is symmetric, among solutions to this linear system there are "antisymmetric" solutions such that $\overline{p}_1' = \overline{p}_2' = -\overline{p}_1'' = -\overline{p}_2'' = (\overline{p})$ and $\overline{T}' = -\overline{T}'' = (\overline{T})$.

The stability of some solution to the linear system implies that this system is stable on "antisymmetric" solutions.

The system of equations describing "antisymmetric" oscillations of the construction in a neighborhood of the state of equilibrium is as follows:

$$
\begin{aligned}
&m\,\frac{d\overline{u}}{dt} = 0,\\
&(x_1A_1 + x_2A_2)\,\frac{d\overline{p}}{dt} + (x_1B_1 + x_2B_2)\,\frac{d\overline{T}}{dt} + \overline{u} = 0,\\
&(x_1C_1 + x_2C_2)\,\frac{d\overline{p}}{dt} + (x_1D_1 + x_2D_2)\,\frac{d\overline{T}}{dt} + k\overline{T} = 0.
\end{aligned}
$$

We omit the justification of these equations since the proof is an obvious modification of the arguments for obtaining the equations as in Fig. 30.

We obtain the same equation as in the case of the system represented in Fig. 30, but the coefficients A_i, B_i, C_i, D_i are expressed in terms of the derivatives of the equation of state and the derivatives are taken at two distinct points (p_{10}, T_0) and (p_{20}, T_0):

$$
\begin{aligned}
A_i &= \frac{\partial V_i(p,T)}{\partial p}, \quad & C_i &= \frac{\partial E_i(p,T)}{\partial p} + \frac{\partial V_i(p,T)}{\partial p},\\
B_i &= \frac{\partial V_i(p,T)}{\partial T}, \quad & D_i &= \frac{\partial E_i(p,T)}{\partial T} + \frac{\partial V_i(p,T)}{\partial T},
\end{aligned}
$$

where $p = p_{i0}$ and $T = T_0$. This system was used above.

Another construction is represented in Fig. 32. It consists of four equal cylinders each of which is divided into two boxes by some of the pistons Π', Π'', Π''', Π''''. The pistons Π' and Π''' are rigidly joined by a rod $l'l'''$. Therefore, a displacement of one of these pistons leads to the same displacement of some other piston. The pistons Π'' and Π'''' are also joined by a rod $l''l''''$. Boxes are marked by $1'$, $2'$, $1''$ and so on. Boxes $1'$ and $1'''$ are joined by a tube with the piston Π_B. Boxed $1''$ and $1''''$ are also joined by a tube with the piston Π_H. For each of the pistons Π_B, Π_H, Π', Π'', Π''', Π'''' the mass is equal to 1 and for each of the pistons Π_B, Π_H the mass is equal to $2m$. The pistons Π', Π'', Π''', Π'''', together with joining rods, are assumed to be weightless. A heat-conductive window is made in the wall separating boxes $1'$ and $2''$. Therefore, the temperature is the same in boxes $1'$ and $2''$. Similar heat-conducted windows are made between boxes $1''$ and $2'$, $1''''$ and $2'''$, $1'''$ and $2''''$, as is illustrated in Fig. 32. The pistons Π', Π'', Π''', Π'''' are assumed to be not heat-conductive and the heat conductivity coefficien of each piston Π_B, Π_H is assumed to be equal to $k/2$.

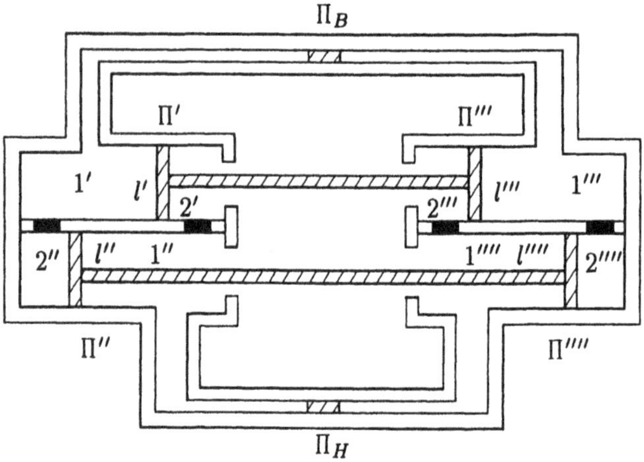

Fig. 32

Each of boxes $1'$, $1''$, $1'''$, $1''''$, together with joining part of the tubes, is occupied by the gas of mass y_1 with the equation of state $V_1 = V_1(p,T)$, $E_1 = E_1(p,T)$, pressure p_{10} and temperature T_0. Each of boxes $2'$, $2''$, $2'''$, $2''''$ is occupied by the gas of mass y_2 with the equation of state $V_2 = V_2(p,T)$, $E_2 = E_2(p,T)$, pressure p_{20}, and the same temperature T_0. It is easy to verify that the system is in equilibrium. We postulate that the state of equilibrium is stable and clarify the necessary conditions on the equation of state.

It suffices to consider only "symmetrically antisymmetric" oscillations in a neighborhood of the state of equilibrium such that $p_1' = p_{10} + \varepsilon\overline{p}$, $p_2' = p_{20} + \varepsilon\overline{p}$, $p_1'' = p_{10} - \varepsilon\overline{p}$, $p_2'' = p_{20} - \varepsilon\overline{p}$, $p_1''' = p_{10} - \varepsilon\overline{p}$, $p_2''' = p_{20} - \varepsilon\overline{p}$, $p_1'''' = p_{10} + \varepsilon\overline{p}$, $p_2'''' = p_{20} + \varepsilon\overline{p}$, $T' = T_0 + \varepsilon\overline{T}$, $T''' = T_0 - \varepsilon\overline{T}$, $T'' = T_0 + \varepsilon\overline{T}$, $T'''' = T_0 - \varepsilon\overline{T}$, $u_B = \varepsilon\overline{u}$, $u_H = -\varepsilon\overline{u}$. The system of ordinary differential equations describing such "symmetrically antisymmetric" oscillations turns out to be exactly the same as required:

$$m\,\frac{d\overline{u}}{dt} = 0,$$

$$(y_1A_1 + y_2A_2)\,\frac{d\overline{p}}{dt} + (y_1B_1 - y_2B_2)\,\frac{d\overline{T}}{dt} = -\overline{u},$$

$$(y_1C_1 - y_2C_2)\,\frac{d\overline{p}}{dt} + (y_1D_1 + y_2D_2)\,\frac{d\overline{T}}{dt} = -k\overline{T}.$$

The values of the coefficients A_i, B_i, C_i, D_i are the same as in the previous system describing oscillations presented in Fig. 31.

Chapter V

Multi-Dimensional Thermodynamically Compatible Conservation Laws

§ 23. Gas Dynamics Equations and Magnetic Hydrodynamics Equations

In Chapter IV, we studied the symmetrization of thermodynamically compatible equations of gas dynamics

$$\begin{aligned}
&\langle 1\rangle \quad \frac{\partial \rho}{\partial t}+\frac{\partial(\rho u_j)}{\partial x_j}=0,\\
&\langle 2\rangle \quad \frac{\partial \rho u_i}{\partial t}+\frac{\partial(\rho u_i u_j+\delta_{ij}\rho^2 E_\rho)}{\partial x_j}=0, \quad i=1,2,3,\\
&\langle 3\rangle \quad \frac{\partial[\rho(E+u_\alpha u_\alpha/2)]}{\partial t}+\frac{\partial[\rho u_j(E+\rho E_\rho+u_\alpha u_\alpha/2)]}{\partial x_j}=0,\\
&\langle 4\rangle \quad \frac{\partial(\rho S)}{\partial t}+\frac{\partial(\rho S u_j)}{\partial x_j}=0.
\end{aligned} \tag{23.1}$$

As was shown, it is possible to write the system obtained from (23.1) by eliminating the last equation in the form of a symmetric hyperbolic system with respect to "integrating factors." Multiplying the equations of the obtained system by these factors and adding the products, we obtain equation ⟨4⟩.

Similar constructions can be made if we choose other equations to eliminate. Further, for such an equation we take the energy equation ⟨3⟩, whereas ⟨4⟩ is included in the main part of the system. We describe the corresponding procedure. We assume that $q_\omega = T = E_S$, $q_0 = E - VE_V - SE_S - \frac{1}{2}(u_1^2 + u_2^2 + u_3^2)$, where the equation of state $E = E(V, S)$, temperature T, and pressure p are connected by the first thermodynamics law $dE = pdV + TdS$. In other words, $p = \rho^2 E_\rho = -E_V$.

For the generating potential we take $L(q_0, q_\omega, u_1, u_2, u_3) = p$. It is easy to check that

$$
\begin{aligned}
dL = d(\rho^2 E_\rho) &= \rho S dE_S + \rho d\left(E - VE_V - SE_S - \frac{u_1^2 + u_2^2 + u_3^2}{2}\right) + \rho u_i du_i \\
&= L_T dT + L_{q_0} dq_0 + L_{u_i} du_i,
\end{aligned}
\tag{23.2}
$$

i.e.,

$$
L_{q_0} = \rho, \quad L_T = \rho S, \quad L_{u_i} = \rho u_i. \tag{23.3}
$$

Therefore, the system

$$
\begin{aligned}
\frac{\partial L_{q_0}}{\partial t} + \frac{\partial (u_k) L_{q_0}}{\partial x_k} &= 0, \\
\frac{\partial L_{u_i}}{\partial t} + \frac{\partial (u_k L)_{u_i}}{\partial x_k} &= 0, \\
\frac{\partial L_{q_\omega}}{\partial t} + \frac{\partial (u_k L)_{q_\omega}}{\partial x_k} &= 0
\end{aligned}
\tag{23.4}
$$

is composed by chosen equalities in (23.1).

Multiplying each equation in (23.4) by q_0, u_i, q_ω respectively and adding, we arrive at the equation

$$
\frac{\partial}{\partial t}(q_0 L_{q_0} + u_i L_{u_i} + q_\omega L_{q_\omega} - L) + \frac{\partial}{\partial x_k} u_k (q_0 L_{q_0} + u_i L_{u_i} + q_\omega L_{q_\omega}) = 0. \tag{23.5}
$$

This equation coincides with ⟨3⟩ in (23.1), i.e., with the energy conservation law. We recommend the reader to check the above assertions by direct computation. It is useful to repeat the material of Section 20.

We consider a slightly more complicated collection of conservation laws (equalities in the divergence form)

$$\begin{aligned}
&\frac{\partial L_{q_0}}{\partial t}+\frac{\partial u_k L_{q_0}}{\partial x_k}=0,\\
&\frac{\partial L_T}{\partial t}+\frac{\partial u_k L_T}{\partial x_k}=0,\\
&\frac{\partial L_{u_i}}{\partial t}+\frac{\partial[(u_k L)_{u_i}-b_i L_{b_k}]}{\partial x_k}=0,\\
&\frac{\partial L_{b_i}}{\partial t}+\frac{\partial[(u_k L)_{b_i}-u_i L_{b_k}]}{\partial x_k}=0.
\end{aligned}\tag{23.6}$$

In these equations, together with the scalars q_0, T and the velocity vector with components u_i, one more vector with components b_i appears. Taking the generating potential in the form $L = p + (b_i b_i)/2$, where $p = \rho^2 E_\rho$, and introducing the notation $T = E_S(\rho, S)$. $q_0 = E + \rho E_\rho - S E_S - (u_i u_i)/2$, we can write (23.6) as follows:

$$\begin{aligned}
&\frac{\partial \rho}{\partial t}+\frac{\partial \rho u_k}{\partial x_k}=0,\\
&\frac{\partial \rho S}{\partial t}+\frac{\partial \rho S u_k}{\partial x_k}=0,\\
&\frac{\partial \rho u_i}{\partial t}+\frac{\partial}{\partial x_k}\left[\rho u_i u_k+p\delta_{ik}+b_i b_k-\frac{b_j b_j}{2}\delta_{ik}\right]=0,\\
&\frac{\partial b_i}{\partial t}+\frac{\partial(u_i b_k-u_k b_i)}{\partial x_k}=0.
\end{aligned}\tag{23.7}$$

This system presents the conservation laws in *ideal magnetohydrodynamics*, where b_i are the components of the intensity of a magnetic field.

The last equality in (23.7) implies that

$$\frac{\partial}{\partial t}\left(\frac{\partial b_i}{\partial x_i}\right)\equiv\frac{\partial}{\partial t}\left(\frac{\partial L_{b_i}}{\partial x_i}\right)=0.$$

Therefore, if at some time moment t_0 it turns out that $\dfrac{\partial b_i}{\partial x_i}=0$ everywhere, then the same equality remains valid for $t > t_0$, i.e., at every next time moment. Due to this fact, we can add to the systems (23.6), (23.7) a compatible equation, a stationary conservation law

$$\frac{\partial b_i}{\partial x_i}=0 \quad \left(\frac{\partial}{\partial x_i}L_{b_i}=0\right).$$

It turns our that this additional law should be used in order to derive the energy conservation law from (23.6). Namely, multiplying the equalities in (23.6) by

q_0, T, u_i, b respectively and adding the products, we add to the result the null sum of products $u_i L_{b_i} \dfrac{\partial}{\partial x_k} L_{b_k}$. Then we obtain the equality

$$\frac{\partial}{\partial t}(q_0 L_{q_0} + T L_T + u_i L_{u_i} + b_i L_{b_i} - L)$$
$$+ \frac{\partial}{\partial x_k}[u_k(q_i L_{q_i} + T L_T + u_i L_{u_i} + b_i L_{b_i} - u_m b_m L_{b_k}] = 0,$$

which in the initial (physical) variables was written as follows:

$$\frac{\partial\left[\rho\left(E + \dfrac{u_1^2 + u_2^2 + u_3^2}{2}\right) + \dfrac{b_1^2 + b_2^2 + b_3^2}{2}\right]}{\partial t}$$

$$+\frac{\partial}{\partial x_k}\left[\rho u_k\left(E + \frac{u_1^2 + u_2^2 + u_3^2}{2} + \frac{p}{\rho}\right) + u_k\,\frac{b_1^2 + b_2^2 + b_3^2}{2} - u_i b_i b_k\right] = 0.$$

§ 24. Equations of Elasticity

To present a more complicated thermodynamically compatible system of conservation laws, we consider the equations of nonlinear elasticity. We consider only smooth solutions, which is essential for the further computation.

We use parameters describing the state of an element of a medium in a infinitesimal neighborhood of some fixed point with coordinates x_1, x_2, x_3 at some time moment t. For the parameters we take elements c_{ij} of the matrix C (distortion), the entropy S, the specific volume V (or the density $\rho = 1/V$), and the components of the velocity u_i, $i = 1, 2, 3$. If the initial state of the medium is obtained by a purely elastic deformation, then the distortion tensor satisfies the compatibility conditions. In particular, the Burgers tensor B vanishes (recall that B is expressed in terms of the matrix $A = C^{-1}$ by formula (10) in Section 12) and the density ρ and the specific volume $V = 1/\rho$ are connected with C by the relations

$$\rho = 1/V = \rho^*/\det C, \quad \det[\rho C] = \rho^* \rho^2,$$

where ρ^* is the same constant for all elements. These conditions are satisfied if the above-mentioned elastic deformation started at the moment when $C = I$ and the density is equal to ρ^*.

At this moment we do not assume that $\rho/\det C = \text{const}$. Also, we do not assume that the compatibility condition holds at the initial time moment. In other words, we assume that the initial state is described by the *effective* elastic distortion C (cf. Chapter II) and is characterized by some arbitrary density distribution. Furthermore, the entropy is not assumed to be constant.

We assume that, starting with some time moment, the further deformation is elastic and the change of the distortion C is described by the equations (cf. Sections 8 and 12):

$$\frac{\partial c_{ij}}{\partial t} + u_0 \frac{\partial c_{ij}}{\partial x_k} - \frac{\partial u_i}{\partial x_k} x_{kj} = 0,$$

so that along each trajectory $\frac{dx_k}{dt} = u_k$ the density ρ changes by the rule

$$\frac{d\rho}{dt} \equiv \frac{\partial}{\partial t} + u_k \frac{\partial \rho}{\partial x_k} = -\rho \frac{\partial u_k}{\partial x_0}$$

and the entropy remains unchanged:

$$\frac{dS}{dt} = \frac{\partial S}{\partial t} + u_k \frac{\partial S}{\partial x_k} = 0.$$

It is convenient to write these relations in the following equivalent way:

$$\frac{\partial \rho}{\partial t} + \frac{\partial(\rho u_k)}{\partial x_k} = 0, \tag{24.1}$$

$$\frac{\partial(\rho c_{ij})}{\partial t} + \frac{\partial(u_k \rho c_{ij})}{\partial x_k} - \rho c_{kj} \frac{\partial u_i}{\partial x_k} = 0, \tag{24.2}$$

$$\frac{\partial \rho S}{\partial t} + \frac{\partial u_k \rho S}{\partial x_k} = 0. \tag{24.3}$$

To describe the evolution of the velocity field, we can use the momentum conservation law (cf. Section 14):

$$\frac{\partial \rho u_i}{\partial t} + \frac{\partial(\rho u_i u_k - \sigma_{ik})}{\partial x_k} = 0. \tag{24.4}$$

At present, we restrict ourselves to the equations with zero right-hand side, i.e., we assume that no distributed mass forces act.

To write the equation of state, we define the dependence of the internal energy E (per unit mass) on the above parameters c_{ij}, V, S defining the state of the medium and, possibly, on a number of some other parameters (for example, $V_0 = 1/\rho_0$). It is clear that V_0 and ρ_0 are the same for a fixed element at all time moments, i.e., they are constant along each trajectory $\frac{dx_i}{dt} = u_i$. Therefore, we have

$$\frac{\partial(\rho/\rho_0)}{\partial t} + \frac{\partial(u_k \rho/\rho_0)}{\partial x_k} = 0. \tag{24.5}$$

Thus, we assume that

$$E = E(c_{11}, c_{12}, \ldots, c_{33}, V, S, V_0). \tag{24.6}$$

We also assume that E is convex with respect to all arguments and $E_V < 0$.

Example 24.1. The required conditions are satisfied by the function

$$E = a_0^2\left(\frac{k_{-1}^2 + k_0^2 + k_1^2}{3(V/V_0)^{2/3}} - 1\right) + c_0^2\left[\frac{e^S}{\varkappa(\varkappa - 1)}\left(\frac{V_0}{V}\right)^{\varkappa - 1} + \frac{V}{\varkappa V_0} - 1\right]$$
$$= a_0^2\Phi(k_1, k_2, k_3, V) + c_0^2\Psi(V, S) \tag{24.7}$$

It is easy to verify that Φ is a convex function of all 4 variables. At the same time,

$$\frac{\Phi_{k_i} + \Phi_{k_j}}{k_i + k_j} > 0, \quad \frac{\Phi_{k_i} - \Phi_{k_j}}{k_i - k_j} > 0, \quad i \neq j.$$

We recall (cf. Section 18) that these inequalities, in view of the convexity of Φ with respect to k_{-1}, k_0, k_1, imply the convexity of Φ with respect to the entries c_{ij} of the matrix C for a fixed V. On the other hand, it is easy to see that $\Psi(V, S)$ is convex with respect to v, S. Therefore, for sufficiently large ratio c_0^2/a_0^2 the total energy E is a convex function of c_{ij}, V, S. We note that, in the case $\rho^* = \rho_0 = 1/V_0 = \text{const}$ (which corresponds to the case of elastic deformations), we have $V/V_0 = k_{-1}k_0k_1$ and $\Phi(k_{-1}, k_0, k_1, V) \geqslant 0$ because of the known inequality between the arithmetic and geometric means.

Remark 24.1. We can indicate a more general example: $E = E(I_1, I_2, I_3)$ is a convex function of $I_1 = k_{-1} + k_0 + k_1$, $I_2 = \sqrt{k_{-1}^2 + k_0^2 + k_1^2}$, V, S with $E_{I_1} > 0$, $E_{I_2} > 0$, $E_V < 0$.

Since E is convex and $E_V < 0$, we obtain the convexity of the total (internal and kinetic) energy u (per unit mass) $U(u_1, u_2, u_3; c_{11}, c_{12}, \ldots, c_{33}, V, V_0, S) = E + \dfrac{u_1^2 + u_2^2 + u_3^2}{2}$ and the inequality $U_V < 0$.

We make the change of variables $\rho = \dfrac{1}{V}$, $\delta = \dfrac{V_0}{V}$, $v_i = \dfrac{u_i}{V} = \rho u_i$, $\widetilde{S} = \dfrac{S}{V} = \rho S$, $f_{ij} = \dfrac{c_{uj}}{V} = \rho c_{ij}$ and choose the potential $W = \dfrac{1}{V}U = \rho U$. Arguing as in Section 20, we can verify that the potential $W(v_1, v_2, v_3, \rho, \delta, \widetilde{S}, f_{11}, f_{12}, \ldots, f_{33})$ is a convex function of its arguments.

Let L be the Legendre potential of W, i.e., $L + v_iW_{v_i} + \rho W_\rho + \delta W_\delta + \widetilde{S}W_{\widetilde{S}} + f_{ij}W_{f_{ij}} - W$. We regard L as a function of $u_i = W_{v_i}$, $n = W_\delta$, $T = W_{\widetilde{S}} = E_S$, $r_{ij} = W_{f_{ij}} = E_{c_{ij}}$, $q_0 = W_\rho = E - VE_V - SE_S - V_0E_{V_0} - c_{ij}E_{c_{ij}} - \dfrac{u_iu_i}{2}$. Moreover,

$$L_{q_0} = \rho,\ L_n = \frac{\rho}{\rho_0},\ L_T = \rho S',\ L_{r_{ij}} = \rho c_{ij},\ L_{u_i} = \rho u_i,\ L = -VE_V. \tag{24.8}$$

In these variables, Eqs. (24.1)–(24.5) describing the evolution of elastic deformations are written in the form

$$\begin{aligned}
&\frac{\partial L_{q_0}}{\partial t} + \frac{\partial (u_k L_{q_0})}{\partial x_k} = 0,\\
&\frac{\partial L_n}{\partial t} + \frac{\partial (u_k L_n)}{\partial x_k} = 0,\\
&\frac{\partial L_T}{\partial t} + \frac{\partial (u_k L_T)}{\partial x_k} = 0,\\
&\frac{\partial L_{r_{ij}}}{\partial t} + \frac{\partial (u_k L_{r_{ij}})}{\partial x_k} - \frac{\partial u_i}{\partial x_k} L_{r_{kj}} = 0,\\
&\frac{\partial L_{u_i}}{\partial t} + \frac{\partial (u_k L_{u_i} - \sigma_{ik})}{\partial x_k} = 0.
\end{aligned} \tag{24.9}$$

To complete the derivation of the required system, we express the elements σ_{ik} of the stress tensor in terms of the above variables and the potential L. To do this, we can use the Murnaghan formulas. However, we choose another way. We try to clarify conditions on σ_{jk} under which (24.9) implies the conservation of energy. In fact, this is also the proof of the Murnaghan formulas which were obtained from the conservation of energy.

Multiplying the equations in (24.9) by the corresponding factors q_0, n, T, r_{ij}, u_i, adding the products, and using the equality

$$q_0 L_{q_0} + n L_n + T L_T + r_{ij} L_{r_{ij}} + u_i L_{u_i} - L = W,$$

we obtain the equality

$$\frac{\partial W}{\partial t} + \frac{\partial (u_k W)}{\partial x_k} - u_i \frac{\partial (\sigma_{ik} + \delta_{ik} L)}{\partial x_k} - r_{ij} L_{r_{kj}} \frac{\partial u_i}{\partial x_k} = 0,$$

where the last two terms can be written as the derivative if

$$\sigma_{ik} = -\delta_{ik} L + r_{ij} L_{r_{kj}}. \tag{24.10}$$

We regard this relation as the required formulation of the Murnaghan rule.

Thus, the equations take the form

$$\begin{aligned}
&\langle 1 \rangle \quad \frac{\partial L_{q_0}}{\partial t} + \frac{\partial (u_k L_{q_0})}{\partial x_k} = 0,\\
&\langle 2 \rangle \quad \frac{\partial L_n}{\partial t} + \frac{\partial (u_k L_n)}{\partial x_k} = 0,\\
&\langle 3 \rangle \quad \frac{\partial L_T}{\partial t} + \frac{\partial (u_k L_T)}{\partial x_k} = 0,\\
&\langle 4 \rangle \quad \frac{\partial L_{r_{ij}}}{\partial t} + \frac{\partial (u_k L_{r_{ij}})}{\partial x_k} - \frac{\partial u_i}{\partial x_k} L_{r_{kj}} = 0,\\
&\langle 5 \rangle \quad \frac{\partial L_{u_i}}{\partial t} + \frac{\partial [(u_k L)_{u_i} - u_i r_{ij} L_{r_{kj}}]}{\partial x_k} = 0.
\end{aligned} \tag{24.11}$$

We proceed by studying the hyperbolicity of the equations. To this end, we extend the system by introducing new unknowns and adding extra equations. Equation ⟨4⟩ in (24.11) can be written in the form

$$\frac{\partial L_{r_{ij}}}{\partial t} + \frac{\partial (u_k L_{r_{ij}} - u_i L_{r_{kj}})}{\partial x_k} + u_i \frac{\partial L_{r_{kj}}}{\partial x_k} = 0.$$

Differentiating both sides of this equality with respect to x_i, we find

$$\frac{\partial}{\partial t}\left(\frac{\partial L_{r_{ij}}}{\partial x_i}\right) + \frac{\partial}{\partial x_i}\left(u_i \frac{\partial L_{r_{kj}}}{\partial x_k}\right) = 0,$$

where we adopt the convention about summation over double indices. Using the notation

$$b_j = \frac{1}{L_{q_0}} \frac{\partial L_{r_{ij}}}{\partial x_i}, \tag{24.12}$$

we arrive at the equations for the vector b which can be written in the following two forms (equivalent because of Eq. ⟨1⟩ in (24.11));

$$\frac{\partial (b_j L_{q_0})}{\partial t} + \frac{\partial (u_k b_j L_{q_0})}{\partial x_k} = 0, \tag{24.13}$$

$$\frac{\partial b_j}{\partial t} + u_k \frac{\partial b_j}{\partial x_k} = 0. \tag{24.14}$$

Now, we can write Eq. ⟨5⟩ in (24.11) in the form

$$\frac{\partial L_{u_i}}{\partial t} + \frac{\partial (u_k L)_{u_i}}{\partial x_k} - L_{r_{kj}} \frac{\partial r_{ij}}{\partial x_k} = r_{ij} \frac{\partial L_{r_{kj}}}{\partial x_k},$$

$$\frac{\partial L_{u_i}}{\partial t} + \frac{\partial (u_k L)_{u_i}}{\partial x_k} - L_{r_{kj}} \frac{\partial r_{ij}}{\partial x_k} = L_{q_0} r_{ij} b_j.$$

Finally, the system of the modified equations (24.11) and (24.14) can be written as follows:

$$\begin{aligned}
&\frac{\partial L_{q_0}}{\partial t} + \frac{\partial M_{q_0}^{(k)}}{\partial x_k} = 0,\\
&\frac{\partial L_n}{\partial t} + \frac{\partial M_n^{(k)}}{\partial x_k} = 0,\\
&\frac{\partial L_T}{\partial t} + \frac{\partial M_T^{(k)}}{\partial x_k} = 0,\\
&\frac{\partial L_{r_{ij}}}{\partial t} + \frac{\partial M_{r_{ij}}^{(k)}}{\partial x_k} - L_{r_{kj}} \frac{\partial u_i}{\partial x_k} = 0,\\
&\frac{\partial L_{u_i}}{\partial t} + \frac{\partial M_{u_i}^{(k)}}{\partial x_k} - L_{r_{kj}} \frac{\partial r_{ij}}{\partial x_k} = L_{q_0} r_{ij} b_j,\\
&\frac{\partial b_j}{\partial t} + u_k \frac{\partial b_j}{\partial x_k} = 0,
\end{aligned} \tag{24.15}$$

It is easy to verify that we can write the system (24.15) in the quasilinear form and the obtained system is symmetric and hyperbolic. To proof this assertion, it is necessary to use the convex dependence of L on unknown arguments.

In the initial variables, the equalities $\frac{\partial L_{r_{ij}}}{\partial x_i} = b_j L_{q_0}$ in (24.12) are written as follows:

$$\frac{\partial \rho c_{ij}}{\partial x_i} = \rho b_j. \tag{24.16}$$

By (24.14), if the vector b is zero at the initial time moment, then it will be zero at other time moments. In particular, this is true if, for example, there were no stresses at the initial time moment and, consequently, $C = I$ and if the density $\rho = \rho^* \det C$ is the same at all points, i.e., $\rho = \rho^* = \text{const}$. Under the same assumptions, all the components b_j vanish and will remain zero during the whole time-interval of elastic deformation. The equality $b = 0$ is similar to the compatibility condition $B = 0$, where B is the Burgers tensor.

We also note that for $b_j = 0$ Eq. (4) in (24.11) can be written as the divergence form of conservation laws:

$$\frac{\partial L_{r_{ij}}}{\partial yt} + \frac{\partial (u_k L_{r_{ij}} - u_i L_{r_{kj}})}{\partial x_k} = 0.$$

§ 25. Structure of Multi-Dimensional Equations

The equations in Sections 23 and 24 are remarkable because they form essentially overdetermined (although compatible) systems. Therefore, we cannot study them within the framework of the scheme proposed in Chapter IV for systematization of conservation laws compatible with thermodynamics. In this scheme, only one additional equation was used and it was expressed as a linear combination of the remaining laws.

Now, we consider a collection of the main conservation laws which was gathered when we tried to systematize various physical examples described in the classical monographs and journals on the theoretical physics. This illustration will be completed by some conservation laws that are a consequence of the main laws.

We introduce the notation: u_i, are the components of velocity, d_i are the components of the intensity of electric field, b_i are the components of the intensity of magnetic field, and j_i are the components of the vector describing the flow of a superfluid material. Tensor parameters r_{kl} describe the stressed and deformed state. Among variables are scalars q_0, n, $q_\omega = T$. A number of parameters describing the medium are denoted by q_γ, where γ is an index or a multi-index. We adopt the convention about summation over repeating indices (the Einstein rule). Our collection looks as follows:

$$
\begin{aligned}
&\langle 1\rangle \quad \frac{\partial L_{q_\gamma}}{\partial t} + \frac{\partial u_k L_{q_\gamma}}{\partial x_k} = 0,\\
&\langle 2\rangle \quad \frac{\partial L_{u_i}}{\partial t} + \frac{\partial}{\partial x_k}\big[(u_k L)_{u_i} - r_{il} L_{r_{kl}} - b_i L_{b_k} - d_i L_{d_k}\\
&\qquad\qquad + j_k L_{j_i} - \delta_{ik} j_\alpha L_{j_\alpha}\big] = 0,\\
&\langle 3\rangle \quad \frac{\partial L_{r_{il}}}{\partial t} + \frac{\partial (u_k L_{r_{il}} - u_i L_{r_{kl}})}{\partial x_k} = 0,\\
&\langle 4\rangle \quad \frac{\partial L_{d_i}}{\partial t} + \frac{\partial (u_k L_{d_i} - u_i L_{d_k} - e_{ikl} b_l)}{\partial x_k} = 0,\\
&\langle 5\rangle \quad \frac{\partial L_{b_i}}{\partial t} + \frac{\partial (u_k L_{b_i} - u_i L_{b_k} + e_{ikl} d_l)}{\partial x_k} = 0, \qquad (25.1)\\
&\langle 6\rangle \quad \frac{\partial L_n}{\partial t} + \frac{\partial (u_k L_n + j_k)}{\partial x_k} = 0,\\
&\langle 7\rangle \quad \frac{\partial L_{j_l}}{\partial t} + \frac{\partial (u_\alpha L_{j_\alpha} + n)}{\partial x_l} = 0.
\end{aligned}
$$

Hereinafter, e_{ikl} is the unit pseudoscalar: $e_{ikl} = +1$ if (ikl) is an even permutation and $e_{ikl} = -1$ if (ikl) is an odd permutation.

In the system (25.1), the number of equations is equal to the number of unknown functions. We indicate some consequence of this system under certain assumptions on the initial data. These consequences turn out to be conservation laws which complement (25.1).

From Eqs. ⟨3⟩–⟨5⟩ of the system (25.1) differentiating with respect to x_i and summarizing, it is easy to obtain the relation

$$
\begin{aligned}
&\frac{\partial}{\partial t}\left(\frac{\partial L_{r_{il}}}{\partial x_i}\right) \equiv \frac{\partial}{\partial t}\operatorname{div} L_{r_l} = 0, \quad l = 1,2,3,\\
&\frac{\partial}{\partial t}\left(\frac{\partial L_{d_i}}{\partial x_i}\right) \equiv \frac{\partial}{\partial t}\operatorname{div} L_d = 0,\\
&\frac{\partial}{\partial t}\left(\frac{\partial L_{b_i}}{\partial x_i}\right) \equiv \frac{\partial}{\partial t}\operatorname{div} L_b = 0.
\end{aligned}
$$

In the same way, from ⟨7⟩ we obtain the equality

$$\frac{\partial}{\partial t}\left(\frac{\partial L_{j_k}}{\partial x_i} - \frac{\partial L_{j_i}}{\partial x_k}\right) = 0,$$

which is written in the vector form as follows:

$$\frac{\partial}{\partial t}\operatorname{rot} L_j = 0.$$

Therefore, preserving the compatibility, we can complete the system (25.1) by stationary conservation laws

$$\operatorname{div} L_{r_l} = 0, \quad \operatorname{div} L_d = 0, \quad \operatorname{div} L_b = 0, \quad \operatorname{rot} L_j = 0, \tag{25.2}$$

provided that these laws hold for the initial data. We write (25.2) in the component form:

$$\begin{aligned} &\frac{\partial L_{r_{il}}}{\partial x_i} = 0, \quad \frac{\partial L_{d_i}}{\partial x_i} = 0, \\ &\frac{\partial L_{b_i}}{\partial x_i} = 0, \quad \frac{\partial L_{j_i}}{\partial x_k} - \frac{\partial L_{j_k}}{\partial x_i} = 0. \end{aligned} \tag{25.3}$$

If the system (25.1) is thermodynamically compatible, this means that there is an additional evolution equation satisfied by the solutions to the system (25.1). For such an additional law for smooth solutions we take the conservation laws which is obtained in the following way. We multiply the equations in (25.1) by the corresponding factors: ⟨1⟩ by q_γ, ⟨2⟩ by u_i, ⟨3⟩ by r_{ij}, ⟨4⟩ by d_i, ⟨5⟩ by b_i, ⟨6⟩ by n, ⟨7⟩ by j_i, add the products and add a combination of stationary laws (25.3):

$$u_i d_i \frac{\partial L_{d_k}}{\partial x_k} + u_i b_i \frac{\partial L_{b_k}}{\partial x_k} - 2u_i j_k \left(\frac{\partial L_{j_i}}{\partial x_k} - \frac{\partial L_{j_k}}{\partial x_i} \right) + r_{il} u_i \frac{\partial L_{r_{kl}}}{\partial x_k} = 0.$$

Further, we should use the transformation which is already used while reducing to the exact differential:

$$\begin{aligned} &q_\gamma \frac{\partial L_{q_\gamma}}{\partial t} + u_i \frac{\partial L_{u_i}}{\partial t} + r_{il} \frac{\partial L_{r_{il}}}{\partial t} + d_i \frac{\partial L_{d_i}}{\partial t} + b_i \frac{\partial L_{b_i}}{\partial t} + n \frac{\partial L_n}{\partial t} + j_k \frac{\partial L_{j_k}}{\partial t} \\ &= \frac{\partial}{\partial t} (q_\gamma L_{q_\gamma} + u_i L_{u_i} + r_{il} L_{r_{il}} + d_i L_{d_i} + b_i L_{b_i} + n L_n + j_k L_{j_k} - L). \end{aligned}$$

To obtain the divergence form for the terms containing the derivatives with respect to spatial variables, we use the following chain of equalities:

$$\begin{aligned} &q_\gamma \frac{\partial u_k L_{q_\gamma}}{\partial x_k} + u_i \frac{\partial (u_k L_{u_i} + \delta_{ik} L - r_{il} L_{r_{kl}} - b_i L_{b_k} - d_i L_{d_k} + j_k L_{j_i} - \delta_{ik} j_\alpha L_{j_\alpha})}{\partial x_k} \\ &+ r_{il} \frac{\partial (u_k L_{r_{il}} - u_i L_{r_{kl}})}{\partial x_k} + u_i r_{il} \frac{\partial L_{r_{kl}}}{\partial x_k} + d_i \frac{\partial (u_k L_{d_i} - u_i L_{d_k} - e_{ikl} b_l)}{\partial x_k} \\ &+ u_i d_i \frac{\partial L_{d_k}}{\partial x_k} + b_i \frac{\partial (u_k L_{b_i} - u_i L_{b_k} + e_{ikl} d_l)}{\partial x_k} + u_i b_i \frac{\partial L_{b_k}}{\partial x_k} + n \frac{\partial (u_k L_n + j_k)}{\partial x_k} \\ &+ j_k \frac{\partial (u_\alpha L_{j_\alpha} + n)}{\partial x_k} - u_i j_k \left(\frac{\partial L_{j_i}}{\partial x_k} - \frac{\partial L_{j_k}}{\partial x_i} \right) \end{aligned}$$

$$= q_\gamma \frac{\partial u_k l_{q_\gamma}}{\partial x_k} + u_i \frac{\partial u_k L_{u_i}}{\partial x_k} + r_{il} \frac{\partial u_k L_{r_{il}}}{\partial x_k} + d_i \frac{\partial u_k L_{d_i}}{\partial x_k} + b_i \frac{\partial u_k L_{b_i}}{\partial x_k} + n \frac{\partial u_k L_n}{\partial x_k}$$

$$+ j_i \frac{\partial u_k L_{j_i}}{\partial x_k} + u_k \frac{\partial L}{\partial x_k} - u_i \frac{\partial}{\partial x_k}(r_{il} L_{r_{kl}} + b_i L_{b_k} + d_i L_{d_k} - j_k L_{j_i} + \delta_{ik} j_\alpha L_{j_\alpha})$$

$$- (r_{il} L_{r_{kl}} + d_i L_{d_k} + b_i L_{b_k} - j_k L_{j_i} + j_\alpha L_{j_\alpha} \delta_{ik}) \frac{\partial u_i}{\partial x_k} - d_i \frac{\partial e_{ikl} b_l}{\partial x_k}$$

$$+ b_i \frac{\partial e_{ikl} d_l}{\partial x_k} + n \frac{\partial j_k}{\partial x_k} + j_k \frac{\partial n}{\partial x_n}$$

$$= \frac{\partial}{\partial x_k} u_k (q_\gamma L_{q_\gamma} + u_i L_{u_i} + r_{il} L_{r_{il}} + d_i L_{d_i} + b_i L_{b_i} + n L_n + j_i L_{j_i})$$

$$- u_k \left(L_{q_\gamma} \frac{\partial q_\gamma}{\partial x_k} + L_{u_i} \frac{\partial u_i}{\partial x_k} + L_{r_{il}} \frac{\partial r_{il}}{\partial x_k} + L_{d_i} \frac{\partial d_i}{\partial x_k} + L_{b_i} \frac{\partial b_i}{\partial x_k} + L_n \frac{\partial n}{\partial x_k} \right.$$

$$\left. + L_{j_i} \frac{\partial j_i}{\partial x_k} \right) + u_k \frac{\partial L}{\partial x_k} - \frac{\partial}{\partial x_k} u_i (r_{il} L_{r_{kl}} + b_i L_{b_k} + d_i L_{d_k} - j_k L_{j_i} + \delta_{ik} j_\alpha L_{j_\alpha})$$

$$+ \frac{\partial}{\partial x_k} (e_{ikl} b_i d_l + n j_k).$$

We obtain the equality

$$\frac{\partial}{\partial t} (q_\gamma L_{q_\gamma} + u_i L_{u_i} + r_{il} L_{r_{il}} + d_i L_{d_i} + b_i L_{b_i} + n L_n + j_i L_{j_i} - L)$$
$$+ \frac{\partial}{\partial x_k} [u_k (q_\gamma L_{q_\gamma} + u_i L_{u_i} + r_{il} L_{r_{il}} + d_i L_{d_i} + b_i L_{b_i} + n L_n)$$
$$+ u_\alpha (j_k L_{j_\alpha} - r_{\alpha\beta} L_{r_{k\beta}} - d_\alpha L_{d_k} - b_\alpha L_{b_k}) + j_k n + e_{k\alpha\beta} d_\alpha b_\beta] = 0. \quad (25.4)$$

In all the considered examples, this equation is the differential form of the energy conservation law. The equations in the above collection (25.1), (25.3) (and their generalizations) are called "thermodynamically compatible equations."

To conclude the section, we indicate some more complicated equaitons similar to (25.1) which are useful in applications. Namely, we write some special right-hand sides for some equations.

Namely, we replace the zero right-hand side with $-n/\tau$ in Eq. $\langle 6 \rangle$ and insert the right-hand side $-q_\gamma/\tau_\gamma$ for $\gamma \neq \omega$ in Eq. $\langle 1 \rangle$. In the case $\gamma = \omega$, for the right-hand side we can take

$$\frac{1}{q_\omega} \left(\frac{n^2}{\tau} + \sum_{\gamma \neq \omega} \frac{q_\gamma^2}{\tau} \right).$$

Then we obtain the following equations:

$$\begin{aligned} &\frac{\partial L_{q_\omega}}{\partial t} + \frac{\partial(u_k L_{q_\omega})}{\partial x_k} = \frac{1}{q_\omega}\left(\frac{n^2}{\tau} + \sum_{\gamma\neq\omega}\frac{q_\gamma^2}{\tau_\gamma}\right), \\ &\frac{\partial L_{q_\gamma}}{\partial t} + \frac{\partial(u_k L_{q_\gamma})}{\partial x_k} = -\frac{q_\gamma}{\tau_\gamma} \quad (\gamma \neq \omega), \\ &\frac{\partial L_n}{\partial t} + \frac{\partial(u_k L_n + j_k)}{\partial x_k} = -\frac{n}{\tau}, \end{aligned} \tag{25.5}$$

where $q_\omega > 0$, $\tau > 0$, $\tau_\gamma > 0$. In applications, q_ω is the temperature T and L_{q_ω} is the entropy.

These right-hand sides simulate relaxation processes leading to the increase of entropy because the right-hand side of the first equation of (25.5) is positive.

It is important that these right-hand sides do not contradict the validity of the conservation of energy (25.4), which can be easily verified.

§ 26. Symmetric Hyperbolicity in the Three-Dimensional Case

The thermodynamically compatible system (25.1), (25.3) can be reduced to the quasilinear form in which the matrices at the derivatives with respect to t and those with respect to x_i are symmetric. To this end, it is necessary to transform some of the equations. To Eq. ⟨2⟩ of the system (25.1) we add the following combination of stationary laws (25.3):

$$r_{i\alpha}\frac{\partial L_{r_{k\alpha}}}{\partial x_k} + d_i\frac{\partial L_{d_k}}{\partial x_k} + b_i\frac{\partial L_{b_k}}{\partial x_k} + j_k\left(\frac{\partial L_{j_k}}{\partial x_i} - \frac{\partial L_{j_i}}{\partial x_k}\right) = 0,$$

to Eq. ⟨3⟩ we add the expression $u_i\dfrac{\partial L_{r_{il}}}{\partial x_k} = 0$, and to Eqs. ⟨4⟩, ⟨5⟩, ⟨7⟩ we add the expressions

$$u_i\frac{\partial L_{d_k}}{\partial x_k} = 0, \quad u_i\frac{\partial L_{b_k}}{\partial x_k} = 0, \quad u_i\left(\frac{\partial L_{j_k}}{\partial x_i} - \frac{\partial L_{j_i}}{\partial x_k}\right) = 0$$

respectively. We obtain the system equivalent to the system (25.1):

$$\begin{aligned} &\frac{\partial L_{q_\gamma}}{\partial t} + \frac{\partial(u_k L)_{q_\gamma}}{\partial x_k} = 0, \\ &\frac{\partial L_{u_i}}{\partial t} + \frac{\partial(u_k L)_{u_i}}{\partial x_k} - L_{r_{k\alpha}}\frac{\partial r_{i\alpha}}{\partial x_k} - L_{d_k}\frac{\partial d_i}{\partial x_k} - L_{b_i}\frac{\partial L_{b_k}}{\partial x_k} + L_{j_i}\frac{\partial j_k}{\partial x_k} - L_{j_\alpha}\frac{\partial j_\alpha}{\partial x_k} = 0, \\ &\frac{\partial L_{r_{ij}}}{\partial t} + \frac{\partial(u_k L)_{r_{ij}}}{\partial x_k} - L_{r_{kj}}\frac{\partial u_i}{\partial x_k} = 0, \end{aligned}$$

$$\begin{aligned}
&\frac{\partial L_{d_i}}{\partial t}+\frac{\partial (u_k L)_{d_i}}{\partial x_k}-L_{d_k}\frac{\partial u_i}{\partial x_k}-e_{ikl}\frac{\partial b_l}{\partial x_k}=0,\\
&\frac{\partial L_{b_i}}{\partial t}+\frac{\partial (u_k L)_{b_i}}{\partial x_k}-L_{b_k}\frac{\partial u_i}{\partial x_k}-e_{ikl}\frac{\partial d_l}{\partial x_k}=0,\\
&\frac{\partial L_n}{\partial t}+\frac{\partial (u_k L)_n}{\partial x_k}+\frac{\partial j_k}{\partial x_k}=0,\\
&\frac{\partial L_{j_l}}{\partial t}+\frac{\partial (u_k L)_{j_l}}{\partial x_k}+L_{j_\alpha}\frac{\partial u_\alpha}{\partial x_l}+L_{j_l}\frac{\partial u_k}{\partial x_k}+\frac{\partial n}{\partial x_l}=0.
\end{aligned}\tag{26.1}$$

Let us verify that the system (26.1) can be written in the quasilinear form with symmetric matrices of coefficients at the derivatives. To this end, we introduce the generating functions (potentials)

$$\begin{aligned}
&M^k = u_k L + e_{kil} b_i d_l + n j_k,\\
&\Phi^m = u_\alpha r_{\alpha m},\quad \Psi^m = e_{m\alpha\beta} u_\alpha j_\beta,\quad \Theta^1 = u_\alpha d_\alpha,\quad \Theta^2 = u_\alpha b_\alpha.
\end{aligned}\tag{26.2}$$

Using these potentials, it is easy to show that the system (26.1) reduces to the following form:

$$\begin{aligned}
&\frac{\partial L_V}{\partial t}+\frac{\partial M^k_V}{\partial x_k}-L_{r_{km}}\frac{\partial \Phi^m_V}{\partial x_k}-L_{d_k}\frac{\partial \Theta^1_V}{\partial x_k}-L_{b_k}\frac{\partial \Theta^2_V}{\partial x_k}\\
&-e_{mk\beta}L_{j_\beta}\frac{\partial \Psi^m_V}{\partial x_k}=0,
\end{aligned}\tag{26.3}$$

where all the unknown functions are included in the vector-valued function

$$V=(q_1,\dots,u_1,u_2,u_3,r_{11},\dots,r_{33},d_1,d_2,d_3,b_1,b_2,b_3,n,j_1,j_2,j_3)^T$$

and L_{VV} denote teh matrix with entries $L_{V_\alpha V_\beta}$. It is obvious that in (26.3), the matrices of coefficients at $\dfrac{\partial}{\partial t}$ and $\dfrac{\partial}{\partial x_k}$ are symmetric. These matrices are combinations of the matrices of the second-order derivatives of the generating potentials (26.2):

$$L_{VV}\frac{\partial V}{\partial t}+(M^k_{VV}-L_{r_{km}}\Phi^m_{VV}-L_{d_k}\Theta^1_{VV}-L_{b_k}\Theta^2_{VV}-e_{mk\beta}L_{j_\beta}\Psi^m_{VV})\frac{\partial V}{\partial x^k}=0. \tag{26.4}$$

We assume that the potential L is a convex function of its arguments (components of V). Then the matrix of the second-order derivatives L_{VV} is positive definite. In this case, we have a symmetric hyperbolic system.

We note that our symmetrization is essentially based on the homogeneity of the conservation laws under consideration, i.e., the right-hand sides vanish. Each case of a nonzero right-hand side requires additional consideration.

Appendix

S. K. Godunov[1)]

Structure of Thermodynamically Compatible Systems

§ 1. Mathematical Aspects

In Chapter V, we presented a rather large collection of various overdetermined but compatible systems of equations of mathematical physics. Some of the equations looked like conservation laws and allow us to formulate the law of entropy growth. Therefore, we say that such systems are "thermodynamically compatible."

For some of the examples in the previous chapter we indicated a rule for constructing a special linear combination of equations in such a way that these linear combinations form a quasilinear symmetric hyperbolic system. We recall that symmetric hyperbolic systems have a fine property, namely, for sufficiently smooth initial data there exists a solution to the system and this solution has

---

[1)] This chapter was written for the English edition. It contains the very recent results obtained during the preparation of the English translation of this book.

This work is partially supported by the Russian Foundation of Fundamental Research (grant no. 01-01-00766).

derivatives. If the initial data are more smooth, then the solution has higher smoothness.

We note that the existence theorem is local, i.e., we can construct a solution only in some neighborhood of the initial data only for a bounded time-interval.

The number of equations of a symmetric hyperbolic system is equal to the number of unknown functions. The additional relations in the overdetermined system can be regarded as the integrals of the system.

Now we try to systematize examples similar to those considered in the previous chapter. We first consider rather special symmetric hyperbolic systems. Namely, we consider equations for which the Cauchy problem is solvable. We choose special systems possessing integrals, which allows us, under certain conditions on the initial data, to assert that some conservation laws are valid on solutions of the system.

We construct systems in such a way that the resulting system is Galilei-invariant. However, we cannot justify this for the equations of electrodynamics. Therefore, we have to modify them. We hope that this nonordinary approach attracts the attention of physicist who could suggest some other variant of the problem or confirm our mathematical model by physical arguments.

Similarly, not all of our versions of equations used for the description of elastic and elastic-plastic processes are identical to the equations considered earlier (cf. Section 5), although they are sources of these equations.

Perhaps, the complete systematization of thermodynamically compatible equations becomes possible when the relativistic invariance is expanded to more general cases. Such problems are extensively discussed in the literature, but we do not touch this question.

In Section 2, we consider the "simplest" thermodynamically compatible systems and clarify if these systems are Galilei-invariant. Later, we construct more complicated systems on the basis of these simplest models.

In Sections 2 and 3, we consider the case where the unknown functions can be only scalar functions or three-dimensional vectors. In Section 4, we shortly recall some fundamental facts of the theory of orthogonal representations of the rotation group $SO(3)$ and the group $O(3)$ of orthogonal transformations including reflections. We will use these facts in Section 5 when we construct invariant systems of equations.

Section 6 presents the adapted rules of constructing invariants that are formed by quantities that are transformed by orthogonal representations of the groups $SO(3)$ and $O(3)$. These rules were introduced in a more general case by Yu. B. Rumer in his book "Theory of Spinors" published in 1936. Unfortunately, this book has been published only in Russian and is now a rarity. Rumer meant to use these rules for the study of chemical valency. In this book, we use them to describe the dependence of the "generating potential" L on unknown functions.

The presented material was first based on the papers by Godunov and Gordienko[2] As this book is being prepared, some new aspects appeared. In particular, one of intermediate versions was presented in the paper [3].

Hopefully, this chapter contains ideas and approaches that can be useful for mathematical modelling of processes in continuous media.

§ 2. The Simplest Galilei-Invariant Thermodynamically Compatible Systems

To clarify the structure of the class of thermodynamically compatible systems, some representatives of which were studied in the previous chapter, we begin with the simplest "basic" systems. Later, we will construct more complicated systems by adding special terms to these "basic" objects. These additional terms should subject to some conditions that provide that our modified system remains in the class of thermodynamically compatible system, i.e., the matrix of coefficients in the quasilinear form of the equations should by symmetric and thereby the system will be hyperbolic. On the other hand, we should guarantee that the conservation laws of mass, momentum, energy and, in addition, the thermodynamical identity expressing the growth of entropy, are valid after our modifications. Moreover, the modified systems should be invariant under the Galilei change of coordinates.

To clarify the main ideas, we first consider equations in which the unknown functions are either scalar functions or 3-dimensional vectors. It should be noted that this restriction, in particular, means that equations of elasticity are not included in the class under consideration because the deformation tensor cannot be described by only three-dimensional vectors and scalars. But we will return to equations of elasticity in Section 5 after recalling some facts of the representation theory which are necessary for our approach to the study of equations of elasticity.

2.1. In this section, we consider the simplest equations of the form

$$\frac{\partial L_{q_0}}{\partial t} + \frac{\partial (u_k L_{q_0})}{\partial x_k} = 0,$$

[2] S. K. Godunov and V. M. Gordienko, *The simplest Galilean-invariant and thermodynamically consistent conservation laws*, Prikl. Mekh. Tekh. Fiz. **43** (2002), no. 1, 3–16; English transl., J. Appl. Mech. Tech. Phys. **43** (2002), no. 1, 1–12; *Complicated structures of Galilean-invariant conservation laws*, Prikl. Mekh. Tekh. Fiz. **43** (2002), no. 2, 3–21; English transl., J. Appl. Mech. Tech. Phys. **43** (2002), no. 2, 175–189.

[3] S. K. Godunov, On correct mathematical description of processes in continuous media with Maxwell viscosity, *Nonlinear Problems in Mathematical Physics and Related Topics II. In Honor of Professor O. A. Ladyzhenskaya*, International Mathematical Series (www.wkap.nl/prod/s/IMAT), Kluwer / Plenum Publishers, 2002, pp. 193–200.

$$\frac{\partial L_a}{\partial t}+\frac{\partial(u_k L_a)}{\partial x_k}=-\alpha,$$
$$\frac{\partial L_b}{\partial t}+\frac{\partial(u_k L_b)}{\partial x_k}=-\beta,$$

$$\frac{\partial L_{p_i}}{\partial t}+\frac{\partial(u_k L_{p_i})}{\partial x_k}=-\pi_i, \tag{2.1}$$
$$\frac{\partial L_{r_i}}{\partial t}+\frac{\partial(u_k L_{r_i})}{\partial x_k}=-\rho_i,$$

$$\frac{\partial L_{u_i}}{\partial t}+\frac{\partial(u_k L)_{u_i}}{\partial x_k}=0,$$
$$\frac{\partial L_T}{\partial t}+\frac{\partial(u_k L_T)}{\partial x_k}=\frac{a\alpha+b\beta+\cdots+p_i\pi_i+r_i\rho_i+\ldots}{T}.$$

The first equation governs the conservation of mass (so, the right-hand side is zero, and the variable q_0 is a scalar). The variables $a, b, \ldots$ are also scalars; they describe the internal state of a medium governed by the above equations. We assume that the right-hand sides $\alpha, \beta, \ldots$ are sufficiently smooth functions of variables $q_0, a, b, \ldots, p_i, r_i, \ldots, u_i, T$ that govern rates of dissipative processes in a medium. The variables p_i, $r_i, \ldots$ are components of three-dimensional vectors which, as well as $a, b \ldots$, describe the inner state of the medium. The right-hand sides $\pi_i, \rho_i \ldots$ express dissipative source terms. The vector with components u_i is the velocity of the motion of the medium and the equations

$$\frac{\partial L_{u_i}}{\partial t}+\frac{\partial(u_k L)_{u_i}}{\partial x_k}=0$$

express the momentum conservation law (therefore, the right-hand sides of these equations are zero because we assume no exterior mass forces for our system). The scalar variable T is the temperature of the medium, and the last equation expresses the law of the growth (or nondecrease) of entropy (here, we consider only one-temperature media and assume that $T > 0$). The increase of entropy is justified by the condition

$$(a\alpha+b\beta+\cdots+p_i\pi_i+r_i\rho_i+\ldots)/T \geqslant 0.$$

This inequality is valid under a suitable choice of $a, \beta, \ldots, \pi_i, \rho_i \ldots$.

We consider the system in this general form (2.1). One can obtain some particular system after an appropriate choice of the right-hand sides and explicit expression for the thermodynamical potential L:

$$L = L(q_0, a, b \ldots; p_i, r_i, \ldots, u_i, T) \tag{2.2}$$

which "generates" the system. We already discussed such systems and now we shortly describe them.

On solutions to systems of the form (2.1), some identity (the conservation of energy) holds. To justify this identity, one can multiply each of the equations of the system by the corresponding multiplier $q_0, a, b, \ldots, p_i, r_i, \ldots, u_i, T$, and add the products. After some computations (we have repeatedly done this, beginning from Section 12, Chapter IV), in the new notation

$$E = q_0 L_{q_0} + aL_a + bL_b + \cdots + p_i L_{p_i} + r_i L_{r_i} + \cdots + u_i L_{u_i} + TL_T - L \quad (2.3)$$

for the Legendre transform of the generating potential L, we obtain the equality

$$\frac{\partial E}{\partial t} + \frac{\partial [u_k(E+L)]}{\partial x_k} = 0, \quad (2.4)$$

which can be interpreted as the "conservation of energy."

2.2. We enumerate the coordinates x_i and components $u_i, p_i, r_i \ldots$ of the unknown vector-valued functions by subscript $i = -1, 0, 1$ (not by $i = 1, 2, 3$ as before). This notation will be useful, when we use the facts and results of the representation theory (cf. Section 4) of the group $O(3)$ of orthogonal transformations of the three-dimensional space. We now shortly explain how this group, whose elements are rotations of the three-dimensional space and reflections with respect to the origin, acts on the above equations.

Each rotation is given by some orthogonal real 3×3-matrix $P(P^T P = I)$ with positive determinant $\det P = +1$. Such matrices form the group $SO(3)$.

The transformation $P \in SO(3)$ replaces the coordinates x_{-1}, x_0, x_1 of a point x with the new coordinates y_{-1}, y_0, y_1 by the formula

$$y = \begin{pmatrix} y_{-1} \\ y_0 \\ y_1 \end{pmatrix} = P \begin{pmatrix} x_{-1} \\ x_0 \\ x_1 \end{pmatrix} = Px.$$

We can return to the "old" coordinates by the inverse matrix $P^{-1} = P^T$:

$$x = P^{-1} y = P^T y.$$

If the coordinates x_{-1}, x_0, x_1 are transformed by $P \in SO(3)$, then an M-dimensional vector q should be transformed by a linear transformation $\Omega(P)$ associated with the block-diagonal matrix

$$\Omega q = \begin{bmatrix} 1 & & & & & & & & \\ & 1 & & & & & & & \\ & & \cdots & & & & & & \\ & & & 1 & & & 0 & & \\ & & & & P & & & & \\ & & & & & \cdots & & & \\ & & & & & & P & & \\ & 0 & & & & & & P & \\ & & & & & & & & 1 \end{bmatrix} q, \quad (2.5)$$

where the diagonal blocks P describe a transformation of vector component u_i, p_i, $r_i, \ldots$ of the vector q. The unit diagonal element means that the scalar components $q_0, d, b, \ldots, T$ of the vector q remain unchanged under this transformation. It is obvious that $\Omega = \Omega(P)$ is also an orthogonal $M \times M$-matrix. A correspondence $P \to \Omega(P)$ is called an *orthogonal representation of the rotation group* if with every matrix $P \in SO(3)$ we can associate an orthogonal $M \times M$-matrix $\Omega = \Omega(P)$ $(\Omega\Omega^T = I_M)$ such that

$$\Omega(I_3) = I_M, \quad \Omega(P_2P_1) = \Omega(P_2)\Omega(P_1). \tag{2.6}$$

It is obvious that the representation (2.5) satisfies these conditions. By construction, it splits into representations in subspaces. Each of these internal representations either is described in some basis by the matrix P or is a scalar representation (scalars remain unchanged under rotations). More complicated representations will be considered in Section 4. For our goal in this section it suffices to deal with representations of the form (2.5).

2.3. Let us explain what we understand under a transformation $\Omega(P)$ of the unknown vector-valued function $q = q(t, x_{-1}, x_0, x_1)$ that is induced by the change of spatial variables

$$\begin{pmatrix} y_{-1} \\ y_0 \\ y_1 \end{pmatrix} = P \begin{pmatrix} x_{-1} \\ x_0 \\ x_1, \end{pmatrix}, \quad \begin{pmatrix} x_{-1} \\ x_0 \\ x_1 \end{pmatrix} = P^{-1} \begin{pmatrix} y_{-1} \\ y_0 \\ y_1 \end{pmatrix}.$$

We assume that

$$\Omega(P)q(t, x_{-1}, x_0, x_1) = p = p(t, y_{-1}, y_0, y_1)$$

or, shortly,

$$p = p(t, y) = \Omega(P)q(t, P^{-1}y). \tag{2.7}$$

For any scalar function

$$g(t, y) = g(t, y_{-1}, y_0, y_1) = g(t, P_{-1j}x_j P_{0j}x_j, P_{1j}x_j)$$

the formulas for the gradient of g in the original coordinates and new ones are as follows:

$$\frac{\partial g}{\partial y} = \begin{pmatrix} g_{y_1} \\ g_{y_0} \\ g_{y_1} \end{pmatrix} = P \begin{pmatrix} g_{x_{-1}} \\ g_{x_0} \\ g_{x_1} \end{pmatrix} = P\frac{\partial g}{\partial x}.$$

If the velocity vector $u = [u_{-1}, u_0, u_1]^T$ with components $u_i(t, x_{-1}, x_0, x_1)$ goes to the vector with components v_i, then $v(t, y) = Pu(t, P^{-1}y)$; moreover, the inner product $\left(v, \dfrac{\partial g}{\partial y}\right)$ coincides with $\left(u, \dfrac{\partial g}{\partial x}\right)$:

$$\left(v, \frac{\partial g}{\partial y}\right) = \left(Pu, P\frac{\partial g}{\partial x}\right) = \left(P^*Pu, \frac{\partial g}{\partial x}\right) = \left(u, \frac{\partial g}{\partial x}\right).$$

Under the representations described by transformations (2.5), (2.7), the products $u_k L_{q_0}$, $u_k L_a$, $u_k L_b$, $u_k L_T, \ldots$ become $v_k L_{q_0}$, $v_k L_a$, $v_k L_b$, $v_k L_T, \ldots$ and their derivatives

$$\frac{\partial}{\partial x_k}(u_k L_{q_0}), \quad \frac{\partial}{\partial x_k}(u_k L_a), \quad \frac{\partial}{\partial x_k}(u_k L_b), \quad \frac{\partial}{\partial x_k}(u_k L_T)$$

go to the expressions

$$\frac{\partial}{\partial y_k}(v_k L_{q_0}), \quad \frac{\partial}{\partial y_k}(v_k L_a), \quad \frac{\partial}{\partial y_k}(v_k L_b), \quad \frac{\partial}{\partial y_k}(v_k L_T),$$

i.e., they are invariant (scalar). For example,

$$\frac{\partial}{\partial x_k}[u_k(t,x)L_{q_0}(t,x)] = \frac{\partial}{\partial y_k}[v_k(t,y)L_{\tilde{q_0}}(t,y)],$$

where $\tilde{q_0}(t,y) = q(t, P^{-1}x)$. Hereinafter, we assume that $L(q_0, a, b, r_i, p_i, \ldots, u_i, T)$ is a scalar and is invariant under transformations $\Omega(P)$ of its arguments. We also suppose that L can be expressed as a function of (u,u), (r,r), (p,p), (u,r), (u,p), (r,p), of the determinants of coordinates of triples of vectors

$$\begin{vmatrix} u_{-1} & p_{-1} & r_{-1} \\ u_0 & p_0 & r_0 \\ u_1 & p_1 & r_1 \end{vmatrix}$$

and of all scalars q_0, a, b, $T, \ldots$. Under these assumptions, the equations

$$\begin{aligned} &\frac{\partial L_{q_0}}{\partial t} + \frac{\partial(u_k L_{q_0})}{\partial x_k} = 0, \\ &\frac{\partial L_a}{\partial t} + \frac{\partial(u_k L_a)}{\partial x_k} = -\alpha, \\ &\frac{\partial L_b}{\partial t} + \frac{\partial(u_k L_b)}{\partial x_k} = -\beta, \\ &\frac{\partial L_T}{\partial t} + \frac{\partial(u_k L_T)}{\partial x_k} = \frac{a\alpha + b\beta + p_i\pi_i + r_i\rho_i + \ldots}{T}. \end{aligned}$$

take the form

$$\begin{aligned} &\frac{\partial L_{q_0}}{\partial t} + \frac{\partial(v_k L_{q_0})}{\partial y_k} = 0, \\ &\frac{\partial L_a}{\partial t} + \frac{\partial(v_k L_a)}{\partial y_k} = -\alpha, \\ &\frac{\partial L_b}{\partial t} + \frac{\partial(v_k L_b)}{\partial y_k} = -\beta, \\ &\frac{\partial L_T}{\partial t} + \frac{\partial(v_k L_T)}{\partial y_k} = \frac{a\alpha + b\beta + p_i\pi_i + r_i\rho_i \ldots}{T}. \end{aligned}$$

i.e., these equations preserve the form under any rotations of coordinates.

If $p_i(t,x)$ is a vector-valued function, then the vectors with components $p_i(t,x)$, $\frac{\partial}{\partial x_k}(u_k L_{p_i})$ (the summation is taken over k) have the following components after rotation:

$$\widehat{p}_i(t,y) = p_i(t, P^{-1}y), \qquad \frac{\partial}{\partial y_k}(v_k L_{\widehat{p}_i}),$$

and the derivatives $\frac{\partial L_{p_i}}{\partial t}$ take the form $\frac{\partial}{\partial t} L_{\widehat{p}_i}(t,y)$. Consequently, in these coordinates, the equations

$$\frac{\partial L_{p_i}}{\partial t} + \frac{\partial (u_k L_{p_i})}{\partial x_k} = -\pi_i$$

keep the same form

$$\frac{\partial L_{\widehat{p}_i}}{\partial t} + \frac{\partial (v_k L_{\widehat{p}_i})}{\partial y_k} = -\widehat{\pi}_i(t,y) = -P\pi_i(t, P^{-1}\, y).$$

Therefore, we can assert that the form of Eqs. (2.1) is invariant under rotations of spatial coordinates. This assertion is valid if the generating potential L is invariant.

2.4. Our next goal is to clarify the conditions under which the basic simplest system (2.1) is Galilei-invariant not only under translations of coordinates with constant velocity vector.

If we already establish the invariance of the form of equations under rotations, it suffices to clarify the invariance conditions for the passage from t, x_{-1}, x_0, x_1 to t, y_{-1}, y_0, y_1 ($y_j = x_j - U_j t$, $U_j = \text{const}$). Moreover, the components of the velocity vector $u_j(t, x_{-1}, x_0, x_1)$ should be replaced with $v_j(t, y_{-1}, y_0, y_1) = u_j(t, x_{-1} + U_{-1}t, x_0 + U_0 t, x_1 + U_1 t) - U_j$. For our purposes, it is necessary to require some additional conditions on the generating potential L.

We assume that

$$L = \Lambda(q_0 + u_i u_i/2, a, b, \ldots, p_{-1}, p_0, p_1, r_{-1}, r_0, r_1, \ldots, T)$$

where the scalar function Λ of scalar variables $q_0 + \frac{u_i u_i}{2}, a, b, \ldots, T$ and vector variables $p, r, \ldots$ is invariant under rotations. Obviously, the generating potential itself is also invariant.

At the same physical point with fixed coordinates x_k and moving coordinates $y_j = x_j + U_j t$, the corresponding values of all scalar and vector-valued functions $q_0, a, b, \ldots, T^0, p, r, \ldots$ should be the same at some time moment t. Only the components u_j and v_j in the representation of the velocity vector are different.

Introduce the notation

$$\begin{aligned}\widetilde{L}\ &(q_0, v_{-1}, v_0, v_1, a, b, \ldots, p_\cdot, r, \ldots, T) \\ &= L(q_0, v_{-1} + U_{-1}, v_0 + U_0, v_1 + U_1, a, b, \ldots, p, r, \ldots, T).\end{aligned}$$

After the passage to the moving coordinates, equations

$$\frac{\partial F}{\partial t} + \frac{\partial G_k}{\partial x_k} = -f$$

become

$$\frac{\partial F}{\partial t} + \frac{\partial (G_k - U_k F)}{\partial y_k} = -f.$$

Moreover, at the same physical point, we have $\widetilde{L}_{q_0} = L_{q_0}$, $\widetilde{L}_{v_k} = L_{u_k}$, $\widetilde{L}_a = L_a, \ldots, \widetilde{L}_T = L_T$. Therefore, the equalities

$$\begin{aligned}
&\frac{\partial L_{q_0}}{\partial t} + \frac{\partial (u_k L_{q_0})}{\partial x_k} = 0, \\
&\frac{\partial L_a}{\partial t} + \frac{\partial (u_k L_a)}{\partial x_k} = -\alpha, \\
&\frac{\partial L_\beta}{\partial t} + \frac{\partial (u_k L_b)}{\partial x_k} = -\beta, \\
&\frac{\partial L_{p_i}}{\partial t} + \frac{\partial (u_k L_{p_i})}{\partial x_k} = -\pi_i, \\
&\frac{\partial L_{r_i}}{\partial t} + \frac{\partial (u_k L_{r_i})}{\partial x_k} = -\rho_i, \\
&\frac{\partial L_T}{\partial t} + \frac{\partial (u_k L_T)}{\partial x_k} = \frac{a\alpha + b\beta + p_i \pi_i + r_i \rho_i + \ldots}{T}
\end{aligned}$$

also take some new form. We demonstrate this by the following typical examples:

$$\begin{aligned}
&\frac{\partial L_{q_0}}{\partial t} + \frac{\partial (u_k L_{q_0} - U_k L_{q_0})}{\partial y_k} = 0, \\
&\frac{\partial L_a}{\partial t} + \frac{\partial (u_k L_a - U_k L_a)}{\partial y_k} = -\alpha, \\
&\cdots\cdots\cdots\cdots\cdots\cdots\cdots\cdots\cdots \\
&\frac{\partial L_T}{\partial t} + \frac{\partial (u_k L_T - U_k L_T)}{\partial y_k} = \frac{a\alpha + b\beta + p_i \pi_i + r_i \rho_i + \ldots}{T}.
\end{aligned}$$

After the passage to the components v_k and the generating potential $\widetilde{L}$, they differ from the initial equations only by the notation:

$$\begin{aligned}
&\frac{\partial \widetilde{L}_{q_0}}{\partial t} + \frac{\partial (v_k \widetilde{L}_{q_0})}{\partial y_k} = 0, \\
&\frac{\partial \widetilde{L}_a}{\partial t} + \frac{\partial (v_k \widetilde{L}_{q_0})}{\partial y_k} = -\alpha,
\end{aligned}$$

$$\dots\dots\dots\dots\dots\dots\dots\dots\dots\dots\dots\dots$$

$$\frac{\partial \widetilde{L}_T}{\partial t} + \frac{\partial (v_k \widetilde{L}_T)}{\partial y_k} = \frac{a\alpha + b\beta + p_i\pi_i + r_i\rho_i + \dots}{T}.$$

The equations describing momentum conservation are transformed in a similar way. The equalities

$$\frac{\partial L_{u_i}}{\partial t} + \frac{\partial (u_k L)_{u_i}}{\partial x_k} = 0,$$

which can be also written as follows:

$$\frac{\partial L_{u_i}}{\partial t} + \frac{\partial (u_k L_{u_i} + \delta_{ik} L)}{\partial x_k} = 0$$

become

$$\frac{\partial L_{u_i}}{\partial t} + \frac{\partial [(u_k L_{u_i} + \delta_{ik} L) - U_k L_{u_i}]}{\partial y_k} = 0$$

and in some suitable notation take the final form

$$\frac{\partial \widetilde{L}_{v_i}}{\partial t} + \frac{\partial (v_k \widetilde{L}_{v_i} + \delta_{ik} \widetilde{L})}{\partial y_k} = 0.$$

We note that the generating potential $\widetilde{L}$ is not written in the required form : $\widetilde{L} \neq \Lambda(q_0 + v_i v_i/2 + \dots)$. To restore it, we replace q_0 with a new unknown function

$$Q_0 = q_0 + u_{-1}U_{-1} + u_0 U_0 + u_1 U_1 - \frac{1}{2}(U_{-1}^2 + U_0^2 + U_1^2).$$

It is clear that

$$\begin{aligned} &\widetilde{L}(q_0, v_{-1}, v_0, v_1, a, b, \dots, p, r, \dots, T) \\ &= \Lambda\left(q_0 + \frac{(v_i + U_i, v_i + U_i)}{2} a, \dots, T \right) \\ &= \Lambda\left(Q_0 + \frac{v_i v_i}{2}, a, b, \dots, p, r, \dots, T \right) \end{aligned}$$

Hence we again can use our postulate. (Note that the difference $Q_0 - q_0$ remains unchanged under rotations because it is expressed in terms of the invariants (u, U), (U, U).) Assuming that $\widetilde{L}$ is a function of $q_0, v_{-1}, v_0, v_1, a, b, \dots$, and the function L (with the same values) is expressed in terms of $Q_0, v_{-1}, v_0, v_1, a, b, \dots$, it is easy to verify the equalities

$$\widetilde{L}_{q_0} = L_{Q_0}, \quad \widetilde{L}_{v_i} = L_{v_i} + U_i L_{Q_0}, \quad \hat{L}_a = L_a, \quad \hat{L}_{p_i} = L_{p_i}, \dots, L_T = \widetilde{L}_T. \quad (2.8)$$

Using (2.8), we can write the equation expressing the conservation of mass

$$\frac{\partial \hat{L}_{q_0}}{\partial t} + \frac{\partial v_k \hat{L}_{q_0}}{\partial y_k} = 0$$

in the standard form

$$\frac{\partial L_{Q_0}}{\partial t} + \frac{\partial v_k L_{Q_0}}{\partial y_k} = 0.$$

The remaining equations, except for the momentum conservation law, also preserve the form. Using the notation (2.8), we find

$$\frac{\partial L_{v_i}}{\partial t} + \frac{\partial (v_k L_{v_i} + \delta_{ik} L)}{\partial y_k} + U_i \left[\frac{\partial L_{Q_0}}{\partial t} + \frac{\partial (v_k L_{Q_0})}{\partial y_k} \right] = 0.$$

It is obvious that the last terms in these equality vanish (the conservation of mass) and can be eliminated. After that, the equations for momentum again take the standard form.

We complete our review of the "basic" equations which will be modified in the following section in order to obtain a new thermodynamically compatible system. In particular, we consider the case where the unknowns are transformed by representations of the rotation group more complicated than vector or scalar ones.

§ 3. Methods of Constructing Equations

The "simplest" thermodynamically compatible systems in Section 2 are conservation laws and form a symmetric hyperbolic system, i.e., the Cauchy problem is well posed.

Now, we start to describe more complicated thermodynamically compatible systems. Equations will be constructed from equations of some "simplest" system by adding some terms in such a way that the obtained system also belongs to the class of thermodynamically compatible system.

But we do not use the explicit form of conservation laws. In order to represent the resulting system as a collection of conservation laws, we consider combinations of the derivatives of unknown functions as new variables. Then, using some additional equaitons, we obtain a symmetric hyperbolic system that is a consequence of the original equations.

3.1. We begin by considering the "simplest" system ($L = L(q_0, u, d, b, T)$)

$$\frac{\partial L_{q_0}}{\partial t} + \frac{\partial (u_k L_{q_0})}{\partial x_k} = 0, \tag{1}$$

$$
\begin{aligned}
&\langle 2\rangle \qquad \frac{\partial L_{d_i}}{\partial t} + \frac{\partial (u_k L_{d_i})}{\partial x_k} = -j_i, \quad j_i = j_i(q_0, u, d, b, T),\\
&\langle 3\rangle \qquad \frac{\partial L_{b_i}}{\partial t} + \frac{\partial (u_k L_{b_i})}{\partial x_k} = 0,\\
&\langle 4\rangle \qquad \frac{\partial L_T}{\partial t} + \frac{\partial (u_k L_T)}{\partial x_k} = -\frac{d_i j_i}{T},\\
&\langle 5\rangle \qquad \frac{\partial L_{u_i}}{\partial t} + \frac{\partial (u_k L)_{u_i}}{\partial x_k} = 0
\end{aligned}
\tag{3.1}
$$

for unknown scalars q_0, T and three vectors d, b u with components d_i, b_i, u_i $(i = -1, 0, 1)$. As usual, u_i denotes the components of the velocity vector. Hence the last equality is a formal expression of the momentum conservation law. Equation $\langle 1\rangle$ governs the mass conservation, whereas $\langle 4\rangle$ is the compensating entropy equation.

The equality governing conservation of energy is a linear combination of the above equations with the corresponding coefficients q_0, d_i, b_i, T, u_i. It takes the form

$$\frac{\partial E}{\partial t} + \frac{\partial u_k (E + L)}{\partial x_k} = 0,$$

$$E = q_0 L_{q_0} + u_i L_{u_i} + d_i L_{d_i} + b_i L_{b_i} + T L_T - L.$$

We modify the equations containing the derivatives of L_{b_i} and L_{u_i} with respect to t. Namely, we add some terms to the left-hand sides of these equations:

$$\frac{\partial L_{b_i}}{\partial t} + \frac{\partial (u_k L_{b_i})}{\partial x_k} - L_{b_k} \frac{\partial u_i}{\partial x_k} = 0,$$

$$\frac{\partial L_{u_i}}{\partial t} + \frac{\partial (u_k L)_{u_i}}{\partial x_k} - L_{b_k} \frac{\partial b_i}{\partial x_k} - \theta b_i L_{q_0} = 0 \quad (\theta = \text{const}).$$

It is obvious that these equations remain invariant under orthogonal transformations, in particular, under rotations. These equations also remain unchanged after passing to the coordinates moving with respect to the initial coordinates with constant velocity because the additional terms contain only the derivatives of the components u_i of velocity.

We note that the hyperbolicity of the original system follows from the possibility to write it in the quasilinear form with symmetric coefficient matrices at the derivatives $\dfrac{\partial}{\partial x_k}$ (under the assumption that the generating potential L is convex). To preserve the hyperbolicity of the modified system, we should add some terms to these matrices such that their coefficients form symmetric matrices at the derivatives of b_i and u_i with respect to x_l. We write these additional terms:

$$-\left[\begin{array}{ccc|ccc} 0 & 0 & 0 & L_{b_k} & 0 & 0 \\ 0 & 0 & 0 & 0 & L_{b_k} & 0 \\ 0 & 0 & 0 & 0 & 0 & L_{b_k} \\ \hline L_{b_k} & 0 & 0 & 0 & 0 & 0 \\ 0 & L_{b_k} & 0 & 0 & 0 & 0 \\ 0 & 0 & L_{b_k} & 0 & 0 & 0 \end{array}\right] \frac{\partial}{\partial x_k} \begin{pmatrix} b_{-1} \\ b_0 \\ b_1 \\ u_{-1} \\ u_0 \\ u_1 \end{pmatrix}.$$

Therefore, under our modifications, we again have a symmetric hyperbolic system.

The situation with the conservation of energy is more complicated. To obtain an energy equation as a linear combination of modified equations, we should add the sum

$$-b_i L_{b_k} \frac{\partial u_i}{\partial x_k} - u_i \left(L_{b_k} \frac{\partial b_i}{\partial x_k} + \theta b_i L_{q_0} \right) = -\frac{\partial}{\partial x_k}(u_i b_i L_{b_k}) + u_i b_i \frac{\partial L_{b_k}}{\partial x_k} - \theta u_i b_i L_{q_0},$$

which is not written in the divergence form, i.e., it is not represented as a conservation law. Before we explain how to overcome this obstacle, we study these modified equations more carefully. Note that it is convenient to write the equations

$$\frac{\partial L_{b_i}}{\partial t} + \frac{\partial (u_k L_{b_i})}{\partial x_k} - L_{b_k} \frac{\partial u_i}{\partial x_k} = 0$$

as follows:

$$\frac{\partial L_{b_i}}{\partial t} + \frac{\partial (u_k L_{b_i} - u_i L_{b_k})}{\partial x_k} + u_i \frac{\partial L_{b_k}}{\partial x_k} = 0,$$

which implies

$$\frac{\partial}{\partial t}\left(\frac{\partial L_{b_k}}{\partial x_k}\right) + \frac{\partial}{\partial x_i}\left[u_i \left(\frac{\partial L_{b_k}}{\partial x_k}\right)\right] = 0.$$

Therefore, taking into account this fact, from the conservation of mass

$$\frac{\partial L_{q_0}}{\partial t} + \frac{\partial (u_i L_{q_0})}{\partial x_k} = 0$$

it is easy to obtain the equality

$$\frac{\partial}{\partial t}\left[\frac{1}{L_{q_0}}\left(\frac{\partial L_{b_k}}{\partial x_k}\right)\right] + u_i \frac{\partial}{\partial x_i}\left[\frac{1}{L_{q_0}}\left(\frac{\partial L_{b_k}}{\partial x_k}\right)\right] = 0$$

which means that $\left(\frac{\partial L_{b_k}}{\partial x_k}\right)/L_{q_0}$ is constant along the trajectories $\frac{dx_i}{dt} = u_i$. Therefore, if the equality $\frac{\partial L_{b_k}}{\partial x_k} = \theta L_{q_0}$ holds at the initial time moment, then it remains valid for all subsequent time moments. In other words, our system can be completed with one more *compatible* equation

$$\frac{\partial L_{b_k}}{\partial x_k} = \theta L_{q_0}.$$

In this case, the terms added to the conservation of energy are of divergence form:

$$-\frac{\partial}{\partial x_k}(u_i b_i L_{b_k}) + u_i b_i \frac{\partial L_{b_k}}{\partial x_k} - \theta u_i b_i L_{q_0} = -\frac{\partial}{\partial x_k}(u_i b_i L_{b_k}).$$

Moreover, the equations containing the derivatives of L_{u_i} with respect to t can be written in the divergence form and express the momentum conservation law. Indeed, we can assert that

$$-L_{b_k}\frac{\partial b_i}{\partial x_k} = -L_{b_k}\frac{\partial b_i}{\partial x_k} - b_i\frac{\partial L_{b_k}}{\partial x_k} + \theta b_i L_{q_0} = -\frac{\partial(b_i L_{b_k})}{\partial x_k} + \theta b_i L_{q_0}$$

and write the equality

$$\frac{\partial L_{u_i}}{\partial t} + \frac{\partial(u_k L)_{u_i}}{\partial x_k} - L_{b_k}\frac{\partial b_i}{\partial x_k} - \theta b_i L_{q_0} = 0$$

as follows:

$$\frac{\partial L_{u_i}}{\partial t} + \frac{\partial}{\partial x_k}[(u_k L)_{u_i} - b_i L_{b_k}] = 0.$$

Thus, we can conclude that the system (3.1) can be written as a quasilinear symmetric hyperbolic system compatible with the additional equation

$$\frac{\partial L_{b_k}}{\partial x_k} = \theta L_{q_0}. \tag{3.2}$$

To assert the (local) existence of a sufficiently smooth solution to this system, it suffices to know that the initial data are smooth and satisfy the additional equation. If we want to obtain a solution of higher smoothness, it is reasonable to require that the initial data are of higher smoothness (In fact, we can overcome this obstacle if we use some specific function spaces containing a solution, as well as the initial data. But we do not discuss this in detail.)

The system (3.1), together with the additional equation (3.2), can be written as a collection of equalities of divergence form (conservation laws)

$$
\begin{aligned}
&\frac{\partial L_{q_0}}{\partial t} + \frac{\partial (u_k L_{q_0})}{\partial x_k} = 0, \\
&\frac{\partial L_{d_i}}{\partial t} + \frac{\partial (u_k L_{d_i})}{\partial x_k} = -j_i, \quad j_i = j_i(q_0, u, d, b, T), \\
&\frac{\partial L_{b_i}}{\partial t} + \frac{\partial (u_k L_{b_i} - L_{b_k} u_i)}{\partial x_k} = 0, \\
&\frac{\partial L_T}{\partial t} + \frac{\partial (u_k L_T)}{\partial x_k} = \frac{d_i j_i}{T}, \\
&\frac{\partial L_{u_i}}{\partial t} + \frac{\partial [(u_k L)_{u_i} - b_i L_{b_k}]}{\partial x_k} = 0, \\
&\frac{\partial L_{b_k}}{\partial x_k} = \theta L_{q_0}, \\
&\frac{\partial E}{\partial t} + \frac{\partial [u_k (E + L) - u_i b_i L_{b_k}]}{\partial x_k} = 0.
\end{aligned}
\tag{3.3}
$$

Certainly, we could start with these conservation laws whose number is larger than the number of unknown functions and, taking linear combinations, arrive at a symmetric hyperbolic system. It is easy to verify this fact.

3.2. We return to complicated thermodynamically compatible equations. Constructing the systems (3.1) and (3.3), we left the equations

$$
\frac{\partial L_{d_i}}{\partial t} + \frac{\partial (u_k L_{d_i})}{\partial x_k} = -j_i
$$

unchanged, although we could insert additional terms as follows:

$$
\frac{\partial L_{d_i}}{\partial t} + \frac{\partial (u_k L_{d_i})}{\partial x_k} - L_{d_k} \frac{\partial u_i}{\partial x_k} = -j_i \tag{3.4}
$$

and, simultaneously, correct the last equations

$$
\frac{\partial L_{u_i}}{\partial t} + \frac{\partial (u_k L)_{u_i}}{\partial x_k} - L_{b_k} \frac{\partial b_i}{\partial x_k} - L_{d_k} \frac{\partial d_i}{\partial x_k} = 0.
$$

It turns out that, under the above modification, these equations cannot be reduced to the momentum conservation law with zero right-hand side. From (3.4) we obtain the equalities

$$
\frac{\partial}{\partial t}\left(\frac{\partial L_{d_i}}{\partial x_i}\right) + \frac{\partial}{\partial x_k}\left[u_k \left(\frac{\partial L_{d_i}}{\partial x_i}\right) + j_k\right] = 0,
$$

$$
\frac{\partial}{\partial t}\left[\frac{1}{L_{q_0}}\left(\frac{\partial L_{d_i}}{\partial x_k}\right)\right] + u_k \frac{\partial}{\partial x_k}\left[\frac{1}{L_{q_0}}\left(\frac{\partial L_{d_i}}{\partial x_i}\right)\right] + \frac{1}{L_{q_0}} \frac{\partial j_k}{\partial x_k} = 0,
$$

which shows that we cannot expect that the equation $\dfrac{\partial L_{d_k}}{\partial x_k} = 0$ is valid in the case of nonzero "relaxational current" j_k for any initial value of this expression.

If we assume that this assertion is valid, together with the postulate $\dfrac{\partial L_{b_k}}{\partial x_k} = 0$, then the equations

$$\frac{\partial L_{u_i}}{\partial t} + \frac{\partial (u_k L)_{u_i}}{\partial x_k} - L_{b_k} \frac{\partial b_i}{\partial x_k} - L_{d_k} \frac{\partial d_i}{\partial x_k} = 0$$

should be equivalent to the equations

$$\frac{\partial L_{u_i}}{\partial t} + \frac{\partial (u_k L)_{u_i}}{\partial x_k} - L_{b_k} \frac{\partial b_i}{\partial x_k} - b_i \frac{\partial L_{b_k}}{\partial x_k} - L_{d_k} \frac{\partial d_i}{\partial x_k} - d_i \frac{\partial L_{d_k}}{\partial x_k} = 0,$$

$$\frac{\partial L_{u_i}}{\partial t} + \frac{\partial}{\partial x_k} [(u_k L)_{u_i} - b_i L_{b_k} - d_i L_{d_k}] = 0.$$

The last equation can be regarded as a variant of the momentum conservation law. However, we cannot expect that $\dfrac{\partial L_{d_k}}{\partial x_k} = 0$. These obstacles show that we cannot use modifications of the equations

$$\frac{\partial L_{d_i}}{\partial t} + \frac{\partial (u_k L_{d_i})}{\partial x_k} = -j_i.$$

We also note that there are some difficulties arise if we want to obtain the energy conservation law.

3.3. We describe one more method for constructing new variants of systems. We start with the partially modified equations (3.1) with the additional relation (3.2) ($\theta = 0$). The equalities

$$\frac{\partial L_{d_i}}{\partial t} + \frac{\partial (u_k L_{d_i})}{\partial x_u} = -j_i,$$

$$\frac{\partial L_{b_i}}{\partial t} + \frac{\partial (u_k L_{b_i})}{\partial x_k} - L_{b_k} \frac{\partial u_i}{\partial x_k} = 0$$

in (3.1) are replaced with the following:

$$\frac{\partial L_{d_i}}{\partial t} + \frac{\partial (u_k L_{d_i})}{\partial x_u} - \varepsilon_{ikl} \frac{\partial b_l}{\partial x_k} = -j_i,$$

$$\frac{\partial L_{b_i}}{\partial t} + \frac{\partial (u_k L_{b_i})}{\partial x_k} + \varepsilon_{ikl} \frac{\partial d_l}{\partial x_k} - L_{b_k} \frac{\partial u_i}{\partial x_k} = 0$$

($\varepsilon_{ikl} = 0$ if some of i, k, and l are equal, $\varepsilon_{-101} =_{01-1}= \varepsilon_{1-10} = 1$, $\varepsilon_{10-1} = \varepsilon_{0-11} = \varepsilon_{-110} = -1$).

We note that the second line of these equations implies the equality

$$\frac{\partial}{\partial t}\left(\frac{\partial L_{b_i}}{\partial x_i}\right) + \frac{\partial}{\partial x_k}\left[u_k \left(\frac{\partial L_{b_i}}{\partial x_i}\right)\right] = 0$$

which is the same as we obtained before this modification. Therefore, the modified equations, as well as Eqs. (3.1), are compatible with the equality (3.2) ($\theta = 0$) which is used for obtaining the momentum conservation law and conservation of energy. The momentum conservation law does not require any modifications and can be written as follows:

$$\frac{\partial L_{u_i}}{\partial t} + \frac{\partial[(u_k L)_{u_i} - b_i L_{b_k}]}{\partial x_k} = 0.$$

It can be obtained with the help of (3.2) from the equations

$$\frac{\partial L_{u_i}}{\partial t} + \frac{\partial(u_k L)_{u_i}}{\partial x_k} - L_{b_k}\frac{\partial b_i}{\partial x_k} = 0.$$

But it is necessary to modify the equality describing the conservation of energy. Hence we obtain the relation

$$\frac{\partial E}{\partial t} + \frac{\partial[u_k(E+L) - u_i b_i L_{b_k} + \varepsilon_{klm} d_l b_m]}{\partial x_k} = 0.$$

The above modifications allow us to represent some compatible equations as a symmetric hyperbolic system.

It is easy to check that the following equations can be written as a symmetric hyperbolic system:

$$\begin{aligned}
&\frac{\partial L_{q_0}}{\partial t} + \frac{\partial(u_k L_{q_0})}{\partial x_k} = 0,\\
&\frac{\partial L_{d_i}}{\partial t} + \frac{\partial(u_k L_{d_i} - \varepsilon_{ikl} b_l)}{\partial x_k} = -j_i, \quad j_i = j_i(q_0, u, d, b, T),\\
&\frac{\partial L_{b_i}}{\partial t} + \frac{\partial(u_k L_{b_i} + \varepsilon_{ikl} d_l)}{\partial x_k} - L_{b_k}\frac{\partial u_i}{\partial x_k} = 0,\\
&\frac{\partial L_T}{\partial t} + \frac{\partial(u_k L_T)}{\partial x_k} = \frac{d_i j_i}{T},\\
&\frac{\partial L_{u_i}}{\partial t} + \frac{\partial(u_k L)_{u_i}}{\partial x_k} - L_{b_k}\frac{\partial b_i}{\partial x_k} = 0
\end{aligned} \tag{3.5}$$

which is compatible with the additional equation

$$\frac{\partial L_{b_k}}{\partial x_k} = 0$$

which is used in the modification of (3.5) in order to derive the conservation laws

$$\frac{\partial L_{q_0}}{\partial t} + \frac{\partial(u_k L_{q_0})}{\partial x_k} = 0,$$

$$\begin{aligned}
&\frac{\partial L_{d_i}}{\partial t} + \frac{\partial(u_k L_{d_i} - \varepsilon_{ikl} b_l)}{\partial x_k} = -j_i, \\
&\frac{\partial L_{b_i}}{\partial t} + \frac{\partial(u_k L_{b_i} - u_i L_{b_k} + \varepsilon_{ikl} d_l)}{\partial x_k} = 0, \\
&\frac{\partial L_T}{\partial t} + \frac{\partial(u_k L_T)}{\partial x_k} = \frac{d_i j_i}{T}, \\
&\frac{\partial L_{u_i}}{\partial t} + \frac{\partial[(u_k L)_{u_i} - b_i L_{b_k}]}{\partial x_k} = 0, \\
&\frac{\partial E}{\partial t} + \frac{\partial[u_k(E+L) - u_i b_i L_{b_k} + \varepsilon_{klm} d_l b_m]}{\partial x_k} = 0, \\
&E = q_0 L_{q_0} + d_i L_{d_i} + b_i L_{b_i} + T L_T + u_i L_{u_i} - L.
\end{aligned} \tag{3.6}$$

In our opinion, Eqs. (3.5), (3.6) can be used in order to describe the behavior of electromagnetic field in a *conducting* medium in some situations. The question whether such situations exist should be answered by specialists in physics. In our opinion, our motivation of constructing equations allows us to understand deep mathematical reasons which provide, on one hand, the well-posedness of the Cauchy problem and, on the other hand, the validity of conservation laws.

3.4. One more example of the construction of thermodynamically compatible equations can be found in the hydrodynamics of the Newton gravity field. This example is of interest because the equations are not hyperbolic since the gravity potential satisfies the elliptic Poisson equation. We write the corresponding equations ($\gamma \neq 0$)

$$\begin{aligned}
&\langle 1\rangle \qquad \frac{\partial L_{q_0}}{\partial t} + \frac{\partial(u_k L_{q_0})}{\partial x_k} = 0, \\
&\langle 2\rangle \qquad \frac{\partial L_{u_i}}{\partial t} + \frac{\partial(u_k L)_{u_i}}{\partial x_k} = -\gamma L_{q_0}\frac{\partial \varphi}{\partial x_i}, \\
&\langle 3\rangle \qquad -\frac{\partial}{\partial x_k}\left(\frac{\partial \varphi}{\partial x_k}\right) = \gamma L_{q_0}, \\
&\langle 4\rangle \qquad \frac{\partial L_T}{\partial t} + \frac{\partial(u_k L_T)}{\partial x_k} = 0
\end{aligned} \tag{3.7}$$

From $\langle 3\rangle$ we find

$$\frac{\partial}{\partial t}\left(\frac{1}{2}\varphi_{x_k}\varphi_{x_k}\right) = \gamma\varphi\frac{\partial L_{q_0}}{\partial t} + \frac{\partial}{\partial x_k}\left[\varphi\frac{\partial \varphi_t}{\partial x_k}\right],$$

which, in view of $\langle 1\rangle$, can be written as follows:

$$\frac{\partial}{\partial t}\left(\frac{1}{2}\varphi_{x_k}\varphi_{x_k}\right) = -\gamma\varphi\frac{\partial(u_k L_{q_0})}{\partial x_k} + \frac{\partial}{\partial x_k}\left(\varphi\frac{\partial \varphi_t}{\partial x_k}\right). \tag{3.8}$$

To obtain the identity corresponding to the energy conservation, we multiply the equalities ⟨1⟩, ⟨2⟩, ⟨4⟩, in (3.7) by q_i, u_i, T respectively and add (3.8) to their sum. Then we obtain the equality

$$\frac{\partial}{\partial t}\left(E+\frac{1}{2}\varphi_{x_k}\varphi_{x_k}\right)+\frac{\partial}{\partial x_k}[u_k(E+L)]=\frac{\partial}{\partial x_k}\left[\varphi\left(\frac{\partial \varphi_t}{\partial x_k}-\gamma u_k L_{q_0}\right)\right]. \quad (3.9)$$

As usual, $E = q_0 L_{q_0} + u_i L_{u_i} + T L_T - L_0$.

If φ, φ_{tx_k}, $u_k(E+L)$ decrease sufficiently quickly as $r = \sqrt{x_k x_k} \to \infty$, the last equality implies

$$\frac{d}{dt}\left[\int_{-\infty}^{\infty}\int_{-\infty}^{\infty}\int_{-\infty}^{\infty}\left(E+\frac{1}{2}\varphi_{x_k}\varphi_{x_k}\right)dx_{-1}dx_0dx_1\right]=0, \quad (3.10)$$

i.e., the sum of the internal energy $\int\int\int E dx_{-1}dx_0dx_1$ of gas and the energy of gravitation field $\int\int\int \varphi_{x_k}\varphi_{x_k} dx_{-1}dx_0dx_1$ remains constant.

3.5. To conclude the section, we indicate one more example of a rather complicated system of thermodynamically compatible conservation laws. It was initiated by detailed analysis of equations used for the description of processes in a superfluid medium (although it rather differs from these equations) and the study of Romenskii [4)] Comparing with the above examples, this example is more cumbersome.

Consider the following system:

$$\begin{aligned}
&\frac{\partial L_{q_0}}{\partial t}+\frac{\partial (u_k L_{q_0})}{\partial x_k}=0,\\
&\frac{\partial L_{\omega_i}}{\partial t}+\frac{\partial (u_k L_{\omega_i})}{\partial x_k}-\underline{\underline{L_{\omega_k}\frac{\partial u_i}{\partial x_k}}}=0,\\
&\frac{\partial L_n}{\partial t}+\frac{\partial (u_k L_n+\underline{j_k})}{\partial x_k}=-g,\\
&\frac{\partial L_{j_i}}{\partial t}+\frac{\partial (u_k L_{j_i}+\underline{n\delta_{ik}})}{\partial x_k}-\underline{\underline{L_{j_i}\frac{\partial u_k}{\partial x_k}}}+\underline{\underline{L_{j_k}\frac{\partial u_k}{\partial x_i}}}=0,\\
&\frac{\partial L_T}{\partial t}+\frac{\partial (u_k L_T)}{\partial x_k}=\frac{gn}{T},\\
&\frac{\partial L_{u_i}}{\partial t}+\frac{\partial (u_k L)_{u_i}}{\partial x_k}-\underline{\underline{L_{\omega_k}\frac{\partial \omega_i}{\partial x_k}}}+\underline{\underline{L_{j_i}\frac{\partial j_k}{\partial x_k}}}-\underline{\underline{L_{j_k}\frac{\partial j_i}{\partial x_i}}}=\varepsilon_{ilm}j_l L_{\omega_m}.
\end{aligned} \quad (3.11)$$

We recall the changes of the initial system made in this procedure. First, we added the compatible terms j_k, $n\delta_{ik}$ (these terms are underlined) under the

---

[4)] E. I. Romenskii, Thermodynamics and Hyperbolic Systems of Balance Laws in Continuum Mechanics, In: *Godunov Methods. Theory and Applications*, Kluwer Academic / Plenum Publishers, 2001, pp. 745–762.

sign of derivatives $\dfrac{\partial}{\partial x_k}$ in the third and fourth line. Since these terms do not destroy the symmetry, the system remains hyperbolic. For example, the matrix of coefficients at the derivatives with respect to x_0 of the variables n, j_{-1}, j_0, j_1 will contain a new symmetric block

$$\begin{pmatrix} 0 & 0 & 1 & 0 \\ 0 & 0 & 0 & 0 \\ 1 & 0 & 0 & 0 \\ 0 & 0 & 0 & 0 \end{pmatrix}.$$

Second, we added some terms of nondivergence form (they are underlined twice) to the left-hand sides of some equations. After this procedure, the system is again a symmetric hyperbolic system. In particular, the matrix of coefficients at the derivatives with respect to x_0 of the variables ω_{-1}, ω_0, ω_1; j_{-1}, j_0, j_1; u_{-1}, u_0, u_1 will contain the additional block

$$\begin{pmatrix}
0 & 0 & 0 & 0 & 0 & 0 & -L_{\omega_0} & 0 & 0 \\
0 & 0 & 0 & 0 & 0 & 0 & 0 & -L_{\omega_0} & 0 \\
0 & 0 & 0 & 0 & 0 & 0 & 0 & 0 & -L_{\omega_0} \\
0 & 0 & 0 & 0 & 0 & 0 & 0 & -L_{j_{-1}} & 0 \\
0 & 0 & 0 & 0 & 0 & 0 & L_{j_{-1}} & 0 & L_{j_1} \\
0 & 0 & 0 & 0 & 0 & 0 & 0 & -L_{j_1} & 0 \\
-L_{\omega_0} & 0 & 0 & 0 & L_{j_{-1}} & 0 & 0 & 0 & 0 \\
0 & -L_{\omega_0} & 0 & -L_{j_{-1}} & 0 & -L_{j_1} & 0 & 0 & 0 \\
0 & 0 & -L_{\omega_0} & 0 & L_{j_1} & 0\,0 & 0 & 0 &
\end{pmatrix}.$$

Similar symmetric blocks are added to the matrices of coefficients at the derivatives with respect to $\dfrac{\partial}{\partial x_{-1}}$, $\dfrac{\partial}{\partial x_1}$. Only three equations of the system (3.11) have nonzero right-hand sides which are completely determined by the scalar function $g = g(q_0, u, \omega_i, n, j_i, T)$ that describes relaxations. It is natural to assume that $g = n/\tau$, where $\tau > 0$ is the characteristic relaxation time depending on the state of the medium. Owing to this assumption, the right-hand side gn/T in the "entropy" equation is positive. The term $\varepsilon_{ilm} u_i j_l L_{\omega_m}$ added to the right-hand side of the last equation provides the validity of the conservation of energy. We will discuss this in detail below.

The first and second equations of the system (3.11) lead to the relation

$$\frac{\partial}{\partial t}\left[\frac{1}{L_{q_0}}\left(\frac{\partial L_{\omega_k}}{\partial x_k}\right)\right] + u_i \frac{\partial}{\partial x_i}\left[\frac{1}{L_{q_0}}\left(\frac{\partial L_{\omega_k}}{\partial x_k}\right)\right] = 0,$$

which means that this system is compatible with the additional equation

$$\frac{\partial L_{\omega_k}}{\partial x_k} = 0 \tag{3.12}$$

Introducing the notation $\Omega_i = \varepsilon_{ikl}\frac{\partial L_{j_l}}{\partial x_k}$,[5] which imply, in particular, the equality $\frac{\partial \Omega_k}{\partial x_k} = 0$ we can write the equations for L_{j_i} in the form

$$\frac{\partial L_{j_i}}{\partial t} + \frac{\partial(u_k L_{j_k} + n\delta_{ik})}{\partial x_i} - \varepsilon_{ikl} u_k \Omega_l = 0 \tag{3.13}$$

and obtain the following consequences:

$$\frac{\partial \Omega_i}{\partial t} = \varepsilon_{ikl}\frac{\partial u_r \varepsilon_{rls}\Omega_s}{\partial x_k},$$

$$\frac{\partial \Omega_i}{\partial t} + \frac{\partial u_k \Omega_i}{\partial x_k} - \Omega_k \frac{\partial u_i}{\partial x_k} = 0.$$

Comparing with the second line of the system, we find

$$\frac{\partial}{\partial t}\left(L_{\omega_i} - \varepsilon_{iml}\frac{\partial L_{j_l}}{\partial x_m}\right) + \frac{\partial}{\partial x_k}\left[u_k\left(L_{\omega_i} - \varepsilon_{iml}\frac{\partial L_{j_l}}{\partial x_m}\right)\right]$$
$$= \left(L_{\omega_k} - \varepsilon_{kml}\frac{\partial L_{j_l}}{\partial x_m}\right)\frac{\partial u_i}{\partial x_k},$$
$$\frac{\partial}{\partial t}\left(L_{\omega_i} - \varepsilon_{iml}\frac{\partial L_{j_l}}{\partial x_m}\right) + u_k\frac{\partial}{\partial x_k}\left(L_{\omega_i} - \varepsilon_{iml}\frac{\partial L_{j_l}}{\partial x_m}\right)$$
$$= \frac{\partial u_i}{\partial x_k}\left(L_{\omega_k} - \varepsilon_{kml}\frac{\partial L_{j_l}}{\partial x_m}\right) - \frac{\partial u_k}{\partial x_k}\left(L_{\omega_i} - \varepsilon_{iml}\frac{\partial L_{j_l}}{\partial x_m}\right) = 0.$$

Therefore, if $L_{\omega_i} = \varepsilon_{iml}\frac{\partial L_{j_l}}{\partial x_m}$ at the initial time moment, then this equality is preserved along each trajectory $\frac{dx_k}{dt}$. This fact follows from the uniqueness theorem for the system of linear homogeneous ordinary equations.

So, the system (3.11) is compatible with the additional equality

$$L_{\omega_k} = \varepsilon_{kml}\frac{\partial L_{j_l}}{\partial x_m} \quad (L_{\omega_k} = \Omega_k) \tag{3.14}$$

which implies, in particular, the relation (3.12). Using (3.14), we can write the equality (3.13) as follows:

$$\frac{\partial L_{j_i}}{\partial t} + \frac{\partial(u_k L_{j_k} + n\delta_{ik})}{\partial x_i} - \varepsilon_{ikl} u_k L_{\omega_l} = 0 \tag{3.15}$$

[5] Here, ε_{ikl} is the **Levi Civita symbol** which is equal to 0 if all i, k, l are distinct, $\varepsilon_{-101} = \varepsilon_{01-1} = \varepsilon_{1-10} = +1, \varepsilon_{0-11} = \varepsilon_{-110} = \varepsilon_{10-1} = -1$

Before we begin to derive the conservation of energy, it is convenient, based on (3.14) and its consequence (3.12), to write the last equation of the system (3.11) as an equation of the divergence form that should govern the momentum conservation law

$$\frac{\partial L_{u_i}}{\partial t} + \frac{\partial[(u_k L)_{u_i} + j_k L_{j_i} - \omega_i L_{\omega_k} - \delta_{ik} j_m L_{j_m}]}{\partial x_k} = 0. \tag{3.16}$$

Here, we used the specific form of the right-hand side. Multiplying each of the equations in (3.11), except for the last one, by the corresponding multipliers q_0, ω_i, n, j_i, T and adding the sum of these products with the equalities (3.16) multiplied by u_i, we arrive at the conservation of energy

$$\frac{\partial E}{\partial t} + \frac{\partial}{\partial x_k}[u_k(E + L - j_i L_{j_i}) + u_i(j_k L_{j_i} - \omega_i L_{\omega_k}) + n j_k] = 0, \tag{3.17}$$

where E is determined in terms of L with the help of the Legendre transform

$$E = q_0 L_{q_0} + \omega_i L_{\omega_i} + n L_n + j_i L_{j_i} + T L_T + u_i L_{u_i} - L.$$

As usual, we assume that the generating potential L is a convex function of all scalars and vectors describing the behavior of the medium.

For this conservation law of energy the right-hand sides of the equations with respect to $\dfrac{\partial L_n}{\partial t}$, $\dfrac{\partial L_T}{\partial t}$, $\dfrac{\partial L_{u_i}}{\partial t}$ are compensated and, consequently, the law remains valid.

We have presented examples of overdetermined thermodynamically compatible equations in the case where the unknown functions are only scalar or three-dimensional vectors. Of course, this case cannot cover all possible processes in media. In particular, the problems of the theory of elasticity are outside our consideration because, in this case, we should deal with deformation tensor and stress tensor which cannot be determined only by scalar and vectors. To apply our approach to more complicated situations, we need more information about representations of the rotation group.

§ 4. Some Facts of the Theory of Representations of Orthogonal Transformations of Three-Dimensional Space

Constructing systems of differential equations, we should keep in mind that the form of these equations is independent of the choice of orthogonal coordinates in the spaces where the values of the unknown functions are contained at each time moment. In the examples in Sections 2 and 3, the role of such functions was played by either scalars or three-dimensional vectors. In this section, we introduce new notions which allow us to generalize the above examples and construct systems with respect to unknown vector-valued functions of other dimensions.

Each orthogonal transformation of the three-dimensional space is determined by some orthogonal 3×3-matrix Q ($Q^T Q = I_3$). All such matrices form a group denoted by $O(3)$. The letter "O" abbreviates "orthogonal" and the digital "3" indicates the dimension. This group contains the subgroup $SO(3)$ of "special" (indicated by the letter "S") matrices, i.e., orthogonal matrices P with positive determinant ($P^T P = I_3, \det P = 1$). Any element $Q \in O(3)$ can be represented as $Q = P$ or as $Q = -P$, where $P \in SO(3)$. We first recall some facts concerning representations of the *rotation group* $SO(3)$.

The transformation $P \in SO(3)$ sends the coordinates x_{-1}, x_0, x_1 of a coordinate vector x to the coordinates y_{-1}, y_0, y_1 by the formula

$$y \equiv \begin{pmatrix} y_{-1} \\ y_0 \\ y_1 \end{pmatrix} = P \begin{pmatrix} x_{-1} \\ x_0 \\ x_1 \end{pmatrix} \equiv Px.$$

The "old" coordinates can be found in terms of the "new" ones with the help of the inverse matrix $P^{-1} = P^T = P^*$:

$$x = P^{-1} y = P^T y.$$

After the transformation P, a field of some vector-valued function, for example, the field of velocity vector $u = u(x,t)$,

$$u(x,t) = \begin{pmatrix} u_{-1}(x_{-1}, x_0, x_1, t) \\ u_0(x_{-1}, x_0, x_1, t) \\ u_1(x_{-1}, x_0, x_1, t) \end{pmatrix}$$

is determined by a vector-valued function $v = v(y,t)$ computed in terms of $u(x,t)$ by the formula

$$v(y,t) = Pu(P^{-1}y, t).$$

If P consists of two subsequent transformations P_1 and P_2, i.e., $P = P_2 P_1$, then

$$V(y,t) = P_2 P_1 u(P_1^{-1} P_2^{-1} y, t).$$

It is reasonable to suppose that some other vector-valued function $q = q(x,t)$ of some dimension (not necessarily of dimension 3) is transformed under rotations of coordinates in a similar way so that, after a rotation P of coordinates, it becomes the vector-valued function

$$p = p(x,t) = p(P^{-1}y, t) = \Omega(P) q(P^{-1}y, t),$$

where $\Omega(P)$ is some matrix corresponding to transformation of coordinates of the vector q which can be found from P. A correspondence $P \to \Omega(P)$, called a representation, is denoted by $\Omega(P) = T_P$.

By definition, any representation should satisfy the following conditions:

$$T_{I_3} = I_M$$

($M \times M$ is the size of the matrix Q and M is the dimension of the vectors q and p),

$$T_{P_2P_1} = T_{P_2} \cdot T_{P_1}.$$

In other words, $\Omega(I_3) = I_M$, $\Omega(P_2P_1) = \Omega(P_2)\cdot\Omega(P_1)$. The matrix representation $\Omega(P)$ realizes the representation $P \to \Omega(P)q(P^{-1}y,t)$ in the space of M-dimensional vector-valued functions.

It turns our that any representation of the rotation group can be divided into simpler irreducible representations which already cannot be simplified.

We say that a representation realized by matrices $\Omega(P)$ is reducible if there exists a similar transformation $W(\det W \neq 0)$ such that

$$W\Omega(P)W^{-1} = \begin{pmatrix} \Omega'(P) & & \\ & \Omega''(P) & 0 \\ & 0 & \ddots \end{pmatrix}.$$

Here, W is the same for all P and the square matrix blocks $\Omega'(P)$, $\Omega''(P)$, ... describe a representation of the same group. If these blocks do not admit further decomposition by a similar procedure, then the representations associated with these matrix blocks are said to be *irreducible.*

It turns out that the dimension M of irreducible representations of the group $SO(3)$ may be only odd numbers: $M = 1$, $M = 3$, $M = 5$, ... The value N such that $M = 2N + 1$ is called the *weight* of this representation. (In the quantum mechanics, the so-called representations of half-integer weight N are also used; these representations are realized by complex $M \times M$-matrices, where $M = 2N + 1$ is even. But two matrices $\pm\Omega(P)$ are associated with each P. We do not consider this in detail.) It turns out that the irreducible representation is determined by its weight N (or, which is the same, by the dimension $2N + 1$). Namely, taking suitable W in order to simplify $\Omega(P)$, we can reach the situations where each diagonal block $\Omega(P)$, $\Omega''(P)$, ... coincides with one of the following matrices:

$$\Omega^{(0)}(P),\ \Omega^{(1)}(P),\ \Omega^{(2)}(P),\ldots,\ \Omega^{(N)}(P),\ \ldots.$$

(The same block $\Omega^{(k)}(P)$ can be placed into the diagonal several times.) All the matrices of this list are orthogonal and there are explicit formulas for them.

As is well known, each rotation $P \in SO(3)$ can be written as the product of rotations $P = P_0(\psi)P_{-1}(\theta)P_0(\varphi)$ around the coordinate axes:

$$P_0(\varphi) = \begin{pmatrix} \cos\varphi & 0 & \sin\varphi \\ 0 & 1 & 0 \\ -\sin\varphi & 0 & \cos\varphi \end{pmatrix},$$

$$P_{-1}(\theta) = \begin{pmatrix} 1 & 0 & 0 \\ 0 & \cos\theta & -\sin\theta \\ 0 & \sin\theta & \cos\theta \end{pmatrix}.$$

Since, by the definition of a representation, we have the equality

$$\Omega(P) = \Omega(P_0(\psi))\Omega(P_{-1}(\theta))\Omega(P_0(\varphi)),$$

it suffices to have formulas for the matrices

$$\Omega_0^{(N)}(\varphi) \equiv \Omega^{(N)}(P_0(\varphi)), \quad \Omega_{-1}^{(N)}(\theta) \equiv \Omega^{(N)}(P_{-1}(\theta))$$

whose entries will be denoted by $\Omega^{(N)}_{0,n_1,n_2}(\varphi)$, $\Omega^{(N)}_{-1,n_1,n_2}(\theta)$. The number of rows and the number of columns run all integers not larger than N by modulo $(-N \leqslant n_i \leqslant N)$.

We write formulas for these matrix entries.[6)]

For the matrix $\Omega_0^{(N)}(\varphi)$:

$$\Omega^{(N)}_{0,0,0}(\varphi) = 1,$$
$$\Omega^{(N)}_{0,0,0}(\varphi) = \cos n\varphi,$$
$$\Omega^{(N)}_{0,-n,n} = \sin n\varphi, \quad n = \pm 1, \pm 2, \pm N,$$

and $\Omega^{(N)}_{0,n,k}(\varphi) = 0$ in the remaining cases.

For the matrix $\Omega_{-1}^{(N)}(\theta)$ $(\cos\theta = \mu)$:

$$\Omega^{(N)}_{-1,\pm n,\pm k}(\theta) = \frac{(-1)^{N-k}}{2^N(1-\mu^2)^{k/2}}\sqrt{\frac{(N+k)!}{(N-k)!(N-n)!(N+n)!}}$$
$$\times \{(\frac{1-\mu}{1+\mu})^{\frac{n}{2}}\frac{d^{N-k}}{d\mu^{N-k}}[(1+\mu)^{N+n}(1-\mu)^{N-n}]$$
$$\pm (-1)^n(\frac{1+\mu}{1-\mu})^{\frac{n}{2}}\frac{d^{N-k}}{d\mu^{N-k}}[(1+\mu)^{N-n}(1-\mu)^{N+n}]\}, \quad n,k \geqslant 1.$$

$$\Omega^{(N)}_{-1,0,0}(\theta) = \frac{(-1)^N}{2^N \cdot N!}\frac{d^N}{d\mu^N}(1-\mu^2)^N,$$

$$\Omega^{(N)}_{-1,0,k}(\theta) = -\frac{(-1)^{N-k}}{2^N \cdot N!}\sqrt{\frac{2\cdot(N+k)!}{(N-k)!}}\frac{1}{(1-\mu^2)^{k/2}}\frac{d^{N-k}}{d\mu^{N-k}}(1-\mu^2)^N, \quad k \geqslant 1,$$

$$\Omega^{(N)}_{-1,n,0}(\theta) = -\frac{(-1)^N}{2^N \cdot N!}\sqrt{\frac{2(N+n)!}{(N-n)!}}\frac{1}{(1-\mu^2)^{n/2}}\frac{d^{N-n}}{d\mu^{N-n}}(1-\mu^2)^N, \quad n \geqslant 1.$$

We note that the above formulas imply the equality $\Omega^{(1)}(P) = P$.

We shortly discuss irreducible representations of the orthogonal group $O(3)$. As was mentioned, any element $Q \in O(3)$ can be represented as $Q = P$ or $Q = -P$, where $P \in SO(3)$. Therefore, if we are given an irreducible representation T_P of weight N of the group $SO(3)$, then in order to determine the representation of the group $O(3)$, it suffices to define the transformation

---

[6)] Cf. V. M. Gordienko, *Sib. Math. J.*, 2002.

$T_{(-I_3)}$. There are only two possibilities: $T_{(-I_3)} = \pm I_{2N+1}$. Thus, for each weight N (for the dimension $2N+1$), there exist two different irreducible representations of weight N of the group $O(3)$. A quantity transformed in accordance with an irreducible representation of the group $O(3)$ of weight N is called a *vector* if $T_{(-I_3)} = (-I_{2N+1})^N$ and a *pseudovector* if $T_{(-I_3)} = (-I_{2N+1})^{N+1}$.

We return to representations of the rotation group $SO(3)$. The formulated assertion about the reducibility of representations can be explained as follows. Any space of the action of a vector representation of the group $SO(3)$ can be divided into subspaces that are invariant under this representation. In each of these subspaces, we can choose a canonical basis such that, in this basis, the representation is realized with the help of the above orthogonal matrices. In particular, in three-dimensional invariant subspaces, representations are realized by the transformation P itself.

We illustrate the above arguments by one important example. Consider $(2K+1)\cdot(2L+1)$-dimensional vector space elements of which have components a_{kl} marked by two subscripts k and l, where k and l, run all integers and satisfy the inequalities $-K \leqslant k \leqslant K$, $-L \leqslant l \leqslant L$. It is convenient to arrange the components as entries of a rectangle matrix A so that k is the number of rows and l is the number of columns of this matrix:

$$A = \begin{pmatrix} a_{-K(-L)} & a_{-K(-L+1)} & \cdots & \cdots & a_{-K(L)} \\ a_{(-K+1)(-L)} & a_{(-K+1)(-L+1)} & \cdots & \cdots & a_{(-K+1)(-L)} \\ \vdots & \vdots & & & \vdots \\ \vdots & \vdots & & & \vdots \\ a_{K(-L)} & a_{K(-L+1)} & \cdots & \cdots & a_{K(L)} \end{pmatrix}$$

We introduce a representation T_P of the rotation group by the following rule for transformation of the above matrices regarded as vectors of the $(2K+1)(2L+1)$-dimensional space:

$$T_P A \equiv \Omega^{(K)}(P)\cdot A\cdot[\Omega^{(L)}(P)]^T.$$

This representation is called the *Kronecker product* of the representations with weights K and L. It turns out that this representation is reducible and it can be decomposed into irreducible weights $|K-L|$, $|K-L|+1$, $|K-L|+2, \ldots,$ $K+L$. There are no equivalent representations in this list.

An irreducible representation of weight $N(|K-L| \leqslant N \leqslant K+L)$ is realized in this $2N+1$-dimensional space whose canonical basis is formed by some special $(2K+1)\times(2L+1)$-matrices denoted by $G^n_{N[K,L]}$ $(-N \leqslant n \leqslant N)$. These matrices are called the *Clebsch-Gordan matrices*, and entries of these matrices $G^{n[k,l]}_{N[K,L]}$ $(-K \leqslant k \leqslant K,\ -L \leqslant l \leqslant L)$ are called the *Clebsch-Gordan coefficients*.

Each matrix of the subspace where a representation of weight N acts, can be written as a linear combination of Clebsch–Gordan matrices

$$\sum_{n=-N}^{N} z_n^{(N)} G^n_{N[K,L]}.$$

Under a transformation of this representation

$$\Omega^{(K)}(P)\left[\sum_{n=-N}^{N} z_n^{(N)} G^n_{N[K,L]}\right]\left[\Omega^{(L)}(P)\right]^T = \sum_{n=-N}^{N} \hat{z}_n^{(N)} G^n_{N[K,L]}$$

the coefficients of this linear combination are transformed by the formula

$$\hat{z}^{(N)} = \Omega^{(N)}(P)\cdot z^{(N)},$$

$$(\hat{z}_m^{(N)} = \sum_{n=-N}^{N} \Omega_{mn}^{(N)}(P) z_n^{(N)}).$$

The matrices $G^n_{N[KL]}$ satisfy the condition

$$\sum_{k,l} G^{n_1[k,l]}_{N_1[K,L]}\cdot G^{n_2[k,l]}_{N_2[K,L]} = \operatorname{tr}\left\{G^{n_1}_{N_1[K,L]}\cdot [G^{n_2}_{N_2[K,L]}]^T\right\} = \delta_{N_1N_2}\cdot\delta_{n_1n_2},$$

which can be considered as the orthonormalization condition in the space of $(2K+1)\times(2L+1)$-matrices A, B, ..., with respect to the inner product

$$(A,B) = \operatorname{tr}\{A\cdot[B]^T\}.$$

The common number $\sum\limits_{N_j=|K-L|}^{K+L}(2N_j+1) = (2K+1)(2L+1)$ of matrices $G^{n_j}_{N_j[KL]}$ coincides with the number of entries of a matrix of this size. Therefore, the Clebsch–Gordan matrices can be regarded as basis elements for the space of rectangle matrices. The coefficients of the representation

$$A = \sum_{n,m} w_n^{(N)} G^n_{N[K,L]}$$

can be found by the formula

$$w_n^{(N)} = \operatorname{tr}\left\{G^n_{N[K,L]}\cdot A^T\right\}.$$

Definition 4.1. By the *product of weight* N of a vector $u^{(K)}$ of weight K and a vector $v^{(L)}$ of weight L with components $u_k^{(K)}$, $v_l^{(L)}$ $(-K\leqslant k\leqslant K, -L\leqslant l\leqslant L)$ we mean the vector $w^{(N)} = [u^{(K)}\times v^{(L)}]^{(N)}$ with components $w_n^{(N)}$ $(-N\leqslant n\leqslant N)$ computed by the formula

$$w_n^{(N)} = (u^{(K)}, G^n_{N[K,L]}v^{(L)}) \equiv [u^{(K)}\times v^{(L)}]_n^{(N)}.$$

If the vectors $u^{(K)}$ and $v^{(L)}$ are transformed by the matrices $\Omega^{(K)}(P)$ and $\Omega^{(L)}(P)$ corresponding to representations of the group $SO(3)$, then $w^{(N)}(P)$ is transformed with the help of $\Omega^{(N)}(P)$:

$$\Omega^{(N)}(P)w^{(N)} = [\Omega^{(K)}(P)u^{(K)} \times \Omega^{(L)}(P)v^{(L)}]^{(N)}.$$

The product $[u^{(K)} \times v^{(L)}]^{(N)}$ has meaning if $|K-L| \leqslant N \leqslant K+L$.

Consider an example. The matrices $G^n_{1[1,1]}$ have the form

$$G^{-1}_{1[1,1]} = \begin{pmatrix} 0 & 0 & 0 \\ 0 & 0 & \frac{1}{\sqrt{2}} \\ 0 & -\frac{1}{\sqrt{2}} & 0 \end{pmatrix},$$

$$G^{0}_{1[1,1]} = \begin{pmatrix} 0 & 0 & -\frac{1}{\sqrt{2}} \\ 0 & 0 & 0 \\ \frac{1}{\sqrt{2}} & 0 & 0 \end{pmatrix},$$

$$G^{1}_{1[1,1]} = \begin{pmatrix} 0 & \frac{1}{\sqrt{2}} & 0 \\ -\frac{1}{\sqrt{2}} & 0 & 0 \\ 0 & 0 & 0 \end{pmatrix}.$$

Therefore, $[u^{(1)} \times v^{(1)}]^{(1)}$ coincides up to a factor $1/\sqrt{2}$ with an ordinary vector product of three-dimensional vectors. Similarly, in view of $G^{0}_{0[1,1]} = \dfrac{1}{\sqrt{3}} I_3$, it is easy to see that $[u^{(1)} \times u^{(1)}]^{(0)}$ coincides up to the factor $1/\sqrt{3}$ with the classical inner product.

Assuming that $K = 1$ and replacing the vector $u^{(1)}$ with the operator vector, i.e., with the gradient

$$\nabla^{(1)} = \begin{pmatrix} \frac{\partial}{\partial x_{-1}} \\ \frac{\partial}{\partial x_0} \\ \frac{\partial}{\partial x_1} \end{pmatrix},$$

we can define the invariant differential operations

$$D^{(-)}_{(L)} v^{(L)} = [\nabla^{(1)} \times v^{(L)}]^{(L-1)},$$
$$D^{(0)}_{(L)} v^{(L)} = [\nabla^{(1)} \times v^{(L)}]^{(L)},$$
$$D^{(+)}_{(L)} v^{(L)} = [\nabla^{(1)} \times v^{(L)}]^{(L+1)},$$

which can be regarded as a generalization of the classical operations div, curl, grad because they can be applied not only to vectors and scalars, but also to vector-valued functions that are transformed by irreducible representations of the rotation group with any weight.

Let us write the matrix representations of these operators:

$$D^{(-)}_{(L)} = \sqrt{\frac{2L-1}{3}}\frac{\partial}{\partial x_i}G^i_{1[L-1,L]},$$

$$D^{(0)}_{(L)} = -\sqrt{\frac{2L+1}{3}}\frac{\partial}{\partial x_i}G^i_{1[L,L]},$$

$$D^{(+)}_{(L)} = \sqrt{\frac{2L+3}{3}}\frac{\partial}{\partial x_i}G^i_{1[L+1,L]}.$$

Now, we write various relations connected the Clebsch–Gordan matrices and their entries, the Clebsch–Gordan coefficients:

$$G^{0[\lambda,l]}_{0[L,L]} = \frac{\delta_{\lambda l}}{\sqrt{2L+1}},$$

$$G^{k[l,m]}_{K[L,M]} = \sqrt{\frac{2K+1}{2M+1}}G^{m[k,l]}_{M[K,L]} = (-1)^{K+L+M}\sqrt{\frac{2K+1}{2L+1}}G^{l[k,m]}_{L[K,M]}.$$

Some of these relations were used in representations of $D^{(0)}_{(L)}$ and $D^{(\pm)}_{(L)}$.

We also write the matrices $G^j_{2[1,1]}$:

$$G^{-2}_{2[1,1]} = \begin{pmatrix} 0 & 0 & -\frac{1}{\sqrt{2}} \\ 0 & 0 & 0 \\ -\frac{1}{\sqrt{2}} & 0 & 0 \end{pmatrix}, \quad G^{-1}_{2[1,1]} = \begin{pmatrix} 0 & \frac{1}{\sqrt{2}} & 0 \\ \frac{1}{\sqrt{2}} & 0 & 0 \\ 0 & 0 & 0 \end{pmatrix},$$

$$G^{0}_{2[1,1]} = \begin{pmatrix} -\frac{1}{\sqrt{6}} & 0 & 0 \\ 0 & \sqrt{\frac{2}{3}} & 0 \\ 0 & 0 & 0-\frac{1}{\sqrt{6}} \end{pmatrix},$$

$$G^{1}_{2[1,1]} = \begin{pmatrix} 0 & 0 & 0 \\ 0 & 0 & \frac{1}{\sqrt{2}} \\ 0 & \frac{1}{\sqrt{2}} & 0 \end{pmatrix}, \quad G^{2}_{2[1,1]} = \begin{pmatrix} \frac{1}{\sqrt{2}} & 0 & 0 \\ 0 & 0 & 0 \\ 0 & 0 & -\frac{1}{\sqrt{2}} \end{pmatrix}.$$

§ 5. The Clebsch–Gordan Coefficients

Using facts of the representation theory (cf. Section 4), we describe how to construct a model system of thermodynamically compatible conservation laws. In this case, the unknown functions are not necessarily scalars or three-dimensional vectors. Among unknown functions there are 5-dimensional vector-valued functions that are transformed under rotations according to irreducible representations of weight 2.

As in all the cases considered in Section 3, we first choose a standard "simplest" system

$$\begin{aligned} &\frac{\partial L_{q_0}}{\partial t} + \frac{\partial (u_k L_{q_0})}{\partial x_k} = 0, \\ &\frac{\partial L_n}{\partial t} + \frac{\partial (u_k L_n)}{\partial x_k} = -g, \\ &\frac{\partial L_{h_j}}{\partial t} + \frac{\partial (u_k L_{h_j})}{\partial x_k} = -f_j, \\ &\frac{\partial L_T}{\partial t} + \frac{\partial (u_k L_T)}{\partial x_k} = \frac{ng + h_j f_j}{T}, \\ &\frac{\partial L_{u_i}}{\partial t} + \frac{\partial (u_k L)_{u_i}}{\partial x_k} = 0. \end{aligned} \tag{5.1}$$

Here, the first equation expresses the mass conservation law, the last equation is the momentum conservation law, and the equation before is the compensating entropy equation. As usual, u_i $(i = -1, 0, 1)$ are components of the velocity vector and T is the temperature (it is assumed to be positive; $T > 0$). The parameter n is also a scalar characteristic of a medium (for example, it can characterize the porosity of the medium which can be compressed under the action of presser arising as a result of deformations). The 5-dimensional vector with components h_j $(-2 \leqslant j \leqslant +2)$ can describe, for example, the internal state of a deformed medium where the stress value on different area elements, passing through a fixed point depend on the direction of these elements.

The generating potential L is a function of the form

$$L = \Lambda\left(q_0 + \frac{u_i u_i}{2}, n, h, T\right).$$

It is convex with respect to q_0, n, h, and T. As was done in Section 2, it is possible to show that the system (5.1) with this potential is Galilei-invariant if, in addition to the assumptions made in Section 2, we require that, under rotations P $(P^T P = I_3, \det\ P > 0)$, the vector-valued function h is transformed by the rule

$$h(x,t) \to \Omega^{(2)}(P) h(P^{-1}x, t),$$

where $\Omega^{(2)}(P)$ is a matrix corresponding to a representation of weight 2 in the canonical basis, and that the generating potential is invariant under such transformations.

Below, we will not change the generating potential L but add new terms that are invariant under rotations and do not contain the velocity vector (but they may contain its derivatives). Then the Galilei-invariance is preserved. We first consider the 3rd line of the system (5.1):

$$\frac{\partial L_{h_j}}{\partial t} + \frac{\partial (u_k L_{h_j})}{\partial x_k} = -f_j. \tag{5.2}$$

There are 5 such equations ($-2 \leqslant j \leqslant 2$).

We propose to add new terms to the left-hand sides of the equations of the 3rd and 5th lines. These new terms (they are underlined) contain the derivatives of the unknown vector-functions u and h transformed with the help of the Clebsch–Gordan matrices:

$$\begin{aligned}
&\frac{\partial L_{q_0}}{\partial t} + \frac{\partial (u_k L_{q_0})}{\partial x_k} = 0, \\
&\frac{\partial L_n}{\partial t} + \frac{\partial u_k L_n}{\partial x_k} = -g = -\frac{n}{\tau_0}, \\
&\frac{\partial L_{h_j}}{\partial t} + \frac{\partial u_k L_{h_j}}{\partial x_k} \underline{- \frac{\partial}{\partial x_k} G^{k[ji]}_{1[21]} u_i} = -f_j = -\frac{h_j}{\tau} \\
&\frac{\partial L_T}{\partial t} + \frac{\partial (u_k L_T)}{\partial x_k} = \frac{ng + h_j f_j}{T} \equiv \frac{Q}{T} \geqslant 0, \\
&\frac{\partial L_{u_i}}{\partial t} + \frac{\partial (u_k L_{u_i} + \delta_{ik} L)}{\partial x_k} - \frac{\partial}{\partial x_k} G^{k[ij]}_{1[12]} h_j = 0.
\end{aligned} \tag{5.3}$$

Introduce the notation

$$\sigma_{ik} = -\delta_{ik} L + G^{k[ij]}_{1[12]} h_j = -\delta_{ik} L + \sqrt{\frac{5}{3}} h_j G^{j[ik]}_{2[11]} = \sigma_{ki}.$$

Then the last equation in (5.3) takes the form which is usual for the momentum conservation law in elasticity

$$\frac{\partial L_{u_i}}{\partial t} + \frac{\partial (u_k L_{u_i} - \sigma_{ik})}{\partial x_k} = 0. \tag{5.4}$$

Taking into account this analogy, we note that the system we construct will not coincide with the equations of nonlinear elasticity (cf. the previous section) even for $g = 0$ and $f = 0$. However, it is reasonable to expect that with the help of Eqs. (5.3) it is possible to simulate high-rate deformations with sufficiently high accuracy.

Multiplying the first four lines in (5.3) by the corresponding factors q_0, n, h, T and adding to the sum of the products the sum of the equalities (5.3) multiplied by u_i, we arrive at the conservation of energy

$$\frac{\partial E}{\partial t} + \frac{\partial (u_k E - u_i \sigma_{ik})}{\partial x_k} = 0, \tag{5.5}$$

where, as usual, $E = q_0 L_{q_0} + n L_n + h_j L_{h_j} + u_i L_{u_i} + T L_T - L$. We again emphasize that in order to keep the Galilei invariance of (5.3) and the invariance of the conservation law (5.5) which is a consequence of (5.3), the generating potential L should be a function of the form

$$L = \Lambda \left[q_0 + \frac{u_i u_i}{2}, n, h_j, h_j, \det(h_j G^j_{2[11]}), T \right]$$

(We note that $h_j h_j$, $\det(h_j G^j_{2[11]})$ are invariant under rotations.)

It is convenient to consider

$$E = \rho\left\{\mathcal{E}[\rho, S, n, h_j h_j, \det(h_j G^j_{2[11]})] + \frac{u_i u_i}{2}\right\}.$$

Introducing the energy of state in such a way, we arrive at the parametric representation of the generating function L and its arguments q_0 and T:

$$\begin{aligned}
&q_0 = \mathcal{E} + \rho\mathcal{E}_\rho - \mathcal{E}_S - n\mathcal{E}_n - h_j\mathcal{E}_{h_j} - u_i u_i/2,\\
&L = \rho^2\mathcal{E}_\rho, \quad T = \mathcal{E}_S.
\end{aligned}$$

Moreover, $\rho = L_{q_0}$, $\rho u_i = L_{u_i}$, $\rho S = L_T$.

Now, we describe one more system where the Clebsch–Gordan coefficients are used. This system is obtained from the above-mentioned system (cf. Section 24)

$$\begin{aligned}
&\frac{\partial \rho c_{ij}}{\partial t} + \frac{\partial u_k \rho c_{ij}}{\partial x_k} - \frac{\partial u_i}{\partial x_k}\rho c_{kj} = 0,\\
&\frac{\partial(\rho/\rho_0)}{\partial t} + \frac{\partial}{\partial x_k}(u_k\rho/\rho_0) = -\varphi,\\
&\frac{\partial \rho S}{\partial t} + \frac{\partial u_k \rho S}{\partial x_k} = \frac{Q}{T},\\
&\frac{\partial \rho u_i}{\partial t} + \frac{\partial(\rho u_i u_k - \sigma_{ik})}{\partial x_k} = 0,
\end{aligned} \tag{5.6}$$

where $\rho = \sqrt{\det \rho C}/\rho_0^{2/3} = D/\rho_0^{2/3}$. The variables ρc_{ij}, ρS, $\rho/_0$, ρu_i are independent variables and the number of independent variables coincides with the number of equations. We indicate the dependence of σ_{ik} on the independent variables ρc_{ij}, ρ/ρ_0, ρS later. We emphasize that the density ρ is not regarded as an independent variable but is expressed in terms of ρ and the entries ρc_{ij} of the matrix ρC by some formula which allows us to derive from the first and second equations in (5.6) the following additional relations:

$$\begin{aligned}
&\frac{\partial(D/\rho_0^{3/2})}{\partial t} + \frac{\partial(u_k D/\rho_0^{3/2})}{\partial x_k} = 0,\\
&\frac{\partial(D/\rho_0^{5/2})}{\partial t} + \frac{\partial(u_k D/\rho_0^{5/2})}{\partial x_k} = -\varphi.
\end{aligned}$$

The first equation is the continuity equation

$$\frac{\partial \rho}{\partial t} + \frac{\partial(u_k\rho)}{\partial x_k} = 0 \tag{5.7}$$

which is valid for solutions to the system (5.6). However, it is not included in the system (5.6). From the above system we also obtain the relation

$$\frac{\partial \rho_0}{\partial t} + u_k\frac{\partial \rho_0}{\partial x_k} = \frac{\rho_0^2\varphi}{\rho}$$

which describes the change of ρ_0 depending on the dissipative relaxation, the rate of which is determined in terms of the right-hand side $-\varphi$ of the second equation in (5.6). Using (5.7), one can verify that the system (5.6) is invariant under the passage to a new coordinate system moving relative to the original coordinate system with constant speed. This fact is essential for our purpose because the equations which will be constructed cannot be studied within the framework of the above scheme.

We consider such media for which the system (5.6) is invariant under rotation, which is convenient if we pass to new unknown functions transformed by irreducible representations. In particular, we deal with scalars $gd = \rho/\rho_0$, $\sigma = \rho S$, T_0 and vectors with components η_i, $\nu_i = \rho u_i$ $(i = 1, 2)$ transformed by representation of weight 1 and vectors with component ζ_j $(j = -2, -1, 0, 1, 2)$ transformed by representations of weight 2. Moreover, ξ_0, η_i, ζ_j are parameters of the representation of the matrix ρC:

$$\rho c_{km} = \xi_0 G^{0[km]}_{0[11]} + \eta_i G^{i[km]}_{1[11]} + \zeta_j G^{j[kk]}_{2[11]}. \tag{5.8}$$

Using the new variables, we can write the system (5.6) as follows:

$$\begin{aligned}
&\frac{\partial \xi_0}{\partial t} + \frac{\partial(u_k \xi_0)}{\partial x_k} - \xi_0 \frac{\partial}{\partial x_k} G^{0[ks]}_{0[11]} G^{0[si]}_{0[11]} u_i = 0,\\
&\frac{\partial \eta_a}{\partial t} + \frac{\partial(u_k \eta_a)}{\partial x_k} - \eta_b \frac{\partial}{\partial x_k} G^{b[ks]}_{1[11]} G^{a[si]}_{1[11]} u_i = 0,\\
&\frac{\partial \zeta_j}{\partial t} + \frac{\partial(u_k \zeta_j)}{\partial x_k} - \zeta_l \frac{\partial}{\partial x_k} G^{l[ks]}_{2[11]} G^{j[si]}_{2[11]} u_i = 0,\\
&\frac{\partial \sigma}{\partial t} + \frac{\partial(u_k \sigma)}{\partial x_k} = \frac{q}{T},\\
&\frac{\partial \delta}{\partial t} + \frac{\partial(u_k \delta)}{\partial x_k} = -\varphi,\\
&\frac{\partial \nu_i}{\partial t} + \frac{\partial(\nu_i u_i - \sigma_{ik})}{\partial x_k} = 0.
\end{aligned} \tag{5.9}$$

We explain our arguments for the equations on the third line of (5.9). Using (5.8), we distinguish relations in the first line of (5.6) that are transformed by representations of weight 2:

$$\frac{\partial}{\partial t} \zeta_j G^{j[ai]}_{2[11]} + \frac{\partial}{\partial x_k}(u_k \zeta_j G^{j[ai]}_{2[11]}) - \frac{\partial u_i}{\partial x_k} \zeta_l G^{l[ak]}_{2[11]} = 0.$$

Multiplying from the left this equality by $G^{l[ia]}_{2[11]}$, summarizing with respect to i, and taking into account the orthogonality of the Clebsch–Gordan matrices, we obtain the required third line in (5.9).

Let the total internal energy of a volume element of the medium under consideration is invariant under rotation

$$F = \rho E + \frac{\rho u_i u_i}{2} = \rho E + \frac{\nu_i \nu_i}{2\rho},$$

where E and ρ are represented as invariants of scalar and vector parameters $\delta, \sigma, \xi_0, \eta_i, \zeta_j$ so that $F_{\nu_i} = u_i$. We introduce the generating thermodynamical potential L as the Legendre transformation of F:

$$L = \delta F_\delta + \sigma F_\sigma + \xi_0 F_{\xi_0} + \eta_i F_{\eta_i} + \zeta_j F_{\zeta_j} + \nu_i F_{\nu_i} - F \tag{5.10}$$

and denote

$$T = F_\sigma, \quad n = F_\delta, \quad m_0 = F_{T_0}, \quad r_i = F_{\eta_i}, \quad h_j = F_{\zeta_j}, \quad u_i = F_{\nu_i}. \tag{5.11}$$

Then

$$\begin{aligned} &\delta = \frac{\rho}{\rho_0} = L_n, \quad \sigma = \rho S = L_T, \quad \nu_i = \rho u_i = L_{u_i}, \\ &\xi_0 = L_{m_0}, \quad \eta_i = L_{r_i}, \quad \zeta_j = L_{h_j} \end{aligned} \tag{5.12}$$

and the system (5.9) can be written as follows:

$$\begin{aligned} &\frac{\partial L_n}{\partial t} + \frac{\partial (u_k L_n)}{\partial x_k} = -\varphi, \\ &\frac{\partial L_T}{\partial t} + \frac{\partial (u_k L_T)}{\partial x_k} = \frac{Q}{T}, \\ &\frac{\partial L_{m_0}}{\partial t} + \frac{\partial (u_k L_{m_0})}{\partial x_k} - L_{m_0} \frac{\partial}{\partial x_k} G^{0[ks]}_{0[11]} G^{0[si]}_{0[11]} u_i = 0, \\ &\frac{\partial L_{r_a}}{\partial t} + \frac{\partial (u_k L_{r_a})}{\partial x_k} - L_{r_b} \frac{\partial}{\partial x_k} G^{b[ks]}_{11[11]} G^{a[si]}_{11[11]} u_i = 0, \\ &\frac{\partial L_{h_j}}{\partial t} + \frac{\partial (u_k L_{h_j})}{\partial x_k} - L_{h_l} \frac{\partial}{\partial x_k} G^{l[ks]}_{2[11]} G^{j[si]}_{2[11]} u_i = 0, \\ &\frac{\partial L_{u_i}}{\partial t} + \frac{\partial (u_k L_{u_i} - \sigma_{ik})}{\partial x_k} = 0. \end{aligned} \tag{5.13}$$

From (5.13) and the continuity equation we find

$$\begin{aligned} &\left(\frac{\partial}{\partial t} + u_k \frac{\partial}{\partial x_k}\right) \frac{1}{\rho} \left[\frac{\partial L_{m_0}}{\partial x_k} G^{0[is]}_{0[11]}\right] = 0, \\ &\left(\frac{\partial}{\partial t} + u_k \frac{\partial}{\partial x_k}\right) \frac{1}{\rho} \left[\frac{\partial L_{r_k}}{\partial x_i} G^{k[is]}_{1[11]}\right] = 0, \\ &\left(\frac{\partial}{\partial t} + u_k \frac{\partial}{\partial x_k}\right) \frac{1}{\rho} \left[\frac{\partial L_{h_j}}{\partial x_i} G^{j[is]}_{2[11]}\right] = 0, \end{aligned}$$

which provides us with the compatibility of our system and the additional equality

$$\frac{\partial L_{m_0}}{\partial x_k} G^{0[ki]}_{0[11]} + \frac{\partial L_{r_a}}{\partial x_k} G^{a[ki]}_{1[11]} + \frac{\partial L_{h_j}}{\partial x_k} G^{j[ki]}_{2[11]} = 0 \tag{5.14}$$

or, shortly,

$$[\nabla \times L_{m_0}]^{(1)} - [\nabla \times L_{r_a}]^{(1)} + [\nabla \times L_{h_j}]^{(1)} = 0$$

which, in turn, can be written as follows:

$$\frac{\partial \rho c_{kj}}{\partial x_k} = 0. \tag{5.15}$$

As usual, we assume that $\varphi = n/r$ and $q/T = n^2/(\tau T)$ in (5.13), which guarantees that the entropy increases under relaxation of the loading density. Multiplying the equations in (5.13) by the corresponding factors n, T, m_0, r_0, h_j, u_i and adding the results with (5.14) multiplied by u_i, we arrive at the following additional conservation law simulating the energy conservation law:

$$\frac{\partial F}{\partial t} + \frac{\partial (u_k F - u_i \sigma_{ik})}{\partial x_k} = 0 \tag{5.16}$$

if

$$\sigma_{ik} = -\delta_{ik} L + L_{m_0} G^{0[is]}_{0[11]} G^{0[sk]}_{0[11]} m_0 + L_{r_a} G^{a[is]}_{1[11]} G^{b[sk]}_{1[11]} r_b$$
$$+ L_{h_j} G^{j[is]}_{2[11]} G^{l[sk]}_{2[11]} h_l. \tag{5.17}$$

We do not discuss here the rigorous derivation of formula (5.16) for the stress tensor, the symmetric hyperbolicity of the equations, and conditions under which the thermodynamical potential is convex. These questions require further investigations.

§ 6. Orthogonal Invariants

In the above constructions of thermodynamically compatible models, we used the generating potential L invariant under rotations of coordinates. In this section, we discuss rules for constructing various invariants under rotations starting with a given collection of vectors $v_{(1)}, v_{(2)}, \ldots, v_{(s)}, \ldots, v_{(R)}$, where each of these vectors is transformed by an irreducible representation of the corresponding weight N_s under rotations.

We begin by constructing basic polynomial invariants B_r. The invariant B is constructed from these basic invariants in the form

$$B = B(B_1, B_2, \ldots).$$

The following important question arises: How to provide the convex dependence of this invariant on all its vector arguments? But we do not discuss this question here. We consider the rule for constructing independent invariants that are homogeneous polynomials of given degree $\sigma(s)$ in the components $v_{n(s)}$ $(-N_s \leqslant n(s) \leqslant N_s)$ of the vectors $v_{(s)}$. This rule was first presented by Yu. B. Rumer in his book "Theory of Spinors" published only in Russian in 1936.

Let $\sigma(s)$ be degrees of the components $v_{n(s)}$ of the vectors $v_{(s)}$ of weight N_s (of dimension $2N_s + 1$) in the polynomial invariant B_r. We divide the construction of B_r in several steps.

STEP 1. On a circle, we place (in some order) R various points F_1, F_2, ... F_R, where R is the number of the vectors v_s. From each point F_i, we draw $2N_s$ segments of curves inside the circle (N_s is the weight of the vector V_s corresponding to the point F_i).

STEP 2. We join the curves by arcs (generally, curvilinear) lying inside the circle. We consider only those (admissible) joins when arcs do not intersect. All the curves should be used. It is obvious, that after the choice of joins, we can make arcs shorter by rectifying them so that they do not intersect. We explain the last requirement: if points F_i and F_j are joined by several arcs, then they will coincide after rectification. We allow such situations. We assume that the number of joining chords (after rectification) is equal to the number of arcs joining these points.

STEP 3. We introduce an orientation on the circle and, as a consequence, the order of location of the points F_1, F_2, ..., F_R. With every chord (constructed at Step 2) joining points F_i and F_j, where F_j is the first before the second, we associate the bilinear binomial $\xi_i\eta_j - \xi_j\eta_i$. If k_{ij} is the number of such chords, then with the pair F_iF_j we associate the product of binomials corresponding to the chord, i.e., the polynomial $(\xi_i\eta_j - \xi_j\eta_i)^{k_{ij}}$.

STEP 4. We construct the product of all polynomials associated with the pairs of points

$$P(\xi_1, \eta_1; \xi_2, \eta_2; \dots ; \xi_R, \eta_R) = \text{const} \prod_{1 \leqslant i < j \leqslant R} (\xi_i\eta_j - \xi_j\eta_i)^{k_{ij}}.$$

We can assume that "const" is equal to 1, although sometimes (cf. the examples below) it is more convenient to choose it in other way.

It is easy to check that the common degree of variables ξ_i, η_i in the constructed polynomial is equal to $2N_i$, the number of curves started from the point F_i.

STEP 5. We choose basis polynomials of the form

$$h_n^{(N)}(\xi, \eta) = \begin{cases} h_{-|n|}^{(N)} = i^{N+1}\sqrt{\frac{(2N+1)!}{2(N+n)!(N-n)!}}(\xi^{N+|n|}\eta^{N-|n|} - (-1)^n\xi^{N-|n|}\eta^{N+|n|}) \\ h_0^{(N)} = i^N\sqrt{\frac{(2N+1)!}{N!}}\xi^N\eta^N \\ h_{-|n|}^{(N)} = i^{N+1}\sqrt{\frac{(2N+1)!}{2(N+n)!(N-n)!}}(\xi^{N+|n|}\eta^{N-|n|} + (-1)^n\xi^{N-|n|}\eta^{N+|n|}) \end{cases}$$

Using these basic polynomials, we write P as follows:

$$P=\sum b^{(N_1,N_2,\dots,N_R)}_{(n_1,n_2,\dots,n_r)}\prod_{i=1}^{i=R} h^{(N_i)}_{n_i}(\xi_i\eta_j).$$

STEP 6 (final). Using the coefficients determined at Step 5, we explicitly express the invariant in terms of the components $v^{(N_i)}_{n_i}$ of the vectors $v_{(s)}$:

$$P=\sum b^{(N_1,N_2,\dots,N_R)}_{(n_1,n_2,\dots,n_r)}\prod_{\substack{i=1\\ -N_i\leqslant n_i\leqslant N_i}}^{i=R} v^{(N_i)}_{n_i}.$$

Although $h^{(N)}_n(\xi,\eta)$ are complex, all the coefficients $b^{(N_1,N_2,\dots,N_R)}_{(n_1,n_2,\dots,n_r)}$ are real.

We could obtain more invariants if we assume that intersecting chords are also admissible for joining points at the circle. However, such invariants can be represented as linear combinations of the constructed invariants. In the above construction, only the simplest quadratic invariants, i.e., the sums of squares of the components $\sum\limits_{n=-N}^{N}[v^{(N)}_n]^2$ of the vectors $v^{(N)}$ are not constructed. These sums are invariant because we consider only orthogonal representations of the group $SO(3)$. One can prove that any representation of this group can be described by orthogonal matrices with respect to some special basis.

In the above-mentioned book "Theory of Spinors" by Rumer, for the basis the monomials $\xi^{N+n}\eta^{N-n}$ were used instead of $h^{(N)}_n$. Moreover, we constructed invariants in the case of representation of half-integer weights of the group $SU(2)$. For our purposes the basis consisting of $h^{(N)}_n$ (this basis was proposed by Gordienko (cf. footnote [6])) is more preferable. This basis was used for computing matrix elements of orthogonal representations of the group $SO(3)$ in Section 4.

We indicate some examples of construction of Rumer graphs.

Example 6.1. Let $N_1=N_2=2$, $N_3=N_4=1$, and let the points are located according to the increase of their numbers. We draw 4 curves starting at the point F_1 and 4 curves starting at the point F_2. Also, we draw two curves starting at the point 3 and two curves starting at the point 4. There are the following possibilities of joining these curves by nonintersecting arcs.

1st. Between points of each pairs (F_2,F_3), (F_3,F_4), (F_4,F_1) there is one arc and between the points F_1 and F_2 there are three arcs.

2nd. Between points of each pairs (F_1,F_2), (F_2,F_3), (F_4,F_1) there are two arcs.

3rd. There are 4 arcs between the point F_1 and F_2, and there are two arcs between the points F_3 and F_4. The polynomials $P(\xi_1,\eta_1;\xi_2,\eta_;\xi_3,\eta_3;\xi_4,\eta_4)$ have the form

$$(\xi_2\eta_3 - \xi_3\eta_2)(\xi_3\eta_4 - \xi_4\eta_3)(\xi_4\eta_1 - \xi_1\eta_4)(\xi_1\eta_2 - \xi_2\eta_1)^3,$$
$$(\xi_1\eta_2 - \xi_2\eta_1)^2(\xi_2\eta_3 - \xi_3\eta_2)^2(\xi_4\eta_1 - \xi_1\eta_4)^2,$$
$$(\xi_1\eta_2 - \xi_2\eta_1)^4(\xi_3\eta_4 - \xi_4\eta_3)^2.$$

Using the third way, we obtain a polynomial that can be represented as the product of $(\xi_1\eta_2 - \xi_2\eta_1)^4$ and $(\xi_3\eta_3 - \xi_4\eta_3)^2$. With each of these factors we associate an invariant; moreover, the first one depends only on the vectors $V^{(1)}$ and $V(2)$, whereas the second invariant depends on the vectors $V(3)$ and $V(4)$. Each of these invariants could be constructed by placing on the circle only one pair of points (either F_1 and F_2 or F_3 and F_4). The final invariant is the product of these invariants.

To avoid such situations where invariants are decomposed into independent invariant factors, it is necessary to exclude nonconnected graphs.

Example 6.2. Let $N_1 = N_3 = 1$, $N_2 = N_4 = 2$ (the weights are the same as in Example 6.1 but are ordered in other way). We recommend the reader to compare the invariants with the invariants in Example 6.1.

Example 6.3. The number of points is 3, and N_1, N_2, N_3 are arbitrary integers subject to the triangle inequality. Then the polynomial P takes the form

$$P = \sqrt{\frac{(2N_3)!(2N_1+1)!(2N_2+1)!}{(N_1+N_2+N_3+1)!(N_1+N_2-N_3)!(N_3+N_1-N_2)!(N_3-N_1+N_2)!}}$$
$$\times (\xi_1\eta_2 - \xi_2\eta_1)^{N_1+N_2-N_3}(\xi_3\eta_1 - \xi_1\eta_3)^{N_3+N_1-N_2}(\xi_2\eta_3 - \xi_3\eta_2)^{N_2+N_3-N_1}.$$

Here, the coefficient differs from 1. Under this choice, the coefficients of the representation

$$P = \sum G^{n_3[n_1,n_2]}_{N_3[N_1,N_2]} \; h^{n_3}_{N_3}(\xi_3, \eta_3) h^{n_1}_{N_1}(\xi_1, \eta_1) \; h^{n_2}_{N_2}(\xi_2, \eta_2)$$

coincide with the Clebsch–Gordan coefficients described in Section 4.

References

1. **F. Chalot, T. J. R. Hughes, and F. Shakib**, Symmetrization of conservation laws with entropy for high-temperature hypersonic computations, *Comput. Syst. Sci. Eng.* **1** (1990), no. 2-4, 495–521.

2. **V. N. Dorovskii, A. M. Iskoldskii, and E. I. Romenskii**, Dynamics of flash heating of metal by current and electric explosion of conductors [in Russian], Prikl. Mekh. Tekh. Fiz. (1983), no. 4, 10–25.

3. **K. O. Friedrichs**, Conservation equations and the laws of motion in classical physics, *Commun. Pure Appl. Math.* **31** (1978), no. 1, 123-131.

4. **K. O. Friedrichs and P. D. Lax**, Systems of conservation laws with a convex extension, *Proc. Nat. Acad. Sci. USA* **68** (1971), no. 8, 1686–1688.

5. **S. K. Godunov**, *Elements of Continuum Mechanics* [in Russian], Moscow: "Nauka," 1978.

6. **S. K. Godunov**, Lois de conservation et integrales d'energie des equations hyperboliques, *Lect. Notes Math.* **1270**, 1986, pp. 135–149.

7. **S. K. Godunov and E. I. Romenskii**, *Elements of Continuum Mechanics and Conservation Laws* [in Russian], Novosibirsk: Scientific Books (IDMI), 1998.

8. **L. A. Merzhievskii, A. D. Resnyanskii, and V. M. Titov**, Strength phenomena in the inverse cumulation [in Russian], *Dokl. AN SSSR* **290** (1986), no. 6, 1310–1314.

9. **I. Mueller and T. Ruggeri**, *Extended Thermodynamics*, Springer Tracts in Natural Philosophy, **37**, New York: Springer-Verlag, 1993.

10. **M. Sever**, Estimate of the time rate of entropy dissipation for systems of conservation laws, *J. Differ. Equations* **130** (1996), no. 1, 127–141.

Subject Index

www.ingramcontent.com/pod-product-compliance
Lightning Source LLC
LaVergne TN
LVHW012304141025
823511LV00003B/195
* 9 7 8 0 3 0 6 4 7 7 3 5 5 *